Introduction to Smooth Ergodic Theory
Second Edition

GRADUATE STUDIES
IN MATHEMATICS **231**

Introduction to Smooth Ergodic Theory
Second Edition

Luís Barreira
Yakov Pesin

AMS
AMERICAN
MATHEMATICAL
SOCIETY
Providence, Rhode Island

EDITORIAL COMMITTEE
Matthew Baker
Marco Gualtieri
Gigliola Staffilani (Chair)
Jeff A. Viaclovsky
Rachel Ward

Nonsequential material taken from *Nonuniform Hyperbolicity: Dynamics of Systems with Nonzero Lyapunov Exponents*, by Luis Barreira and Yakov Pesin, Copyright ©2007 Luis Barreira and Yakov Pesin. Reprinted with the permission of Cambridge University Press.

2020 *Mathematics Subject Classification*. Primary 37D25, 37C40.

For additional information and updates on this book, visit
www.ams.org/bookpages/gsm-231

Library of Congress Cataloging-in-Publication Data
Names: Barreira, Luís, 1968- author. | Pesin, Ya. B., author.
Title: Introduction to smooth ergodic theory / Luís Barreira, Yakov Pesin.
Description: Second edition. | Providence, Rhode Island : American Mathematical Society, [2023] | Series: Graduate studies in mathematics, 1065-7339 ; volume 231 | Includes bibliographical references and index.
Identifiers: LCCN 2022050423 | ISBN 9781470470654 (hardcover) | ISBN 9781470473075 (paperback) | ISBN 9781470473068 (ebook)
Subjects: LCSH: Ergodic theory. | Topological dynamics. | AMS: Dynamical systems and ergodic theory – Dynamical systems with hyperbolic behavior – Nonuniformly hyperbolic systems (Lyapunov exponents, Pesin theory, etc.). | Dynamical systems and ergodic theory – Smooth dynamical systems: general theory – Smooth ergodic theory, invariant measures.
Classification: LCC QA611.5 .B37 2023 | DDC 515/.39–dc23/eng20230117
LC record available at https://lccn.loc.gov/2022050423

Copying and reprinting. Individual readers of this publication, and nonprofit libraries acting for them, are permitted to make fair use of the material, such as to copy select pages for use in teaching or research. Permission is granted to quote brief passages from this publication in reviews, provided the customary acknowledgment of the source is given.

Republication, systematic copying, or multiple reproduction of any material in this publication is permitted only under license from the American Mathematical Society. Requests for permission to reuse portions of AMS publication content are handled by the Copyright Clearance Center. For more information, please visit www.ams.org/publications/pubpermissions.

Send requests for translation rights and licensed reprints to reprint-permission@ams.org.

© 2023 by the authors. All rights reserved.
Printed in the United States of America.

∞ The paper used in this book is acid-free and falls within the guidelines established to ensure permanence and durability.
Visit the AMS home page at https://www.ams.org/

10 9 8 7 6 5 4 3 2 1   28 27 26 25 24 23

Contents

Preface to the Second Edition — ix

Preface to the First Edition — xiii

Part 1. The Core of the Theory

Chapter 1.   Examples of Hyperbolic Dynamical Systems — 3
- §1.1.   Anosov diffeomorphisms — 4
- §1.2.   Anosov flows — 9
- §1.3.   Hyperbolic sets — 14
- §1.4.   The Smale–Williams solenoid — 20
- §1.5.   The Katok map of the 2-torus — 22
- §1.6.   Area-preserving diffeomorphisms with nonzero Lyapunov exponents on surfaces — 33
- §1.7.   A volume-preserving flow with nonzero Lyapunov exponents — 39
- §1.8.   A slow-down of the Smale–Williams solenoid — 43

Chapter 2.   General Theory of Lyapunov Exponents — 45
- §2.1.   Lyapunov exponents and their basic properties — 45
- §2.2.   The Lyapunov and Perron irregularity coefficients — 50
- §2.3.   Lyapunov exponents for linear differential equations — 54
- §2.4.   Forward and backward regularity. The Lyapunov–Perron regularity — 66
- §2.5.   Lyapunov exponents for sequences of matrices — 73

Chapter 3. Cocycles over Dynamical Systems — 81
- §3.1. Cocycles and linear extensions — 82
- §3.2. Lyapunov exponents and Lyapunov–Perron regularity for cocycles — 87
- §3.3. Examples of measurable cocycles over dynamical systems — 93

Chapter 4. The Multiplicative Ergodic Theorem — 97
- §4.1. Lyapunov–Perron regularity for sequences of triangular matrices — 98
- §4.2. Proof of the Multiplicative Ergodic Theorem — 105
- §4.3. Normal forms of cocycles — 110
- §4.4. Regular neighborhoods — 115

Chapter 5. Elements of the Nonuniform Hyperbolicity Theory — 119
- §5.1. Dynamical systems with nonzero Lyapunov exponents — 120
- §5.2. Nonuniform complete hyperbolicity — 129
- §5.3. Regular sets — 132
- §5.4. Nonuniform partial hyperbolicity — 139
- §5.5. Hölder continuity of invariant distributions — 141

Chapter 6. Lyapunov Stability Theory of Nonautonomous Equations — 147
- §6.1. Stability of solutions of ordinary differential equations — 148
- §6.2. Lyapunov absolute stability theorem — 153
- §6.3. Lyapunov conditional stability theorem — 158

Chapter 7. Local Manifold Theory — 163
- §7.1. Local stable manifolds — 164
- §7.2. An abstract version of the Stable Manifold Theorem — 168
- §7.3. Basic properties of stable and unstable manifolds — 178

Chapter 8. Absolute Continuity of Local Manifolds — 187
- §8.1. Absolute continuity of the holonomy map — 189
- §8.2. A proof of the Absolute Continuity Theorem — 197
- §8.3. Computing the Jacobian of the holonomy map — 203
- §8.4. An invariant foliation that is not absolutely continuous — 205

Chapter 9. Ergodic Properties of Smooth Hyperbolic Measures — 207
- §9.1. Ergodicity of smooth hyperbolic measures — 207
- §9.2. Local ergodicity — 216
- §9.3. The entropy formula — 222

Chapter 10. Geodesic Flows on Surfaces of Nonpositive Curvature	237
§10.1. Preliminary information from Riemannian geometry	238
§10.2. Definition and local properties of geodesic flows	240
§10.3. Hyperbolic properties and Lyapunov exponents	242
§10.4. Ergodic properties	248
§10.5. The entropy formula for geodesic flows	253
Chapter 11. Topological and Ergodic Properties of Hyperbolic Measures	257
§11.1. Hyperbolic measures with local product structure	258
§11.2. Periodic orbits and approximations by horseshoes	263
§11.3. Shadowing and Markov partitions	264

Part 2. Selected Advanced Topics

Chapter 12. Cone Techniques	271
§12.1. Introduction	271
§12.2. Lyapunov functions	273
§12.3. Cocycles with values in the symplectic group	277
Chapter 13. Partially Hyperbolic Diffeomorphisms with Nonzero Exponents	279
§13.1. Partial hyperbolicity	280
§13.2. Systems with negative central exponents	283
§13.3. Foliations that are not absolutely continuous	285
Chapter 14. More Examples of Dynamical Systems with Nonzero Lyapunov Exponents	291
§14.1. Hyperbolic diffeomorphisms with countably many ergodic components	291
§14.2. The Shub–Wilkinson map	301
Chapter 15. Anosov Rigidity	303
§15.1. The Anosov rigidity phenomenon. I	303
§15.2. The Anosov rigidity phenomenon. II	311
Chapter 16. C^1 Pathological Behavior: Pugh's Example	317
Bibliography	323
Index	331

Preface to the Second Edition

Smooth ergodic theory lies at the core of the modern theory of dynamics and our book is the only currently available source for graduate students to learn the theory, including basic ideas, methods, and examples. Moreover, smooth ergodic theory has numerous and a still-growing number of applications in various areas of mathematics and beyond. As a result there is a growing number of students interested in some aspects of the theory who either major in areas of mathematics outside of dynamics (probability, statistics, number theory, etc.) or in areas outside mathematics (physics, biology, chemistry, engineering, and economy).

Since the publication of our book in 2013, we received many comments and suggestions from colleagues who taught courses in dynamics using the text as well as questions and comments from students who studied the text, which we took into consideration while working on the second edition. We restructured the book and improved its exposition by revising the proofs of some theorems by adding more details and informal discussions. We also included numerous comments on the notions we introduce, added more exercises, and corrected some typos and minor mistakes.

The second edition contains complete proofs of all major results. Moreover, in an attempt to expand its scope and bring the reader closer to the modern research in dynamics, throughout the book we added many remarks which include comments on the notions and results being presented as well as a brief discussion of some recent advances in the field and statements (usually without proofs) of some new results. Here is a brief account of the changes.

In Chapter 1 we included a new Section 1.3 on uniformly hyperbolic sets which are invariant but not compact or compact but not invariant. The goal is to introduce the reader to the special way of setting up the hyperbolicity conditions, which, in a more general form, are used to define nonuniform hyperbolicity. Further, we added the two new Sections 1.4 and 1.8 on the Smale–Williams solenoid and its slow-down version to discuss some features of hyperbolicity for dissipative dynamical systems which we consider later in Section 11.1. We also wrote a proof of Proposition 1.34, which to the best of our knowledge cannot be found in the literature.

In Chapter 2 we added the two new Sections 2.5.2 and 2.5.3 in which we complete our study of the Lyapunov–Perron regularity for sequences of matrices.

In Chapter 5, Section 5.1.1, we added the important Corollary 5.2 and in Section 5.1.2 (see Remark 5.5) we discuss the crucial role the "deterioration conditions" play in nonuniform hyperbolicity and also introduce the important notion of ε-tempered functions. In Section 5.3 we presented a new Theorem 5.10, which establishes some basic properties of regular sets, and added Section 5.3.3 where the Lyapunov inner product is introduced and its properties are described in Theorem 5.15. We also added the important Remark 5.16 where we compare the weak and strong Lyapunov inner products and charts. Section 5.4 on partial hyperbolicity was extended and improved.

In Chapter 7, Section 7.1.3, we simplified the notations and clarified the exposition. The abstract Stable Manifold Theorem in Section 7.2 is now presented in a greater generality of nonlinear operators acting on Banach spaces. Throughout Section 7.3 we added more details and discussions about local stable and unstable manifolds and we substantially improved and expanded Section 7.3.3 on the graph transform property. We also added a new Section 7.3.5 on stable and unstable manifolds for nonuniformly partially hyperbolic sets.

In Chapter 8 the reader will find a new Sections 8.1.1 on foliation blocks and in Section 8.1.3 a new Theorem 8.7 as well as new Remarks 8.5, 8.8, 8.10, and 8.11, which provide some important additional information to the main result of the section, Theorem 8.4. In Section 8.2 we added more details to the proof of the absolute continuity property and, in particular, included a proof of Lemma 8.13. In the new Remark 8.19 we briefly discuss a construction of local stable and unstable manifolds in the setting of partially hyperbolic sets.

In Chapter 9, Section 9.1, the proof of the main result, Theorem 9.2, on ergodic decomposition of hyperbolic smooth measures is substantially improved and we added a new Section 9.1.1 where we introduce conditional measures on local manifolds along with a new Remark 9.8 where we compare two descriptions of ergodic components, one which comes from Theorem 9.2 and another one, which is related to ergodic homoclinic classes, is given by Theorem 9.10. We also included a proof of Theorem 9.6. The main addition to Section 9.3 is Section 9.3.4 where we present the Ledrappier–Young entropy formula and outline the main ingredients of its proof.

In Chapter 10, after the proof of Theorem 10.11, the reader finds a brief discussion of obstacles to proving ergodicity of geodesic flows on surfaces on nonpositive curvature and a new Theorem 10.12 which establishes ergodicity under some additional assumption.

Chapter 11 is entirely new. It contains three sections where we discuss various properties of general hyperbolic measures including the local product structure, existence of periodic points, and approximations by horseshoes. We also describe the important Sinai–Ruelle–Bowen measures and present some recent results on their existence. We complete the section by outlining a crucial new result on constructing Markov partitions and symbolic dynamics for nonuniformly hyperbolic diffeomorphisms.

All of this resulted in approximately 50 new pages.

Acknowledgements. It is a pleasure to thank several colleagues who have helped us in various ways: Snir Ben Ovadia for drafting Section 11.3 on shadowing and Markov partitions and for useful discussions about this topic and related topics on weak and strong Lyapunov inner products and charts; Sebastian Burgos for useful comments on the construction of the slow-down of the Smale–Williams solenoid in Section 1.8; Anton Gorodetsky for useful comments on Section 1.3; and Federico Rodriguez Hertz for many useful comments on Section 1.6.2 and, particularly, on the proof of Proposition 1.34 as well as on Remark 9.7.

Luís Barreira, Lisboa, Portugal   Yakov Pesin, State College, PA USA

October 2022

Preface to the First Edition

This book is a revised and considerably expanded version of our book *Lyapunov Exponents and Smooth Ergodic Theory* [**11**]. When the latter was published, it became the only source of a systematic introduction to the core of smooth ergodic theory. It included the general theory of Lyapunov exponents and its applications to the stability theory of differential equations, nonuniform hyperbolicity theory, stable manifold theory (with emphasis on absolute continuity of invariant foliations), and the ergodic theory of dynamical systems with nonzero Lyapunov exponents, including geodesic flows. In the absence of other textbooks on the subject it was also used as a source or as supportive material for special topics courses on nonuniform hyperbolicity.

In 2007 we published the book *Nonuniform Hyperbolicity: Dynamics of Systems with Nonzero Lyapunov Exponents* [**13**], which contained an up-to-date exposition of smooth ergodic theory and was meant as a primary reference source in the field. However, despite an impressive amount of literature in the field, there has been until now no textbook containing a comprehensive introduction to the theory.

The present book is intended to cover this gap. It is aimed at graduate students specializing in dynamical systems and ergodic theory as well as anyone who wishes to acquire a working knowledge of smooth ergodic theory and to learn how to use its tools. While maintaining the essentials of most of the material in [**11**], we made the book more student-oriented by carefully selecting the topics, reorganizing the material, and substantially expanding the proofs of the core results. We also included a detailed description of essentially all known examples of conservative systems with nonzero Lyapunov exponents and throughout the book we added many exercises.

The book consists of two parts. While the first part introduces the reader to the basics of smooth ergodic theory, the second part discusses more advanced topics. This gives the reader a broader view of the theory and may help stimulate further study. This also provides nonexperts with a broader perspective of the field.

We emphasize that the new book is self-contained. Namely, we only assume that the reader has a basic knowledge of real analysis, measure theory, differential equations, and topology and we provide the reader with necessary background definitions and state related results.

On the other hand, in view of the considerable size of the theory we were forced to make a selection of the material. As a result, some interesting topics are barely mentioned or not covered at all. We recommend the books [**13, 24**] and the surveys [**12, 77**] for a description of many other developments and some recent work. In particular, we do not consider random dynamical systems (see the books [**7, 68, 75**] and the survey [**69**]), dynamical systems with singularities, including "chaotic" billiards (see the book [**67**]), the theory of nonuniformly expanding maps (see the survey [**76**]), and one-dimensional "chaotic" maps (such as the logistic family; see [**58**]).

Smooth ergodic theory studies the ergodic properties of smooth dynamical systems on Riemannian manifolds with respect to "natural" invariant measures. Among these measures most important are smooth measures, i.e., measures that are equivalent to the Riemannian volume. There are various classes of smooth dynamical systems whose study requires different techniques. In this book we concentrate on systems whose trajectories are hyperbolic in some sense. Roughly speaking, this means that the behavior of trajectories near a given orbit resembles the behavior of trajectories near a saddle point. In particular, to every hyperbolic trajectory one can associate two complementary subspaces such that the system acts as a contraction along one of them (called the stable subspace) and as an expansion along the other (called the unstable subspace).

A hyperbolic trajectory is unstable—almost every nearby trajectory moves away from it with time. If the set of hyperbolic trajectories is sufficiently large (for example, has positive or full measure), this instability forces trajectories to become separated. On the other hand, compactness of the phase space forces them back together; the consequent unending dispersal and return of nearby trajectories is one of the hallmarks of chaos.

Indeed, hyperbolic theory provides a mathematical foundation for the paradigm that is widely known as "deterministic chaos"—the appearance of irregular chaotic motions in purely deterministic dynamical systems. This paradigm asserts that conclusions about global properties of a nonlinear dynamical system with sufficiently strong hyperbolic behavior can be deduced from studying the linearized systems along its trajectories.

The study of hyperbolic phenomena originated in the seminal work of Artin, Morse, Hedlund, and Hopf on the instability and ergodic properties of geodesic flows on compact surfaces (see the survey [**53**] for a detailed description of results obtained at that time and for references). Later, hyperbolic behavior was observed in other situations (for example, Smale horseshoes and hyperbolic toral automorphisms).

The systematic study of hyperbolic dynamical systems was initiated by Smale (who mainly considered the problem of structural stability of hyperbolic systems; see [**107**]) and by Anosov and Sinai (who were mainly concerned with ergodic properties of hyperbolic systems with respect to smooth invariant measures; see [**5, 6**]). The hyperbolicity conditions describe the action of the linearized system along the stable and unstable subspaces and impose quite strong requirements on the system. The dynamical systems that satisfy these hyperbolicity conditions uniformly over all orbits are called Anosov systems.

In this book we consider the weakest (hence, most general) form of hyperbolicity, known as nonuniform hyperbolicity. It was introduced and studied by Pesin in a series of papers [**87**–**91**]. The nonuniform hyperbolicity theory (which is sometimes referred to as Pesin theory) is closely related to the theory of Lyapunov exponents. The latter originated in works of Lyapunov [**78**] and Perron [**86**] and was developed further in [**35**]. We provide an extended excursion into the theory of Lyapunov exponents and, in particular, introduce and study the crucial concept of Lyapunov–Perron regularity. The theory of Lyapunov exponents enables one to obtain many subtle results on the stability of differential equations.

Using the language of Lyapunov exponents, one can view nonuniformly hyperbolic dynamical systems as those systems where the set of points for which *all* Lyapunov exponents are nonzero is "large"—for example, has full measure with respect to an invariant Borel measure. In this case the Multiplicative Ergodic Theorem of Oseledets [**85**] implies that almost every point is Lyapunov–Perron regular. The powerful theory of Lyapunov exponents then yields a profound description of the local stability of trajectories, which, in turn, serves as grounds for studying the ergodic properties of these systems.

Luís Barreira, Lisboa, Portugal   Yakov Pesin, State College, PA USA

February 2013

Part 1

The Core
of the Theory

Chapter 1

Examples of Hyperbolic Dynamical Systems

We begin our journey into smooth ergodic theory by constructing some principal examples of smooth dynamical systems (both with discrete and continuous time), which illustrate various fundamental phenomena associated with uniform as well as nonuniform hyperbolicity. We will first describe some examples of systems which preserve a smooth measure (i.e., a measure which is equivalent to the Riemannian volume). Among them are Anosov diffeomorphisms and flows as well as the Katok map and its continuous-time version. We then discuss an example of a dissipative system—the Smale–Williams solenoid—and outline a construction of Sinai–Ruelle–Bowen measures which are discussed in detail in Section 11.1.

While experts in the field believe that hyperbolic behavior is typical in a sense, it is usually difficult to rigorously establish that a given dynamical system is hyperbolic. In fact, known examples of uniformly hyperbolic diffeomorphisms include only those that act on tori and factors of some nilpotent Lie groups and it is believed—due to the strong requirement that every trajectory of such a system must be uniformly hyperbolic—that those are the only manifolds which can carry uniformly hyperbolic diffeomorphisms. Despite such a shortage of examples (at least in the discrete-time case) the uniform hyperbolicity theory provides great insight into many principal phenomena associated with hyperbolic behavior and, in particular, allows one to obtain an essentially complete description of stochastic behavior of such systems.

In a quest for more examples of dynamical systems with hyperbolic behavior one examines those that are not uniformly hyperbolic but possess "just enough" hyperbolicity to exhibit a high level of stochastic behavior. A "typical" trajectory in these systems is hyperbolic although hyperbolicity is weaker than the one observed in Anosov systems and some trajectories are not hyperbolic at all. This is the case of nonuniform hyperbolicity.

It is believed that nonuniformly hyperbolic volume-preserving systems are generic in some sense in the space of $C^{1+\alpha}$ volume-preserving systems although this remains one of the most challenging problems in the area. In much contrast with Anosov diffeomorphisms, this belief is supported by the fact that every compact manifold of dimension ≥ 2 carries a $C^{1+\alpha}$ volume-preserving nonuniformly hyperbolic diffeomorphism and that every compact manifold of dimension ≥ 3 carries a $C^{1+\alpha}$ volume-preserving nonuniformly hyperbolic flow.

1.1. Anosov diffeomorphisms

We begin with dynamical systems exhibiting hyperbolic behavior in the strongest form. They were introduced by Anosov and are known as Anosov systems (see [**5**] and also [**6**]; for modern expositions of uniform hyperbolicity theory see [**29, 64**]).

To describe the simplest example of such a system, consider the matrix $A = \begin{pmatrix} 2 & 1 \\ 1 & 1 \end{pmatrix}$. It induces a linear transformation T of the two-dimensional torus $\mathbb{T}^2 = \mathbb{R}^2/\mathbb{Z}^2$. The eigenvalues of A are λ^{-1} and λ where $\lambda = (3+\sqrt{5})/2$, and the corresponding eigendirections are given by the orthogonal vectors $(\frac{1+\sqrt{5}}{2}, 1)$ and $(\frac{1-\sqrt{5}}{2}, 1)$ (see Figure 1.1). For each $x \in \mathbb{T}^2$, we denote by

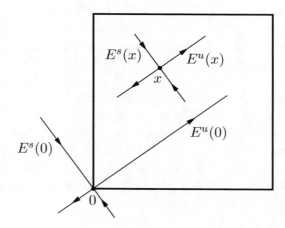

Figure 1.1. A hyperbolic toral automorphism.

1.1. Anosov diffeomorphisms

$E^s(x)$ and $E^u(x)$ the one-dimensional subspaces of the tangent space $T_x\mathbb{T}^2$ (which can be identified with $\mathbb{R}^2$) obtained by translating the eigenlines of A. These subspaces are called, respectively, *stable* and *unstable subspaces* at x in view of the following estimates:

$$\|d_x T^n v\| = \lambda^{-n}\|v\| \quad \text{whenever } v \in E^s(x) \text{ and } n \geq 0,$$
$$\|d_x T^{-n} v\| = \lambda^{-n}\|v\| \quad \text{whenever } v \in E^u(x) \text{ and } n \geq 0.$$

Furthermore, since $E^s(x)$ and $E^u(x)$ are parallel to the eigendirections, they form dT-invariant bundles ($dT = A$), i.e., for every $x \in \mathbb{T}^2$,

$$d_x T E^s(x) = E^s(T(x)) \quad \text{and} \quad d_x T E^u(x) = E^u(T(x)).$$

Note that the origin is a fixed saddle (hyperbolic) point of T and that the trajectory of any point $x \in \mathbb{T}^2$ lies along a hyperbola; see Figure 1.2.

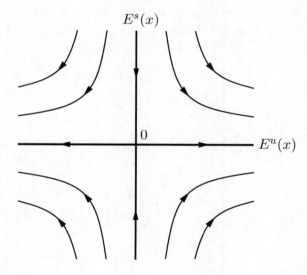

Figure 1.2. The structure of orbits near a hyperbolic fixed point.

Let $\pi\colon \mathbb{R}^2 \to \mathbb{T}^2$ be the canonical projection. For each $x \in \mathbb{T}^2$ the sets $\pi(E^s(x))$ and $\pi(E^u(x))$ are C^∞ curves that are called *global stable* and *unstable curves (manifolds)* at x. They depend C^∞ smoothly on the base point x and form two partitions of $\mathbb{T}^2$.

The above example gives rise to the general notion of Anosov map. Let $f\colon M \to M$ be a diffeomorphism of a compact Riemannian manifold.

We say that f is an *Anosov diffeomorphism* if there are constants $c > 0$, $0 < \lambda < 1 < \mu$ and, for each $x \in M$, subspaces $E^s(x)$ and $E^u(x)$ such that:

(1) *invariant splitting:* for every $x \in M$,

$$T_x M = E^s(x) \oplus E^u(x), \tag{1.1}$$

$$d_x f E^s(x) = E^s(f(x)) \quad \text{and} \quad d_x f E^u(x) = E^u(f(x)); \tag{1.2}$$

(2) *hyperbolicity estimates:*

$$\begin{aligned} \|d_x f^n v\| &\leq c\lambda^n \|v\| \quad \text{whenever } v \in E^s(x) \text{ and } n \geq 0, \\ \|d_x f^{-n} v\| &\leq c\mu^{-n} \|v\| \quad \text{whenever } v \in E^u(x) \text{ and } n \geq 0. \end{aligned} \tag{1.3}$$

Condition (2) justifies calling $E^s(x)$ a *stable subspace* and $E^u(x)$ an *unstable subspace* of $T_x M$. The numbers $\{\lambda, \mu, c\}$ are called *parameters of hyperbolicity*.

If x is a fixed point for f, then the classical Grobman–Hartman and Hadamard–Perron theorems assert that the behavior of orbits in a sufficiently small neighborhood of x imitates the behavior of orbits near a saddle point (see Figure 1.3). More precisely, there exists $\delta > 0$ such that for each $x \in M$ the sets

$$\begin{aligned} V^s(x) &= \{y \in B(x, \delta) : \rho(f^n(y), f^n(x)) \to 0 \text{ as } n \to +\infty\}, \\ V^u(x) &= \{y \in B(x, \delta) : \rho(f^n(y), f^n(x)) \to 0 \text{ as } n \to -\infty\}, \end{aligned} \tag{1.4}$$

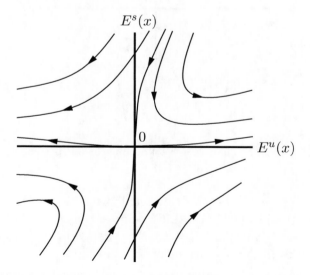

Figure 1.3. Invariant curves near a hyperbolic fixed point.

1.1. Anosov diffeomorphisms

where ρ is the distance in M, are immersed local smooth manifolds for which
$$d_x f V^s(x) = E^s(x) \quad \text{and} \quad d_x f V^u(x) = E^u(x).$$
The manifolds $V^s(x)$ and $V^u(x)$ are called, respectively, *local stable* and *unstable manifolds* at x of size δ. In general they depend only Hölder continuously on x (see Theorem 5.18). If two local stable manifolds intersect, then one of them is a continuation of the other, and hence, they can be "glued" together. Continuing in this fashion we obtain the *global stable manifold* $W^s(x)$ at x. It can also be defined by the formula
$$W^s(x) = \bigcup_{n \geq 0} f^n(V^s(f^{-n}(x)))$$
and it consists of all the points in the manifold whose trajectories converge to the trajectory of x, i.e.,
$$W^s(x) = \{y \in M : \rho(f^n(y), f^n(x)) \to 0 \text{ as } n \to +\infty\}.$$
The global stable leaves form a partition W^s of M called a foliation. More precisely, a partition W of M is called a *continuous foliation of M with smooth leaves* or simply a *foliation* if there exist $\delta > 0$ and $\ell > 0$ such that for each $x \in M$:

(1) the element $W(x)$ of the partition W containing x is a smooth ℓ-dimensional immersed submanifold; it is called the *(global) leaf* of the foliation at x; the connected component of the intersection $W(x) \cap B(x, \delta)$ that contains x is called the *local leaf* at x and is denoted by $V(x)$;

(2) there exists a continuous map $\varphi_x \colon B(x, \delta) \to C^1(D, M)$ (where $D \subset \mathbb{R}^\ell$ is the unit ball) such that for every $y \in B(x, \delta)$ the manifold $V(y)$ is the image of the map $\varphi_x(y) \colon D \to M$.

The function $\varphi_x(y, z) = \varphi_x(y)(z)$ is called the *foliation coordinate chart*. It follows from the definition of foliation that the function $\varphi_x(y, z)$ is continuous in both y and z and is differentiable in z for each $y \in B(x, \delta)$; moreover, the derivative $\frac{\partial}{\partial z}\varphi_x(y, z)$ is continuous in y.

A continuous distribution E on TM is said to be *integrable* if there exists a foliation W of M such that $E(x) = T_x W(x)$ for every $x \in M$.

The stable foliation W^s is an integrable foliation for the stable distribution E^s. It is also invariant under f, i.e.,
$$f(W^s(x)) = W^s(f(x)).$$
We stress that this foliation is Hölder continuous but, in general, not smooth.[1]

---

[1] Recall that a foliation is called *Hölder continuous* (respectively, *smooth*) if for every $x \in M$ the derivative $\frac{\partial}{\partial z}\varphi_x(y, z)$ depends Hölder continuously (respectively, smoothly) on $y \in B(x, \delta)$.

In a similar fashion one can glue unstable local leaves to obtain global unstable leaves. They form an invariant unstable foliation W^u of M that integrates the unstable distribution E^u. The stable and unstable foliations are transverse at every point on the manifold and each of them possesses the absolute continuity property (see Chapter 8). This property of invariant foliations is crucial in proving that an Anosov diffeomorphism of a compact connected smooth manifold is ergodic with respect to a smooth measure.

Recall that an f-invariant measure ν is *ergodic* if for each f-invariant measurable set $A \subset M$ either A or $M \setminus A$ has measure zero. Equivalently, ν is ergodic if and only if every f-invariant (mod 0) measurable function (i.e., a measurable function φ such that $\varphi \circ f = \varphi$ almost everywhere) is constant almost everywhere.

We also recall the notion of Bernoulli automorphism. Let (X, μ) be a Lebesgue space with a probability measure μ. Assume that μ has at most countably many atoms whose union $Y \subset X$ is such that $\mu | (X \setminus Y)$ is metrically isomorphic to the Lebesgue measure on the unit interval $[0, 1]$. One can associate to (X, μ) the two-sided Bernoulli shift $\sigma \colon X^{\mathbb{Z}} \to X^{\mathbb{Z}}$ defined by $(\sigma x)_n = x_{n+1}$, $n \in \mathbb{Z}$, which preserves the convolution $\bigoplus_{\mathbb{Z}} \mu$. A *Bernoulli automorphism* (S, ν) is an invertible (mod 0) measure-preserving transformation which is metrically isomorphic to the Bernoulli shift associated to some Lebesgue space (X, μ).

In the case where X is a compact smooth manifold and S is a diffeomorphism of X preserving a measure ν, we have that S is a Bernoulli automorphism if it is metrically isomorphic to the Bernoulli shift (Σ, σ, μ) where Σ is the space of double-infinite sequences $\omega = (\omega_n)$ with ω_n taking values in a *finite* set $\{1, \ldots, k\}$, σ is the full shift, and μ is the Bernoulli measure on Σ generated by a probability vector $\{p_1, \ldots, p_k\}$.[2]

Theorem 1.1. *Any C^2 Anosov diffeomorphism[3] f of a compact smooth connected Riemannian manifold M preserving a smooth measure μ is ergodic. Furthermore, there is a number $n > 0$ and a subset $A \subset M$ such that:*

(1) $f^k(A) \cap A = \varnothing$ for $k = 1, \ldots, n-1$ and $f^n(A) = A$;

(2) $\bigcup_{k=0}^{n-1} f^k(A) = M \pmod{0}$;

(3) $f^n | A$ is a Bernoulli automorphism.

This theorem is a particular case of Theorems 9.2 and 9.12.

---

[2]To see that, assume $h_\nu(S) > 0$ (otherwise S cannot be a Bernoulli automorphism) and observe that there is a probability vector $\{p_1, \ldots, p_k\}$ (for some $k > 0$) such that $h_\nu(S) = h_\mu(\sigma)$. By Ornstein's theory, S must be metrically isomorphic to σ.

[3]The requirement that f be of class of smoothness C^2 can be relaxed to the requirement that f be of class of smoothness $C^{1+\alpha}$ for some $\alpha > 0$ (i.e., the differential of f is Hölder continuous). It is not known, however, whether this theorem holds for C^1 Anosov diffeomorphisms.

Anosov diffeomorphisms are *structurally stable*. This means that any sufficiently small C^1 perturbation $g\colon M \to M$ of an Anosov diffeomorphism is still an Anosov diffeomorphism and there exists a Hölder homeomorphism $h\colon M \to M$ such that $g \circ h = h \circ f$. In particular, the set of Anosov diffeomorphisms of class C^1 is open in the C^1 topology.

1.2. Anosov flows

Now we consider uniformly hyperbolic dynamical systems with continuous time. It is well known that every smooth vector field $\mathfrak{X}$ on a compact smooth manifold M can be uniquely integrated; i.e., given $x \in M$, there exists a uniquely defined smooth curve $\gamma_x(t)$, $t \in \mathbb{R}$, such that $\gamma_x(0) = x$ and

$$\frac{d}{dt}\gamma_x(t) = \mathfrak{X}(\gamma_x(t)) \tag{1.5}$$

for every t. One can now define a smooth flow $\varphi_t\colon M \to M$ by $\varphi_t(x) = \gamma_x(t)$. It follows readily from (1.5) that $\mathfrak{X}(x) = \frac{d}{dt}(\varphi_t(x))|_{t=0}$.

We say that a flow φ_t is an *Anosov flow* if there exist constants $c > 0$ and $\mu \in (0, 1)$ and, for each $x \in M$, a decomposition

$$T_x M = E^s(x) \oplus E^0(x) \oplus E^u(x)$$

such that:

(1) $E^0(x)$ is the one-dimensional subspace generated by the vector field $\mathfrak{X}(x)$;
(2) $d_x\varphi_t E^s(x) = E^s(\varphi_t(x))$ and $d_x\varphi_t E^u(x) = E^u(\varphi_t(x))$ for $t \in \mathbb{R}$;
(3) for $t \geq 0$,

$$\|d_x\varphi_t v\| \leq c\mu^t \|v\| \quad \text{whenever } v \in E^s(x),$$
$$\|d_x\varphi_{-t} v\| \leq c\mu^t \|v\| \quad \text{whenever } v \in E^u(x).$$

The stable and unstable distributions E^s and E^u integrate to continuous stable and unstable invariant foliations W^s and W^u for the flow. We also have the weakly stable and weakly unstable invariant foliations W^{st} and W^{ut} whose leaves at a point $x \in M$ are

$$W^{st}(x) = \bigcup_{t \in \mathbb{R}} W^s(\varphi_t(x)), \quad W^{ut}(x) = \bigcup_{t \in \mathbb{R}} W^u(\varphi_t(x)).$$

A simple example of an Anosov flow is a *special flow* over an Anosov diffeomorphism. We recall the definition of a special flow. Given a diffeomorphism f of a compact Riemannian manifold M and a smooth positive function h on M, consider the quotient space

$$M_h = \{(x, t) \in M \times \mathbb{R}_0^+ : 0 \leq t \leq h(x)\}/\equiv,$$

where $\equiv$ is the equivalence relation $(x, h(x)) \equiv (f(x), 0)$. The special flow $\varphi_t \colon M_h \to M_h$ with *roof function* h moves a point $(x, 0)$ along $\{x\} \times \mathbb{R}_0^+$ to $(x, h(x))$, then jumps to $(f(x), 0)$ and continues along $\{f(x)\} \times \mathbb{R}_0^+$ (see Figure 1.4). More precisely,

$$\varphi_t(x, \tau) = (f^n(x), \tau'),$$

where the numbers n, τ, and τ' satisfy $0 \leq \tau' \leq h(f^n(x))$ and

$$\sum_{i=0}^{n-1} h(f^i(x)) + \tau' = t + \tau.$$

A special flow with constant roof function is called a *suspension flow*.

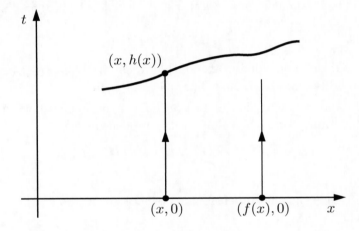

Figure 1.4. The special flow $\varphi_t \colon M_h \to M_h$ with roof function h.

Exercise 1.2. Let f be an Anosov diffeomorphism of a compact Riemannian manifold M and let h be a smooth positive function on M. Show that the special flow φ_t with the roof function h is an Anosov flow on M_h.

Recall that a φ_t-invariant measure ν is *ergodic* if for each φ_t-invariant measurable set $A \subset M$ either A or $M \setminus A$ has measure zero. Equivalently, ν is ergodic if and only if every φ_t-invariant (mod 0) measurable function is constant almost everywhere. Recall also that a flow φ_t is a Bernoulli flow if for each $t \neq 0$ the map φ_t is a Bernoulli automorphism.

An Anosov flow preserving a smooth measure may not be ergodic (just consider a special flow over an Anosov diffeomorphism with constant roof function). We say that a flow φ_t on a compact smooth manifold is *topologically transitive* if for any two nonempty open sets U and V there exists $n \in \mathbb{Z}$ such that $f^n(U) \cap V \neq \varnothing$, and we say that φ_t is *topologically mixing* if for any two nonempty open sets U and V there exists $N \geq 0$ such that for every $n \geq N$ we have that $f^n(U) \cap V \neq \varnothing$.

1.2. Anosov flows

Theorem 1.3. *If an Anosov flow φ_t on a compact smooth connected Riemannian manifold preserves a smooth measure and is topologically transitive, then it is ergodic, and if φ_t is topologically mixing, then it is Bernoulli.*

This result is a corollary of the more general Theorem 9.13 and results in Section 9.2.

Another example of an Anosov flow that stems from Riemannian geometry is the geodesic flow on a compact smooth Riemannian manifold of negative curvature. We will discuss here the simplest case of geodesic flows on surfaces of constant negative curvature, referring the reader to Chapter 10 for a more general case.

We shall use the Poincaré model of the Lobachevsky geometry on the upper-half plane
$$\mathbb{H} = \{z = x + iy \in \mathbb{C} : \operatorname{Im} z = y > 0\}.$$

The line $y = 0$ is called the *ideal boundary* of $\mathbb{H}$ and is denoted by $\mathbb{H}(\infty)$. The Riemannian metric on $\mathbb{H}$ with constant curvature $-k$ is given by the inner product in the tangent space $T_z\mathbb{H} = \mathbb{C}$ defined by
$$\langle v, w \rangle_z = \frac{k}{(\operatorname{Im} z)^2} \langle v, w \rangle,$$
where $\langle v, w \rangle$ is the standard inner product in $\mathbb{R}^2$.

Consider the group $SL(2, \mathbb{R})$ of matrices $A = \begin{pmatrix} a & b \\ c & d \end{pmatrix}$ with real entries and determinant 1 or -1, and define Möbius transformations T_A of $\mathbb{H}$ by
$$T_A(z) = \frac{az+b}{cz+d} \quad \text{and} \quad T_A(z) = \frac{a\bar{z}+b}{c\bar{z}+d}, \tag{1.6}$$
respectively, when $ad - bc = 1$ and $ad - bc = -1$. Clearly, $T_{-A} = T_A$.

Exercise 1.4. Show that the transformations T_A in (1.6) take $\mathbb{H}$ into itself and are isometries.

In fact, the group $G = SL(2, \mathbb{R})/\{\operatorname{Id}, -\operatorname{Id}\}$ is the group of all isometries of $\mathbb{H}$.

Exercise 1.5. Show that the geodesics, i.e., the shortest paths between points in $\mathbb{H}$, are the vertical half-lines and the half-circles centered at points in the axis $\operatorname{Im} z = 0$. More precisely, given $z \in \mathbb{H}$ and $v \in \mathbb{C} \setminus \{0\}$:

(1) if v is parallel to the line $\operatorname{Re} z = 0$, then the geodesic passing through z in the direction of v is the half-line $\{z + vt : t \in \mathbb{R}\} \cap \mathbb{H}$;

(2) if v is not parallel to the line $\operatorname{Re} z = 0$, then the geodesic passing through z in the direction of v is the half-circle centered at the axis $\operatorname{Im} z = 0$ passing through z and with tangent v at that point.

Hint: First consider a path $\gamma\colon [0,1] \to \mathbb{H}$ between ic and id, with $d > c$, and note that its length ℓ_γ satisfies

$$\ell_\gamma = \int_0^\tau \frac{|\gamma'(t)|}{y(t)}\, dt \ge \int_0^\tau \frac{y'(t)}{y(t)}\, dt = \log\frac{d}{c},$$

where $y(t) = \operatorname{Im}\gamma(t)$. Since the vertical path between ic and id has precisely this length, it is in fact the geodesic joining the two points. For the general case, show that given $z, w \in \mathbb{H}$ with $z \ne w$, there exists a Möbius transformation taking the vertical line segment or the arc of the circle centered at the real axis joining z and w to a vertical line segment on the positive part of the imaginary axis and use the former result.

The *geodesic flow* is a flow acting on the unit tangent bundle

$$S\mathbb{H} = \{(z,v) \in \mathbb{H} \times \mathbb{C} : |v|_z = 1\}$$

and is defined as follows. Given $(z,v) \in S\mathbb{H}$, there exists a Möbius transformation T such that $T(z) = i$ and $T'(z)v = i$. It takes the geodesic passing through z in the direction of v into the geodesic ie^t traversing the positive part of the imaginary axis. The geodesic flow $\varphi_t \colon S\mathbb{H} \to S\mathbb{H}$ is given by

$$\varphi_t(z,v) = (\gamma(t), \gamma'(t)),$$

where $\gamma(t) = T^{-1}(ie^t)$.

Exercise 1.6. Show that:

(1) $S\mathbb{H}$ can be identified with G;

(2) $\varphi_t(z,v) = (z,v)g_t$ where $g_t = \begin{pmatrix} e^t & 0 \\ 0 & e^{-t} \end{pmatrix}$ is a one-parameter subgroup;

(3) the Haar measure μ on G is invariant under the geodesic flow and is the volume on SH.

Theorem 1.7. *The geodesic flow on a surface of constant negative curvature is an Anosov flow.*

To explain this result, we describe the global stable manifold through a point $(z,v) \in S\mathbb{H}$. Consider the oriented semicircle $\gamma_{z,v}(t)$ in $\mathbb{H}$ centered at a point on the ideal boundary $\mathbb{H}(\infty)$, which passes through z in the direction of v.[4] Let $\gamma_{z,v}(+\infty) \in \mathbb{H}(\infty)$ be the positive endpoint of the circle. Consider now the collection of all oriented semicircles in $\mathbb{H}$ centered at points in $\mathbb{H}(\infty)$ whose positive endpoints coincide with $\gamma_{z,v}(+\infty)$.[5]

---

[4] That is, the geodesic through z in the direction of v.
[5] This collection also includes the line orthogonal to $\mathbb{H}(\infty)$ at $\gamma_{z,v}(+\infty)$.

1.2. Anosov flows

Exercise 1.8. Show that for any two circles γ_1 and γ_2 in this collection,

$$\rho(\gamma_1(t), \gamma_2(t)) \leq Ce^{\sqrt{-k}t} \tag{1.7}$$

for some $C > 0$ and all $t \geq 0$.

One can show that the circle $c(z, v)$ passing through the point z that is tangent to $\mathbb{H}(\infty)$ at the point $\gamma_{z,v}(+\infty)$ intersects each oriented semicircle in the collection orthogonally. It is called a *horocycle* through z (see Figure 1.5). It follows from (1.7) that the submanifold

$$W^s(z,v) = \big\{(z',v') \in S\mathbb{H} \colon z' \in c(z,v),$$
$$v' \text{ is orthogonal to } c(z,v) \text{ and points inward}\big\}$$

is the global stable leaf through (z, v) (see Figure 1.5).

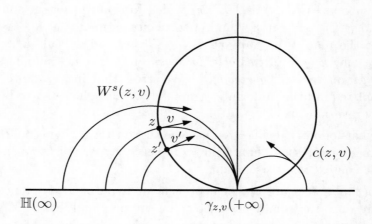

Figure 1.5. Horocycle $c(z,v)$ and global stable leaf $W^s(z,v)$.

A surface M of constant negative curvature is a factor-space $M = \mathbb{H}/\Gamma$ where Γ is a discrete subgroup of G, and any compact surface of genus at least 2 can be obtained in this way. We stress that depending on the choice of the discrete subgroup Γ the factor-space M can be a surface of finite area. The geodesic flow is now defined on the unit tangent bundle SM, and it preserves the Riemannian volume on SM.

The above construction of horocycles can be carried over to general surfaces of negative curvature and an estimate similar to (1.7) can be established. This allows one to extend Theorem 1.7 to surfaces of negative curvature and to study hyperbolic and ergodic properties of geodesic flows.

Theorem 1.9 ([5]). *Let M be a smooth surface of negative curvature $K(x)$ and assume that $K(x) \leq k$ where $k < 0$ is independent of x. The geodesic flow on M is an Anosov flow. Moreover, if the surface has finite area, then the geodesic flow is ergodic and indeed is a Bernoulli flow.*

We emphasize that the study of ergodic properties of geodesic flows on compact surfaces of nonpositive curvature—that lie on the boundary of the metrics of negative curvature—is substantially more complicated (see Chapter 10 where we describe the stable and unstable subspaces and foliations for the geodesic flow in this case). To see this, let us observe that the geodesic flow on the flat torus has only zero Lyapunov exponents.

1.3. Hyperbolic sets

We call an invariant subset $\Lambda \subset M$ *uniformly completely hyperbolic* for a diffeomorphism f if there are constants $c > 0$, $0 < \lambda < 1 < \mu$ and, for each $x \in \Lambda$, subspaces $E^s(x)$ and $E^u(x)$ which form an invariant splitting (see (1.1) and (1.2)) and satisfy the hyperbolicity estimates (1.3). We will also say that the map $f|\Lambda$ is *uniformly hyperbolic*, and we call $E^s(x)$ a *stable subspace* and $E^u(x)$ an *unstable subspace* of T_xM. The numbers $\{\lambda, \mu, c\}$ are the *parameters of hyperbolicity*. We stress that in the above definition we do not require the uniformly completely hyperbolic set Λ to be compact but only invariant.

In what follows we will often use the shorter term "hyperbolic" (instead of the longer expression "uniformly completely hyperbolic") when it is clear from the context what we mean.

Hyperbolic sets are classical objects in hyperbolic dynamics. In the literature they are usually assumed to be compact and invariant but hyperbolicity estimates may be written in somewhat different, albeit equivalent, forms. In our hyperbolicity estimates (1.3) the unstable subspace is viewed as stable for the inverse map. This allows us to study only stable subspaces, since results for unstable ones can be easily obtained by reversing the time. In particular, we do not assume that vectors in the unstable subspace expand with an exponential rate as time goes to $+\infty$ (however, this is indeed the case; see Exercise 1.11). While this may sound like a technical matter, it turns out to be crucial when we move to studying nonuniform hyperbolicity where no exponential expansion along unstable subspaces is expected. In addition, nonuniformly hyperbolic sets, while invariant, are usually not compact. Therefore, we discuss here, as a model case, uniformly hyperbolic sets which are invariant but not compact and with no a priori requirement on the action of the differential along unstable subspaces.

1.3. Hyperbolic sets

Theorem 1.10. *Let Λ be a hyperbolic set for a diffeomorphism f. The following statements hold:*

(1) *for $x \in \Lambda$ the invariant splitting (1.1) and (1.2) is uniquely defined; i.e., if $v \in T_x M$ is such that for all $n \geq 0$,*
$$\|d_x f^n v\| \leq \tilde{c}\tilde{\lambda}^n \|v\| \quad or \quad \|d_x f^{-n} v\| \leq \tilde{c}\tilde{\mu}^{-n} \|v\|$$
for some $\tilde{c} > 0$, $0 < \tilde{\lambda} < 1 < \tilde{\mu}$, then $v \in E^s(x)$ or, respectively, $v \in E^u(x)$;

(2) *the stable and unstable subspaces can be extended to the closure $\overline{\Lambda}$ of Λ so that the set $\overline{\Lambda}$ becomes uniformly completely hyperbolic with the same parameters of hyperbolicity, i.e., estimates (1.3) hold with the same constants c, λ, and μ;*

(3) *the stable and unstable subspaces vary continuously with x on $\overline{\Lambda}$.*

Proof. We prove the first statement. Fix $x \in \Lambda$ and let $v \in T_x M$ be such that for all $n \geq 0$
$$\|d_x f^n v\| \leq \tilde{c}\tilde{\lambda}^n \|v\| \tag{1.8}$$
for some $\tilde{c} > 0$ and $0 < \tilde{\lambda} < 1$. We wish to show that $v \in E^s(x)$. Otherwise, write $v = v^s + v^u$ where $v^s \in E^s(x)$, $v^u \in E^u(x)$, and $v^u \neq 0$. Using (1.2), for any $n \geq 0$ we have that $w = d_x f^n v^u \in E^u(f^n(x))$. By (1.3),
$$\|v^u\| = \|d_x f^{-n} w\| \leq c\mu^{-n}\|w\|$$
implying that
$$\|w\| \geq c^{-1}\mu^n \|v^u\| > 1,$$
for any sufficiently large integer n. On the other hand, since $d_x f^n v = d_x f^n v^s + d_x f^n v^u$, by (1.3) and (1.8),
$$\|w\| = \|d_x f^n v^u\| \leq \|d_x f^n v\| + \|d_x f^n v^s\| \leq c'(\lambda')^n(\|v\| + \|v^s\|) < 1,$$
for all sufficiently large n, leading to a contradiction; here $c' = \max\{c, \tilde{c}\}$ and $\lambda' = \max\{\lambda, \tilde{\lambda}\}$. Let now $v \in T_x M$ be such that for all $n \geq 0$
$$\|d_x f^{-n} v\| \leq c\mu^{-n}\|v\|.$$
Using an argument similar to the above, one can show that $v \in E^u(x)$. This completes the proof of the first statement.

To prove the second statement let $(x_m)_{m \geq 0}$ be a sequence of points in Λ converging to a point $x \in \overline{\Lambda}$. Since the Grassmannian manifold over $\overline{\Lambda}$ is compact, we can choose a subsequence m_k such that the sequence of subspaces $E^s(x_{m_k})$ converges to a subspace $E'(x) \subset T_x M$ and the sequence of subspaces $E^u(x_{m_k})$ converges to a subspace $E''(x) \subset T_x M$. Let us fix $n \geq 0$. Since the hyperbolicity estimates hold at time n for vectors in $E^s(x_{m_k})$ and $E^u(x_{m_k})$ for every m_k, they also hold at time n for vectors in

$E'(x)$ and $E''(x)$, and since n is arbitrary, we obtain that the hyperbolicity estimates hold for the subspaces $E'(x)$ and $E''(x)$.

Now let us fix $\ell \in \mathbb{Z}$ and consider the sequence of points $f^\ell(x_{m_k})$ which converges to the point $f^\ell(x)$. Since Λ is invariant, for every m_k we have subspaces $E^s(f^\ell(x_{m_k}))$ and $E^u(f^\ell(x_{m_k}))$ which clearly converge to some subspaces which we denote by $E'(f^\ell(x))$ and $E''(f^\ell(x))$, respectively. We have that

$$d_x f^\ell(E'(x)) = E'(f^\ell(x)) \quad \text{and} \quad d_x f^\ell(E''(x)) = E''(f^\ell(x)).$$

Observe that

$$\dim E^s(x_{m_k}) = \dim E'(x) =: a, \quad \dim E^u(x_{m_k}) = \dim E''(x) =: b$$

and that $a + b = \dim M$. It remains to show that $E'(x) \cap E''(x) = \{0\}$. Assuming the contrary, we obtain a nonzero vector $v \in E'(x) \cap E''(x)$. For any $n > 0$ write $v = d_x f^n w$ where $w = d_x f^{-n} v \in E'(f^n(x)) \cap E''(f^n(x))$. Since $v \in E'(x)$ and $w \in E''(f^n(x))$, the hyperbolicity estimates imply that for sufficiently large n,

$$\|v\| \le c\lambda^n \|w\| \le c^2 \lambda^n \mu^{-n} \|v\| < \|v\|,$$

leading to a contradiction. In particular, we have that $E'(x) = E^s(x)$ and $E''(x) = E^u(x)$ for every $x \in \overline{\Lambda}$. The second statement now follows. In fact, the above argument and statement (1) imply continuous dependence of the stable and unstable subspaces on $x \in \overline{\Lambda}$. This completes the proof of the theorem. $\square$

Exercise 1.11. Show that for every $x \in \Lambda$ the norm $\|d_x f^n v\|$ grows exponentially with $n \to +\infty$ uniformly over $v \in E^u(x)$, $\|v\| = 1$ and similarly, the norm $\|d_x f^n v\|$ grows exponentially with $n \to -\infty$ uniformly over $v \in E^s(x)$, $\|v\| = 1$. *Hint:* Given $a > 0$, show first that there is $N = N(x, a) > 0$ such that $\|d_x f^n v\| > a$ for every $n > N$ and $v \in E^u(x)$, $\|v\| = 1$; then use continuity of unstable subspaces to show that the number $N(x, a)$ can be chosen to be locally constant on x; finally, apply a similar argument to prove the statement for stable subspaces.

Remark 1.12. Statement (1) of Theorem 1.10 uses invariance of the hyperbolic set Λ in an essential way and may not be true otherwise. This is due to a phenomenon where one can find a set $\Lambda \subset M$ with the property that for every $x \in \Lambda$ there is a direction in $T_x M$ along which the differential of the map acts as an exponential contraction in both positive and negative time (see, for example, [22] and [49] where this phenomenon is observed and discussed). More precisely, consider a diffeomorphism f of a three-dimensional

1.3. Hyperbolic sets

smooth Riemannian manifold M with the following properties:

(1) f has two hyperbolic fixed points p and q for which $\dim E^u(p) = \dim E^s(q) = 2$ and $\dim E^s(p) = \dim E^u(q) = 1$;

(2) the two-dimensional unstable separatrix $W^u(p)$ and the two-dimensional stable separatrix $W^s(q)$ intersect at a point z transversally, that is, the planes $T_z W^u(p)$ and $T_z W^s(q)$ intersect along a line $L \subset T_z M$.

Consider the basis $\{v_1, v_2, v_3\}$ of $T_z M$ where $v_1 \in T_z W^u(p)$, $v_2 \in L$, and $v_3 \in T_z W^s(q)$. Let L_1 and L_3 be lines through vectors v_1 and v_3, respectively. We have two hyperbolic splittings at z: $T_z M = T_z W^u(p) \bigoplus L_3$ and $T_z M = T_z W^s(q) \bigoplus L_1$. Hence, for the singleton set $\Lambda = \{z\}$ statement (1) of Theorem 1.10 fails.

We emphasize that if the set Λ is not invariant, one can modify the hyperbolicity estimates (1.3) to ensure that the hyperbolicity is still uniform, the unstable subspace is still defined as being stable for the inverse map, but that the splitting (1.1) is unique and the stable and unstable subspaces depend continuously on the point. We introduce and study these conditions in Section 5.3; see the estimates (5.25)–(5.27) and Theorem 5.10.

Given numbers λ', μ' satisfying $0 < \lambda < \lambda' < 1 < \mu' < \mu$, one can replace the standard Riemannian metric $\langle \cdot, \cdot \rangle$ on Λ with another metric $\langle \cdot, \cdot \rangle'$ called the *Lyapunov metric*. It is defined in the following way. Given $x \in \Lambda$ and $v, w \in T_x M$, write $v = v^s + v^u$ and $w = w^s + w^u$ where $v^s, w^s \in E^s(x)$ and $v^u, w^u \in E^u(x)$. Then set

$$\langle v^s, w^s \rangle'_x = \sum_{m \geq 0} \langle d_x f^m v^s, d_x f^m w^s \rangle_{f^m(x)} (\lambda')^{-2m},$$

$$\langle v^u, w^u \rangle'_x = \sum_{m \geq 0} \langle d_x f^{-m} v^u, d_x f^{-m} w^u \rangle_{f^{-m}(x)} (\mu')^{2m}, \qquad (1.9)$$

$$\langle v, w \rangle'_x = \langle v^s, w^s \rangle'_x + \langle v^u, w^u \rangle'_x.$$

Exercise 1.13. Show that:

(1) the formulae (1.9) determine a scalar product in $T_x M$;

(2) the subspaces $E^s(x)$ and $E^u(x)$ are orthogonal in the Lyapunov metric.

In Sections 4.3.1 and 7.1 we show how to construct the Lyapunov metric in more general settings of linear cocycles with nonzero Lyapunov exponents over measurable transformations and nonuniformly hyperbolic diffeomorphisms, respectively.

Theorem 1.14. *Let Λ be a hyperbolic set for a diffeomorphism f. The following statements hold:*

(1) *the Lyapunov metric $\langle \cdot, \cdot \rangle'_x$ varies continuously with $x \in \Lambda$;*

(2) *with respect to the Lyapunov metric $\langle \cdot, \cdot \rangle'_x$ the hyperbolicity estimates hold with $c = 1$ and the rates λ, μ replaced with λ', μ', respectively, so that the parameters of hyperbolicity are $\{\lambda', \mu'\}$; more precisely,*

$$\|d_x f v\|'_{f(x)} \leq \lambda' \|v\|'_x \quad \text{for any } v \in E^s(x),$$
$$\|d_x f^{-1} v\|'_{f^{-1}(x)} \leq (\mu')^{-1} \|v\|'_x \quad \text{for any } v \in E^u(x),$$
(1.10)

where $\|\cdot\|'_x$ is the norm generated by the Lyapunov metric;

(3) *the norms $\|\cdot\|_x$ and $\|\cdot\|'_x$ are equivalent, that is, there exists $C > 0$ independent of $x \in \Lambda$ such that*

$$C^{-1} \|v\|_x \leq \|v\|'_x \leq C \|v\|_x \quad \text{for all } v \in T_x M.$$

Proof. The continuity of the Lyapunov metric follows readily from the convergence of the series in (1.9).

For the second statement take $x \in \Lambda$ and $v \in E^s(x)$, and note that

$$(\|d_x f v\|'_{f(x)})^2 = \sum_{m \geq 0} (\|d_x f^{m+1} v\|_{f^{m+1}(x)})^2 (\lambda')^{-2m}$$
$$\leq \sum_{m \geq -1} (\|d_x f^{m+1} v\|_{f^{m+1}(x)})^2 (\lambda')^{-2m}$$
$$= \sum_{k \geq 0} (\|d_x f^k v\|_{f^k(x)})^2 (\lambda')^{-2k} (\lambda')^2 = (\lambda')^2 (\|v\|'_x)^2.$$

This gives the first inequality in (1.10). The second inequality can be obtained by a similar argument.

For property (3) take $x \in \Lambda$ and $v \in T_x M$, and write $v = v^s + v^u$ where $v^s \in E^s(x)$ and $v^u \in E^u(x)$. It follows readily from (1.9) that

$$(\|v\|'_x)^2 = (\|v^s\|'_x)^2 + (\|v^u\|'_x)^2 \geq (\|v^s\|_x)^2 + (\|v^u\|_x)^2$$
$$\geq \frac{1}{2} (\|v^s\|_x + \|v^u\|_x)^2 \geq \frac{1}{2} (\|v\|_x)^2$$

and so $\|v\|'_x \geq \|v\|_x / \sqrt{2}$. Furthermore,

$$(\|v^s\|'_x)^2 \leq \sum_{m \geq 0} c^2 \lambda^{2m} (\lambda')^{-2m} \|v^s\|_x^2 = \frac{c^2}{1 - (\lambda/\lambda')^2} \|v^s\|_x^2$$

and similarly,

$$(\|v^u\|'_x)^2 \leq \frac{c^2}{1 - (\mu'/\mu)^2} \|v^u\|_x^2.$$

1.3. Hyperbolic sets

Therefore,
$$(\|v\|'_x)^2 \le d(\|v^s\|_x^2 + \|v^u\|_x^2),$$
where
$$d = \frac{c^2}{1-(\lambda/\lambda')^2} + \frac{c^2}{1-(\mu'/\mu)^2}.$$
Since
$$\|v\|_x^2 = \|v^s\|_x^2 + \|v^u\|_x^2 + 2\|v^s\|_x\|v^u\|_x \cos\angle(E^s(x), E^u(x))$$
$$\ge \|v^s\|_x^2 + \|v^u\|_x^2,$$
we obtain
$$\frac{1}{\sqrt{2}}\|v\|_x \le \|v\|'_x \le \sqrt{d}\|v\|_x$$
and the two norms are indeed equivalent. $\square$

Given a point x in a hyperbolic set Λ, one can construct local stable $V^s(x)$ and unstable $V^u(x)$ manifolds at x which are tangent to stable and unstable subspaces at x and are given by (1.4). Similarly to the Anosov case, one can then construct global stable $W^s(x)$ and unstable $W^u(x)$ manifolds at x.

A hyperbolic set Λ is called *locally maximal* if there is a neighborhood U of Λ such that if $\Lambda' \subset U$ is an invariant subset, then $\Lambda' \subset \Lambda$. One can show that a hyperbolic set Λ is locally maximal if any of the following conditions hold:

(1) one has $\Lambda = \bigcap_{n \in \mathbb{Z}} f^n(U)$;
(2) there is $\delta > 0$ such that for any two points $x, y \in \Lambda$ with $\rho(x,y) \le \delta$ one has that the local stable manifold $V^s(x)$ and the unstable manifold $V^u(y)$ intersect transversally at a single point $z \in \Lambda$.

For an Anosov diffeomorphism the entire manifold M is a hyperbolic (and, clearly, locally maximal) set.

A locally maximal hyperbolic set Λ is called a *horseshoe* if it is totally disconnected and is such that $f|\Lambda$ is topologically transitive. Horseshoes appear naturally in the following context. Let f be a diffeomorphism of a smooth manifold M and let p and q be two hyperbolic periodic points for f of periods $m(p)$ and $m(q)$, respectively. We say that p and q are *homoclinically related* and write $p \sim q$ if there are $0 \le m < m(p)$ and $0 \le k < m(q)$ such that the global stable (respectively, unstable) manifold at $f^m(p)$ and the global unstable (respectively, stable) manifold at $f^k(q)$ intersect transversally. If $p \sim q$, then the *index* $s(p) = s(q)$ where $s(p)$ is the dimension of the stable manifold at p.

Theorem 1.15 (Smale–Birkhoff Homoclinic Theorem, [107]). *There exists a horseshoe that contains p and q.*

Remark 1.16. The closure H_p of the set of periodic points which are homoclinically related to p is called the *homoclinic class* of p. This is a compact invariant set and it is easy to show that the restriction $f|H_p$ is topologically transitive. This is why homoclinic classes are often viewed in dynamics as "building blocks". They were introduced by Newhouse in [**84**]. A basic hyperbolic set of an Axiom A diffeomorphism[6] gives the simplest example of a homoclinic class, but in general the set H_p can have a much more complicated structure. In particular, it can contain nonhyperbolic periodic points and support nonhyperbolic invariant measures. Moreover, it can contain hyperbolic periodic orbits whose index is different from the index of p as well as hyperbolic periodic points of the same index as p that are not homoclinically related to p; see [**22, 23**] and also Remark 9.8 for a related notion of ergodic homoclinic classes.

1.4. The Smale–Williams solenoid

A compact set $\Lambda \subset X$ is called an *attractor* for f if there exists a neighborhood U of Λ such that $\overline{f(U)} \subset U$ and

$$\Lambda = \bigcap_{n \in \mathbb{N}} f^n(U).$$

The set U is called a *trapping region*. We say that a set Λ is a *hyperbolic attractor* if it is an attractor and also a hyperbolic set.

Exercise 1.17. Show that for any $x \in \Lambda$, we have $V^u(x) \subset \Lambda$ (and hence, $W^u(x) \subset \Lambda$).

A classical example of a hyperbolic attractor is the Smale–Williams solenoid introduced by Smale and Williams (see [**107, 109, 110**]). Consider the solid torus

$$M = S^1 \times \{(x,y) \in \mathbb{R}^2 : x^2 + y^2 \leq 1\}$$

and the map $f \colon M \to M$ defined by

$$f(\theta, x, y) = \bigl(2\theta, \lambda_1 x + a\cos(2\pi\theta), \lambda_2 y + a\sin(2\pi\theta)\bigr) \quad (1.11)$$

for some constants $0 < a < \tfrac{1}{2}$ and $\lambda_1, \lambda_2 \in (0, \min\{a, 1-a\})$.

Exercise 1.18. Using the conditions on a, λ_1, and λ_2 show that $f(M) \subset M$. Also show that the map f is injective and is a diffeomorphism.

The compact f-invariant set $\Lambda = \bigcap_{n \in \mathbb{N}} f^n(M)$ is called a *solenoid* (see Figure 1.6). It is easy to see that $\Lambda = \bigcap_{n \in \mathbb{N}} f^n(\operatorname{int} M)$ implying that Λ is

---

[6] A diffeomorphism $f \colon M \to M$ is said to satisfy *Axiom A* if the set of nonwandering points $NW(f)$ (a point $x \in M$ is said to be *nonwandering* if for any open neighborhood U of x there exists $n \in \mathbb{N}$ such that $f^n(U) \cap U \neq \varnothing$) is hyperbolic and the periodic points are dense in $NW(f)$.

1.4. The Smale–Williams solenoid

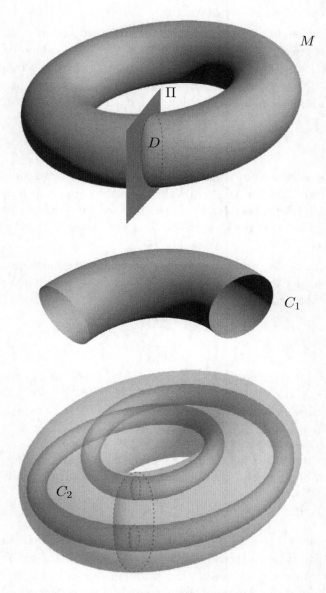

Figure 1.6. The three figures above demonstrate the first step in constructing the Smale–Williams solenoid: (1) a solid torus M is cut by a transverse plane Π along a disk D; (2) M is unfolded into a cylinder C_1; (3) C_1 is extended by 2 and contracted by λ_1, λ_2 in the transverse directions and then is inserted back into M by wrapping around without self-intersections.

an attractor for f. It is also easy to see that Λ is a hyperbolic set for f with $E^s(x)$ being the plane through x which is orthogonal to the solid torus and $E^u(x)$ being the line through x which is orthogonal to this plane. Hence, Λ is a uniformly hyperbolic attractor.

We will construct a special invariant measure which is supported on a hyperbolic attractor Λ (e.g., on the Smale–Williams solenoid). To this end consider a $C^{1+\alpha}$ diffeomorphism f of a compact smooth manifold M and denote by $\mathcal{M}(f, M)$ the space of all probability measures on M which are invariant under f. Consider the distance $d_\mathcal{M}$ in $\mathcal{M}(f, M)$ given by

$$d_\mathcal{M}(\mu, \nu) = \sum_{k=1}^{\infty} \frac{1}{2^k} \left| \int_M \psi_k \, d\mu - \int_M \psi_k \, d\nu \right|,$$

where $\{\psi_k\}_{k \in \mathbb{N}}$ is a dense subset in the unit ball in the space $C^0(M)$ of all continuous functions on M. While the distance defined in this way depends on the choice of the subset $\{\psi_k\}_{k \in \mathbb{N}}$, the topology it generates does not. This topology is called the *weak*-topology*, and the space $\mathcal{M}(f, M)$ is compact in this topology.

Let now m be the normalized volume in the trapping region U. Consider the evolution of m under the dynamics that is the sequence of averages of the push-forward measures

$$\mu_n = \sum_{k=1}^{n-1} f_*^k m. \tag{1.12}$$

Theorem 1.19. *Assume that the map $f|\Lambda$ is topologically transitive. Then the sequence of measures μ_n converges in the weak*-topology to a measure μ which is supported on Λ and is invariant under f; it is ergodic (in fact, has the Bernoulli property).*

The measure μ in Theorem 1.19 is known as *Sinai–Ruelle–Bowen (SRB) measure*; see Section 11.1 for the definition and a proof of this theorem.

1.5. The Katok map of the 2-torus

In this section we describe a construction due to Katok [61] of an area-preserving C^∞ diffeomorphism on the two-dimensional torus $\mathbb{T}^2$ with nonzero Lyapunov exponents. The starting point of our constructions is an area-preserving hyperbolic automorphism, which is then perturbed. The perturbation is not small (otherwise it would be an Anosov diffeomorphism) but is localized in a small neighborhood of a hyperbolic fixed point, say x, and it slows down trajectories, so that the time a given trajectory spends in this neighborhood of x increases. The Poincaré Recurrence Theorem ensures that for almost every trajectory (call them "good") the average increase in time is not significant to affect its hyperbolicity although the expansion and contraction rates get weaker. However, for some "bad" trajectories (which form a set of zero area) the average increase in time is abnormally high and hyperbolicity gets destroyed. Along the way of our constructions we introduce one of our main concepts—the Lyapunov exponents—which measures the asymptotic contraction and expansion rates and distinguishes the "good" trajectories from the "bad" ones—at least some of the values of the Lyapunov exponents for the former are nonzero while all the values of the Lyapunov exponents for the latter are zero.

1.5.1. The slow-down procedure. To effect our construction, consider the matrix $A = \begin{pmatrix} 2 & 1 \\ 1 & 1 \end{pmatrix}$, which induces a linear transformation T of the two-dimensional torus $\mathbb{T}^2 = \mathbb{R}^2/\mathbb{Z}^2$. We shall obtain the desired map by slowing down T near the origin. This construction depends upon a real-valued function ψ, called a *slow-down function*, which is defined on the unit interval $[0, 1]$ and has the following properties:

(1) ψ is a C^∞ function except at the origin;

(2) $\psi(0) = 0$ and $\psi(u) = 1$ for $u \geq r_0$, for some $0 < r_0 < 1$;

(3) $\psi'(u) > 0$ for every $0 < u < r_0$;

(4) the following integral converges:
$$\int_0^1 \frac{du}{\psi(u)} < \infty.$$

The reader can think of the function ψ, which satisfies conditions (1)–(4), as being given near the origin by $\psi_\alpha(u) = u^\alpha$ for some $0 < \alpha < 1$. However, in what follows we will need the function ψ to be "infinitely vertically flat" at the origin.

Let $\lambda = (3+\sqrt{5})/2 > 1$ and $\lambda^{-1} = (3-\sqrt{5})/2 < 1$ be the eigenvalues of the matrix A. In what follows we use the coordinate system (s_1, s_2) centered at zero and obtained from the orthogonal eigendirections of A corresponding to the eigenvalues λ and λ^{-1}. Denote by D_r the closed disk centered at 0 of small radius r, that is,
$$D_r = \{(s_1, s_2) : {s_1}^2 + {s_2}^2 \leq r^2\}.$$
Let us choose numbers $r_2 > r_1 > r_0$ such that
$$T(D_{r_1}) \cup T^{-1}(D_{r_1}) \subset D_{r_2}, \quad D_{r_0} \subset \operatorname{int} T(D_{r_1}). \tag{1.13}$$
Consider the system of differential equations in D_{r_1}
$$\dot{s}_1 = s_1 \log \lambda, \quad \dot{s}_2 = -s_2 \log \lambda. \tag{1.14}$$

Exercise 1.20. Show that the map T is the time-one map of the flow generated by the system of equations (1.14).

We now perturb the system (1.14) to obtain the following system of differential equations in D_{r_2}:

$$\dot{s}_1 = s_1\psi(s_1{}^2 + s_2{}^2)\log\lambda, \quad \dot{s}_2 = -s_2\psi(s_1{}^2 + s_2{}^2)\log\lambda. \qquad (1.15)$$

Let V_ψ be the vector field generated by this system of equations and let g_t be the corresponding local flow. Denote by g the time-one map of the flow. Our choice of the function ψ and numbers r_0, r_1, and r_2 guarantees that the domain of definition of the map g contains the disk D_{r_1} and that $g(D_{r_1}) \subset D_{r_2}$. Furthermore, g is of class C^∞ in $D_{r_1} \setminus \{0\}$ and, in view of (1.13), it coincides with T in some neighborhood of the boundary. Therefore, the map

$$G(x) = \begin{cases} T(x) & \text{if } x \in \mathbb{T}^2 \setminus D_{r_1}, \\ g(x) & \text{if } x \in D_{r_1} \end{cases}$$

defines a homeomorphism of the torus $\mathbb{T}^2$, which is a C^∞ diffeomorphism everywhere except at the origin. It is the map $G(x)$ that is a slow-down of the automorphism T at 0; see Figure 1.7.

The map G preserves the probability measure $d\nu = \kappa_0^{-1}\kappa\,dm$ where m is area and the density κ is a positive C^∞ function that is infinite at 0. It is defined by the formula

$$\kappa(s_1, s_2) = \begin{cases} (\psi(s_1{}^2 + s_2{}^2))^{-1} & \text{if } (s_1, s_2) \in D_{r_1}, \\ 1 & \text{otherwise} \end{cases} \qquad (1.16)$$

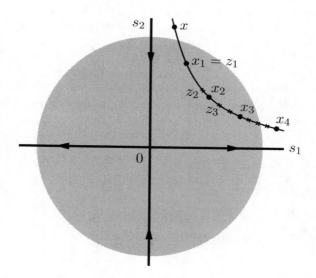

Figure 1.7. The "round" dots show trajectories $x_k = T^k(x)$ of the map T while the "cross" dots show trajectories $z_k = g_1^k(x)$ of the map g_1.

1.5. The Katok map of the 2-torus

and
$$\kappa_0 = \int_{\mathbb{T}^2} \kappa \, dm.$$

Exercise 1.21. Show that the measure ν is G-invariant by verifying that $\operatorname{div}(\kappa v) = 0$ where v is the vector field generated in D_{r_2} by the system of differential equations (1.15).

1.5.2. The Katok map. Our next step is to further perturb the map G so that the new map $G_{\mathbb{T}^2}$ is an area-preserving C^∞ diffeomorphism. We will achieve this by changing the coordinate system in $\mathbb{T}^2$ using a map φ so that $G_{\mathbb{T}^2} = \varphi \circ G \circ \varphi^{-1}$. We define φ in D_{r_1} by the formula

$$\varphi(s_1, s_2) = \frac{1}{\sqrt{\kappa_0(s_1{}^2 + s_2{}^2)}} \left(\int_0^{s_1{}^2 + s_2{}^2} \frac{du}{\psi(u)} \right)^{1/2} (s_1, s_2) \qquad (1.17)$$

and we set $\varphi = \operatorname{Id}$ in $\mathbb{T}^2 \setminus D_{r_1}$.

Exercise 1.22. Show that:

(1) φ is a homeomorphism and is a C^∞ diffeomorphism outside the origin;

(2) $\varphi_* \nu = m$, that is, φ transfers the measure ν into the area m.

Hint: To prove statement (2) first verify that the map $G_{\mathbb{T}^2}$ preserves the probability measure $\varphi_* \nu$ given by $(\varphi_* \nu)(B) = \nu(\varphi^{-1} B)$ where B is a Borel subset of $\mathbb{T}^2$. To show that $\varphi_* \nu = m$ observe that

$$(\varphi_* \nu)(B) = \int_{\varphi^{-1} B} \kappa_0^{-1} \kappa \, dm = \int_B \kappa_0^{-1} (\kappa \circ \varphi^{-1}) |\det d(\varphi^{-1})| \, dm$$

and the desired result would follow if we show that

$$\det d_{(s_1,s_2)} \varphi = \kappa_0^{-1} \kappa(s_1, s_2). \qquad (1.18)$$

To this end write $\varphi(s_1, s_2) = \kappa_0^{-1/2} f(r)(s_1/r, s_2/r)$ where $r = \sqrt{s_1{}^2 + s_2{}^2}$ and $f(r) = \left(\int_0^{r^2} \frac{du}{\psi(u)} \right)^{1/2}$ and then show that

$$d_{(s_1,s_2)} \varphi = \kappa_0^{-1/2} \begin{pmatrix} f'(r)\frac{s_1{}^2}{r^2} + f(r)\frac{r - s_1{}^2/r}{r^2} & f'(r)\frac{s_1 s_2}{r^2} - f(r)\frac{s_1 s_2}{r^3} \\ f'(r)\frac{s_1 s_2}{r^2} - f(r)\frac{s_1 s_2}{r^3} & f'(r)\frac{s_2{}^2}{r^2} + f(r)\frac{r - s_2{}^2/r}{r^2} \end{pmatrix}.$$

Finally, observe that $(f(r)^2)' = 2r/\psi(r^2)$ implying that $f(r)f'(r) = r/\psi(r^2)$, compute $\det d_{(s_1,s_2)} \varphi$, and conclude the equality (1.18).

The behavior of the orbits of $G_{\mathbb{T}^2}$ is sketched in Figure 1.8.

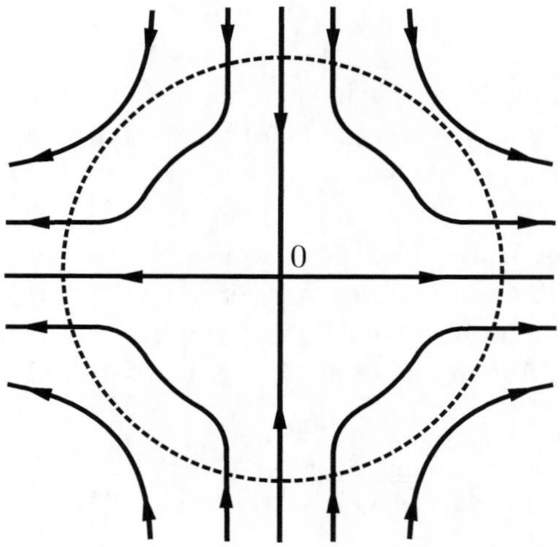

Figure 1.8. Invariant curves for $G_{\mathbb{T}^2}$ near 0.

We now show that $G_{\mathbb{T}^2}$ is infinitely many times differentiable at the origin and hence is a C^∞ diffeomorphism. Observe that the system (1.14) is *Hamiltonian with respect to area m* with the Hamiltonian function $H(s_1, s_2) = s_1 s_2 \log \lambda$. This means (we exploit the fact that the system is two-dimensional) that for every continuous vector field v on D_{r_1} we have

$$\Omega(v, V_H) = dHv,$$

where V_H is the vector field generated by the system (1.14) and Ω is the volume 2-form, i.e., $\Omega = ds_1 \wedge ds_2$. The vector field V_ψ, generated by the system (1.15), is obtained from the vector field V_H by a time change, so it is Hamiltonian with respect to the measure ν with the same Hamiltonian function H, i.e., $\Omega_\psi(v, V_\psi) = dHv$ where $\Omega_\psi = \kappa_0^{-1} \kappa \, ds_1 \wedge ds_2$. Near the origin the map $G_{\mathbb{T}^2}$ is the time-one map generated by the vector field $\varphi_* V_\psi$. Since $\varphi_* \nu = m$, this vector field is Hamiltonian with respect to the area m with the Hamiltonian function

$$H_1(s_1, s_2) = (H \circ \varphi^{-1})(s_1, s_2) = \frac{s_1 s_2 \beta(\sqrt{s_1^2 + s_2^2}) \log \lambda}{s_1^2 + s_2^2},$$

where $\beta(u)$ is the inverse of the strictly increasing function $\gamma(u)$ given by

$$\gamma(u) = \kappa_0^{-1/2} \sqrt{\int_0^u \frac{d\tau}{\psi(\tau)}}. \tag{1.19}$$

Given a C^∞ function β, one can find the function ψ using the relation (1.19).

Exercise 1.23. Show that one can choose the function β such that the function ψ satisfies conditions (1)–(4) in the beginning of this section. *Hint:* Near zero the derivative of β must decrease sufficiently fast.

We call the map $G_{\mathbb{T}^2}$ the *Katok map*.

1.5.3. Hyperbolic structure of the Katok map. We examine the hyperbolic properties of the map $G_{\mathbb{T}^2}$. Since $G_{\mathbb{T}^2} = \varphi \circ G \circ \varphi^{-1}$, it is sufficient to do this for the map G. In particular, we shall show that G is not an Anosov diffeomorphism.

Let $x = (0, s_2) \in D_{r_1}$ be a point on the vertical segment through the origin. Note that
$$g_t(x) = (0, s_2(t)) \to 0 \quad \text{as} \quad t \to \infty,$$
where $s_2(t)$ is the solution of (1.15) with the initial condition $s_2(0) = s_2$. In view of the choice of the function ψ, we obtain that
$$\lim_{t \to +\infty} \frac{\log \rho(g_t(x), 0)}{t} = \lim_{t \to +\infty} \frac{\log |s_2(t)|}{t}$$
$$= \lim_{t \to +\infty} (\log |s_2(t)|)'$$
$$= \lim_{t \to +\infty} (-\psi(s_2(t)^2) \log \lambda) = 0.$$
This implies that
$$\lim_{n \to +\infty} \frac{\log \rho(G^n(x), 0)}{n} = 0.$$
Similar arguments apply to points $x = (s_1, 0) \in D_{r_1}$ on the horizontal segment through the origin, showing that
$$\lim_{n \to +\infty} \frac{\log \rho(G^{-n}(x), 0)}{n} = 0.$$
Thus G is not an Anosov diffeomorphism since otherwise the above limits would be negative.

Exercise 1.24. Show that $d_0 G = \text{Id}$ where Id is the identity map.

For a typical trajectory, the situation is, however, quite different. To study it, we exploit the *cone techniques*—a collection of results aimed at establishing hyperbolic properties of the system. In Chapter 12 the reader can find a detailed exposition of these techniques.

Choose $x \in \mathbb{T}^2$ and define the *stable* and *unstable cones* in $T_x \mathbb{T}^2 = \mathbb{R}^2$ by
$$C^s(x) = \{(v_1, v_2) \in \mathbb{R}^2 : |v_1| \leq |v_2|\},$$
$$C^u(x) = \{(v_1, v_2) \in \mathbb{R}^2 : |v_2| \leq |v_1|\},$$
where v_1 and v_2 are vectors in the eigenlines at x corresponding to λ and λ^{-1}.

Lemma 1.25. *For every $x \neq 0$ we have that*
$$d_x G^{-1} C^s(x) \subset C^s(G^{-1}(x)), \quad d_x G C^u(x) \subset C^u(G(x))$$
and the inclusions are strict.

Proof. We only consider the case of the unstable cones $C^u(x)$ as the other case is completely similar. Clearly, the desired inclusion is true for every x outside the disk D_{r_1}. To establish this property inside the disk, we consider the system of variational equations corresponding to the system (1.15). Recall that given a system of differential equations in $\mathbb{R}^n$,
$$\dot{\mathbf{s}} = F(\mathbf{s}),$$
where $\mathbf{s} = (s_1, \ldots, s_n)$ and $F(\mathbf{s}) = (F_1(s_1, \ldots, s_n), \ldots, F_n(s_1, \ldots, s_n))$, the *system of variational equations* is defined by
$$\dot{\xi} = d_{\mathbf{s}} F \xi,$$
where the (i,j)-entry of the matrix $d_{\mathbf{s}} F$ is $\frac{\partial F_i}{\partial s_j}$. In our case the functions $F_1(s_1, s_2)$ and $F_2(s_1, s_2)$ are given by the right-hand sides of equations (1.15), and hence, the corresponding system of variational equations is
$$\dot{\xi}_1 = \log \lambda((\psi + 2s_1^2 \psi')\xi_1 + 2s_1 s_2 \psi' \xi_2),$$
$$\dot{\xi}_2 = -\log \lambda(2s_1 s_2 \psi' \xi_1 + (\psi + 2s_2^2 \psi')\xi_2).$$
This yields the following equation for the tangent $\eta = \xi_2/\xi_1$:
$$\frac{d\eta}{dt} = -2\log \lambda((\psi + (s_1^2 + s_2^2)\psi')\eta + s_1 s_2 \psi'(\eta^2 + 1)). \quad (1.20)$$
Substituting $\eta = 1$ and $\eta = -1$ in (1.20) gives, respectively,
$$\begin{aligned} \frac{d\eta}{dt} &= -2\log \lambda(\psi + (s_1 + s_2)^2 \psi') \leq 0, \\ \frac{d\eta}{dt} &= 2\log \lambda(\psi + (s_1 - s_2)^2 \psi') \geq 0. \end{aligned} \quad (1.21)$$
Moreover, these inequalities are strict everywhere except at the origin. The desired result follows. $\square$

We define for every $x \neq 0$,
$$E^s(x) = \bigcap_{j=0}^{\infty} d_{G^j(x)} G^{-j} C^s(G^j(x)), \quad E^u(x) = \bigcap_{j=0}^{\infty} d_{G^{-j}(x)} G^j C^u(G^{-j}(x)).$$

Lemma 1.26. $E^s(x)$ *and* $E^u(x)$ *are one-dimensional subspaces of* $T_x \mathbb{T}^2$.

Proof. We only prove the result for $E^u(x)$ as the argument for $E^s(x)$ is quite similar. To show that $E^u(x)$ is one-dimensional, we shall estimate

1.5. The Katok map of the 2-torus

from below the decrease of the angle between two arbitrary lines inside the cone $C^u(x)$ along the whole segment of trajectory of the local flow g_t for $x \in D_{r_2}$.

Let $\eta(t, \alpha_0) = \eta_{s_1^0, s_2^0}(t, \alpha_0)$ be the solution of the differential equation (1.20) with the initial condition $t = 0$ and $\alpha = \alpha_0$ along the trajectory of the flow through the point $(s_1^0, s_2^0) \in D_{r_2}$. We wish to estimate from above the ratio

$$\frac{|\eta(1, \alpha_1) - \eta(1, \alpha_2)|}{|\alpha_1 - \alpha_2|}$$

for all initial conditions α_1, α_2 such that $|\alpha_1| \leq 1$, $|\alpha_2| \leq 1$. To do so, we introduce the function

$$\hat{\eta}(t) = \hat{\eta}_{s_1^0, s_2^0}(t) = \exp\left(-2 \log \lambda \int_0^t (\psi + (s_1^2 + s_2^2)\psi') \, du\right). \quad (1.22)$$

Exercise 1.27. Use equations (1.20) and (1.22) to show that

$$\eta(t, \alpha) = \alpha \hat{\eta}(t) - 2(\log \lambda)\hat{\eta}(t) \int_0^t \frac{s_1 s_2 \psi'(\eta^2(u, \alpha) + 1)}{\hat{\eta}(u)} \, du. \quad (1.23)$$

It follows from (1.21) that the requirement $|\alpha| \leq 1$ implies that $|\eta(t, \alpha)| \leq 1$ for every t and hence by (1.23), we have that

$$\frac{|\eta(t, \alpha_1) - \eta(t, \alpha_2)|}{\hat{\eta}(t)} \leq |\alpha_1 - \alpha_2| + 4 \log \lambda \int_0^t \frac{|s_1 s_2|\psi'|\eta(u, \alpha_1) - \eta(u, \alpha_2)|}{\hat{\eta}(u)} \, du.$$

By the Gronwall inequality,

$$\frac{|\eta(t, \alpha_1) - \eta(t, \alpha_2)|}{\hat{\eta}(t)} \leq |\alpha_1 - \alpha_2| \exp\left(4 \log \lambda \int_0^t |s_1 s_2|\psi' \, du\right).$$

Substituting $t = 1$ and using (1.22), we obtain that

$$|\eta(1, \alpha_1) - \eta(1, \alpha_2)| \leq |\alpha_1 - \alpha_2| \exp\left(-2 \log \lambda \int_0^1 (\psi + \psi'(|s_1| - |s_2|)^2) \, du\right)$$
$$\leq |\alpha_1 - \alpha_2|. \quad (1.24)$$

Let us fix positive numbers a, b and consider the region

$$Q_{\varepsilon_0}^b = \{(s_1, s_2) : |s_1 s_2| \leq \varepsilon_0, |s_1| \leq b, |s_2| \leq b\} \subset D_{r_2}.$$

We wish to estimate the solution of the differential equation (1.20) along the segment of hyperbola

$$\{s_1 s_2 = \varepsilon, \, 0 \leq s_1 \leq b, \, 0 \leq s_2 \leq b\}$$

for all $0 < \varepsilon \leq a$. We consider s_1 as a parameter on the hyperbola. equations (1.15) and (1.20) imply that

$$\frac{d\eta}{ds_1} = -\left(\frac{2}{s_1} + \frac{2(s_1^2 + s_2^2)}{s_1} \frac{\psi'}{\psi}\right)\eta + 2s_2 \frac{\psi'}{\psi}(\eta^2 + 1). \quad (1.25)$$

For $i = 1, 2$ let $\eta(s_1, \varepsilon/b, \eta_i)$ be the solutions of this differential equation with the initial conditions $s_1 = \varepsilon/b$, $\eta = \eta_i$ for $|\eta_i| < 1$. Using equation (1.25), we obtain an equation for the difference $\eta(s_1, \varepsilon/b, \eta_1) - \eta(s_1, \varepsilon/b, \eta_2)$. Then the Gronwall inequality and the requirement that $|\eta(s_1, \varepsilon/b, \eta_i)| < 1$ imply that for every η_1, η_2, $|\eta_1| \leq 1$, $|\eta_2| \leq 1$ we have

$$\left|\eta\left(b, \frac{\varepsilon}{b}, \eta_1\right) - \eta\left(b, \frac{\varepsilon}{b}, \eta_2\right)\right|$$
$$\leq |\eta_1 - \eta_2| \exp\left(-2\int_{\varepsilon/b}^{b}\left(\frac{1}{s_1} + \frac{\psi'}{\psi}\frac{(s_1^2 - \varepsilon)^2}{s_1^3}\right) ds_1\right)$$
$$= |\eta_1 - \eta_2|\frac{\varepsilon^2}{b^4}\exp\left(-2\int_{\varepsilon/b}^{b}\frac{\psi'}{\psi}\frac{(s_1^2 - \varepsilon)^2}{s_1^3} ds_1\right)$$
$$\leq |\eta_1 - \eta_2|\frac{\varepsilon^2}{b^4}.$$

The same inequalities hold in other quadrants, i.e., for $\eta(-b, -\varepsilon/b, \eta_0)$, etc.

Let $\{g^k(x), k = 0, \ldots, n-1\}$ be the segment of the trajectory of the map g through a point x lying inside $Q_{\varepsilon_0}^b$. This means that all these points belong to $Q_{\varepsilon_0}^b$ but the points $g^{-1}(x)$ and $g^n(x)$ do not. Suppose that $x = (s_1^0, s_2^0)$. Then for every η_1, η_2, $|\eta_1| < 1$, $|\eta_2| < 1$ we have that

$$|\eta_{s_1^0, s_2^0}(n, \eta_1) - \eta_{s_1^0, s_2^0}(n, \eta_2)| \leq |\eta_1 - \eta_2|\frac{M(s_1^0 s_2^0)^2}{b^4}$$
$$\leq |\eta_1 - \eta_2|\frac{M\varepsilon^2}{b^4},$$

where M is a constant (independent of ε_0 and b provided that they are sufficiently small). This implies that every angle inside the cone $C^u(x)$ is contracted under the action of dg^n by at least $\mu = M\varepsilon^2/b^4$ times. If we choose ε_0 small enough to ensure that $\mu < 1$, we obtain the desired result for every point x for which the trajectory $g^n(x)$ does not go to the origin.

It remains to show that the conclusion of the lemma holds also for every trajectory converging to the origin. To this end we choose a number $\delta > 0$ and consider the point $x = (\delta, 0) \in D_{r_1}$ whose trajectory $g^n(x)$ goes to zero. On the line $s_2 = 0$, the differential equation (1.25) reduces to the linear differential equation

$$\frac{d\eta}{ds_1} = -\left(\frac{2}{s_1} + \frac{2s_1\psi'}{\psi}\right)\eta.$$

Consider the solution $\eta(\delta, \varepsilon, 1)$ of this equation with the initial condition $s_1 = \varepsilon < \delta$, $\eta = 1$ along the segment $[\varepsilon, \delta]$. For this solution we have

$$\eta(\delta, \varepsilon, 1) = \exp\left(-\int_{\varepsilon}^{1}\left(\frac{2}{s_1} + \frac{2s_1\psi'}{\psi}\right) ds_1\right) \leq \frac{\varepsilon^2}{\delta^2}.$$

1.5. The Katok map of the 2-torus

Therefore, $\eta(\delta, \varepsilon, 1) \to 0$ as $\varepsilon \to 0$ and by the linearity of the equation, we also have $\eta(\delta, \varepsilon, -1) \to 0$ as $\varepsilon \to 0$. The desired result follows. □

Clearly, the subspaces $E^s(x)$ and $E^u(x)$ are transverse to each other, i.e.,
$$T_x \mathbb{T}^2 = E^s(x) \oplus E^u(x),$$
and invariant under the action of the differential, i.e.,
$$d_x G E^s(x) = E^s(G(x)), \quad d_x G E^u(x) = E^u(G(x)).$$

We now wish to show that $E^s(x)$ and $E^u(x)$ are, respectively, stable and unstable subspaces for G. To this end we introduce the crucial notion of Lyapunov exponent which measures the asymptotic rate of growth (or decay) of the length of vectors in $T\mathbb{T}^2$ under $d_x G^n$ as $n \to \infty$. More precisely, given $x \in \mathbb{T}^2$ and $v \in T_x \mathbb{T}^2$, the Lyapunov exponent at the point x of the vector v is given by
$$\chi(x, v) = \limsup_{n \to \infty} \frac{1}{n} \log \|d_x G^n v\|.$$
In the next chapter we shall study properties of the Lyapunov exponents in great detail and we shall see later in the book the important role they play in studying stability of trajectories.

Exercise 1.28. Show that for every point $x \in D_{r_1}$ on the vertical (or horizontal) segment through the origin and every $v \in T_x \mathbb{T}^2$ the Lyapunov exponent $\chi(x, v) = 0$.

For a typical point $x \in \mathbb{T}^2$ the situation is quite different.

Lemma 1.29. *For almost every $x \in \mathbb{T}^2$ with respect to area we have that $\chi(x, v) < 0$ for $v \in E^s(x)$ and $\chi(x, v) > 0$ for $v \in E^u(x)$.*

Proof. By Lemma 1.25, the map G possesses two families of stable and unstable cones, $C^s(x)$ and $C^u(x)$, which are transverse and strictly invariant under dG^{-1} and G, respectively, at every point $x \neq 0$. The desired statement now follows from a deep result by Wojtkowski (see [**111**] and a more general Theorem 12.2): the mere fact that G preserves a measure that is absolutely continuous with respect to area and possesses such cone families implies that for almost every $x \in \mathbb{T}^2$ the Lyapunov exponent $\chi(x, v) < 0$ for $v \in C^s(x)$ and $\chi(x, v) > 0$ for $v \in C^u(x)$.

However, we outline a more direct proof of the lemma that does not use Wojtkowski's result. Fix a point $x \in \mathbb{T}^2 \setminus D_{r_1}$ and a vector $v \in C^u(x)$. Let $n_1 > 0$ be the first moment the trajectory of x under G enters the disk D_{r_1}. There is a number $\lambda > 1$ such that the vector $v_1 = d_x G^{n_1} v$ has norm $\|v_1\| \geq \lambda^{n_1} \|v\|$. Let now $m_1 > n_1$ be the first moment when this trajectory exits D_{r_1}. It follows from (1.24) that the vector $w_1 = d_x G^{m_1}(x)$ has norm

$\|w_1\| \geq \|v_1\|$. Continuing in the same fashion, we will construct a sequence n_k of entries and a sequence m_k of exits of the trajectory $\{G^l(x)\}$ to and from the disk D_{r_1} so that $n_k < m_k < n_{k+1}$ for every $k > 0$. Furthermore, setting
$$v_k = d_x G^{n_k} v, \quad w_k = d_x G^{m_k} v,$$
we have that
$$\|v_{k+1}\| \geq \lambda^{n_{k+1}-m_k}\|w_k\|, \quad \|w_{k+1}\| \geq \|v_{k+1}\|.$$
It follows that given $n > 0$, the length of the vector $\|d_x G^n v\| \geq \lambda^{N(n)}\|v\|$ where $N(n) = \sum_{i=1}^{\ell}(n_{i+1} - m_i)$ is the largest integer such that $n_{\ell+1} \leq n$. This implies that
$$\chi(x, v) \geq \limsup_{n \to \infty} \frac{N(n)}{n} \log \lambda.$$
Since G is area-preserving, by the Birkhoff Ergodic Theorem, we conclude that for almost every $x \notin D_{r_1}$,
$$\lim_{n \to \infty} \frac{N(n)}{n} > 0.$$
This implies that for those x we have that $\chi(x, v) > 0$ for $v \in C^u(x)$. It remains to notice that for every $x \in D_{r_1}$, which does not lie on the stable separatrix, there is $n > 0$ such that $G^n(x) \notin D_{r_1}$. The proof that $\chi(x, v) < 0$ for $v \in C^s(x)$ is similar. $\square$

Lemma 1.29 shows that the length of the vector $\|d_x G^n v\|$ with $v \in E^s(x)$ goes to zero as $n \to \infty$ and that the length of the vector $\|d_x G^n v\|$ with $v \in E^u(x)$ goes to ∞ as $n \to \infty$. In other words $E^s(x)$ is a stable subspace and $E^u(x)$ is an unstable subspace for df. Thus the map G, and hence also $G_{\mathbb{T}^2}$, admits an invariant splitting similar to the one for the hyperbolic automorphism T (except at the origin). However, the contraction and expansion rates along the stable and unstable directions depend nonuniformly in x.

1.5.4. Additional properties of the Katok map. We shall describe some other interesting properties of the map $G_{\mathbb{T}^2}$.

(1) The map $G_{\mathbb{T}^2}$ lies on the boundary of Anosov diffeomorphisms on $\mathbb{T}^2$, i.e., there is a sequence of Anosov diffeomorphisms G_n converging to $G_{\mathbb{T}^2}$ in the C^1 topology.

We outline the proof of this result (see [**61**] for details). Observe that the Katok map $G_{\mathbb{T}^2}$ depends on the choice of the function ψ and so we can write $G_{\mathbb{T}^2} = G_{\mathbb{T}^2}(\psi)$.

Let κ be a real-valued C^∞ function on $[0, 1]$ satisfying:

(a) $\kappa(0) > 0$ and $\kappa(u) = 1$ for $u \geq r_0$, for some $0 < r_0 < 1$;
(b) $\kappa'(u) > 0$ for every $0 < u < r_0$.

Starting with such a function κ, one can repeat the construction in this section and obtain an area-preserving C^∞ diffeomorphism $G_{\mathbb{T}^2}(\kappa)$ of the torus.

Exercise 1.30. Show that $G_{\mathbb{T}^2}(\kappa)$ is an Anosov diffeomorphism.

Let us now choose a function ψ that satisfies conditions (1)–(4) in the beginning of this section. There is a sequence of real-valued C^∞ functions κ_n on $[0, 1]$ that converge uniformly to the function ψ. By Exercise 1.30, every map $G_n = G_{\mathbb{T}^2}(\kappa_n)$ is an area-preserving Anosov diffeomorphism of the torus.

Exercise 1.31. Show that the sequence of maps G_n converges to $G_{\mathbb{T}^2}(\psi)$ in the C^1 topology.

Some other properties of the Katok map $G_{\mathbb{T}^2}$ are (see [**61**] for details):

(2) The map $G_{\mathbb{T}^2}$ is topologically conjugate to T, i.e., there exists a homeomorphism $h\colon \mathbb{T}^2 \to \mathbb{T}^2$ such that $G_{\mathbb{T}^2} \circ h = h \circ T$. Since the automorphism T is topologically mixing, it follows that so is the map $G_{\mathbb{T}^2}$.

(3) Let $W_T^u(x)$ and $W_T^s(x)$ be the projections of eigenlines through x corresponding to the eigenvalues λ and λ^{-1}. They form two smooth transverse invariant foliations that are unstable and stable foliations for T. The curves
$$W_{G_{\mathbb{T}^2}}^u(x) = h(W_T^u(x)), \quad W_{G_{\mathbb{T}^2}}^s(x) = h(W_T^s(x))$$
are smooth and form two continuous everywhere transverse invariant foliations for $G_{\mathbb{T}^2}$ that are tangent to $E^u(x)$ and $E^s(x)$ for $x \in M$, respectively. In particular, the subspaces $E^u(x)$ and $E^s(x)$ depend continuously on x.

(4) $G_{\mathbb{T}^2}$ is ergodic and, in fact, is a Bernoulli diffeomorphism. Ergodicity follows from Theorems 9.2, 9.14, 9.17, and 9.19 and the fact that $G_{\mathbb{T}^2}$ is topologically transitive. The Bernoulli property follows from Theorem 9.12 and the fact that $G_{\mathbb{T}^2}$ is topologically mixing.[7]

1.6. Area-preserving diffeomorphisms with nonzero Lyapunov exponents on surfaces

Following Katok's approach in [**61**], we outline a construction of an area-preserving C^∞ diffeomorphism with nonzero Lyapunov exponents on any surface. Observe that, in particular, this includes the two-dimensional sphere which allows no Anosov diffeomorphisms.

---

[7]Topological mixing (and hence, topological transitivity) follows from property (2): the linear automorphism T, which is topologically conjugate to $G_{\mathbb{T}^2}$, is topologically mixing.

1.6.1. An area-preserving C^∞ diffeomorphism with nonzero Lyapunov exponents on the sphere.

We begin with the automorphism T of the torus $\mathbb{T}^2$ induced by the matrix $A = \begin{pmatrix} 5 & 8 \\ 8 & 13 \end{pmatrix}$ that has four fixed points $x_1 = (0,0)$, $x_2 = (1/2, 0)$, $x_3 = (0, 1/2)$, and $x_4 = (1/2, 1/2)$.

For $i = 1, 2, 3, 4$ consider the disk D_r^i centered at x_i of radius r. We choose numbers r_0, r_1, and r_2 such that the disks $D_{r_k}^i$, $k = 0, 1, 2$, satisfy (1.13) and $D_{r_2}^i \cap D_{r_2}^j = \varnothing$ for $i \neq j$. Choosing the same slow-down function ψ in each disk D_r^i and repeating the arguments in the previous section, we construct a diffeomorphism g_i coinciding with T outside $D_{r_1}^i$. Therefore, the map

$$G_1(x) = \begin{cases} T(x) & \text{if } x \in \mathbb{T}^2 \setminus D, \\ g_i(x) & \text{if } x \in D_{r_1}^i \text{ for some } i = 1, 2, 3, 4, \end{cases}$$

where $D = \bigcup_{i=1}^4 D_{r_1}^i$, defines a homeomorphism of the torus $\mathbb{T}^2$ which is a C^∞ diffeomorphism everywhere except at the points x_i. The Lyapunov exponents of G_1 are nonzero almost everywhere with respect to the measure ν, which coincides with area m in $\mathbb{T}^2 \setminus D$ and is absolutely continuous with respect to m in D with the density function $\kappa(s_1, s_2)$ given in each $D_{r_1}^i$ by (1.16).

Using (1.17) in each disk $D_{r_2}^i$, $r_2 \geq r_1$, we introduce a coordinate change φ_i such that the map

$$\varphi(x) = \begin{cases} \varphi_i(x) & \text{if } x \in D_{r_1}^i \text{ for some } i = 1, 2, 3, 4, \\ x & \text{otherwise} \end{cases} \tag{1.26}$$

defines a homeomorphism of $\mathbb{T}^2$ which is a C^∞ diffeomorphism everywhere except at the points x_i. Repeating arguments of the previous section, it is easy to show that the map $G_2 = \varphi \circ G_1 \circ \varphi^{-1}$ is an area-preserving C^∞ diffeomorphism whose Lyapunov exponents are nonzero almost everywhere.

Using this map, we construct an area-preserving diffeomorphism with nonzero Lyapunov exponents on the sphere S^2. Consider the involution map $I \colon \mathbb{T}^2 \to \mathbb{T}^2$ given by

$$I(t_1, t_2) = (1 - t_1, 1 - t_2).$$

Exercise 1.32. Show that:

(1) the involution I has the points x_i, $i = 1, 2, 3, 4$, as its fixed points;

(2) I commutes with G_2, i.e., $G_2 \circ I = I \circ G_2$.

The factor-space $\mathbb{T}^2 / I$ is homeomorphic to the sphere S^2 (see Figure 1.9) and admits a natural smooth structure induced from the torus everywhere except for the points x_i, $i = 1, 2, 3, 4$. Moreover, the following statement holds.

1.6. Area-preserving diffeomorphisms with nonzero exponents

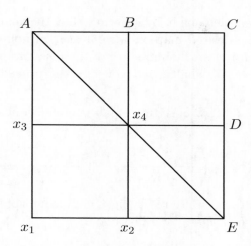

Figure 1.9. Mapping the factor-space $\mathbb{T}^2/I$ to S^2 by making the following identifications: $AB \equiv Ex_2$, $BC \equiv x_2x_1$, $CD \equiv x_1x_3$, $DE \equiv x_3A$, $Ax_4 \equiv Ex_4$, $Bx_4 \equiv x_2x_4$, and $x_3x_4 \equiv Dx_4$.

Proposition 1.33 (Katok [61]). *There exists a map* $\zeta\colon \mathbb{T}^2 \to S^2$ *satisfying:*

(1) ζ *is a double branched covering, is regular (one-to-one on each branch), and is C^∞ everywhere except at the points x_i, $i = 1, 2, 3, 4$, where it branches;*

(2) $\zeta \circ I = \zeta$;

(3) ζ *preserves area, i.e.,* $\zeta_* m = \mu$ *where μ is the area in S^2;*

(4) *there exists a local coordinate system in a neighborhood of each point $\zeta(x_i)$, $i = 1, 2, 3, 4$, in which*

$$\zeta(s_1, s_2) = \left(\frac{s_1^2 - s_2^2}{\sqrt{s_1^2 + s_2^2}}, \frac{2s_1 s_2}{\sqrt{s_1^2 + s_2^2}} \right)$$

in each disk $D_{r_1}^i$.

Finally, one can show that the map $G_{S^2} = \zeta \circ G_2 \circ \zeta^{-1}$ is a C^∞ diffeomorphism which preserves area. It is easy to see that this map has nonzero Lyapunov exponents almost everywhere. It is also ergodic.

1.6.2. An area-preserving C^∞ diffeomorphism with nonzero Lyapunov exponents on any surface. The diffeomorphism G_{S^2} can be used to build an area-preserving C^∞ diffeomorphism with nonzero Lyapunov exponents on any surface. This is a two-step procedure. First, the sphere can be unfolded into the unit disk and the map G_{S^2} can be carried over to a C^∞ area-preserving diffeomorphism g of the disk. The second step is to cut the surface in a certain way so that the resulting surface with the boundary is homeomorphic to the disk. This is a well-known topological construction but we need to complement it by showing that the map g can be carried over to produce the desired map on the surface. In doing so, a crucial fact is that g is the identity on the boundary of the disk and is "sufficiently flat" near the boundary. We shall briefly outline the procedure without going into details.

Set $p_i = \zeta(x_i)$, $i = 1, 2, 3, 4$. In a small neighborhood of the point p_4 we define a map η by

$$\eta(\tau_1, \tau_2) = \left(\frac{\tau_1 \sqrt{1 - \tau_1^2 - \tau_2^2}}{\sqrt{\tau_1^2 + \tau_2^2}}, \frac{\tau_2 \sqrt{1 - \tau_1^2 - \tau_2^2}}{\sqrt{\tau_1^2 + \tau_2^2}} \right). \tag{1.27}$$

One can extend it to an area-preserving C^∞ diffeomorphism η between $S^2 \setminus \{p_4\}$ and the unit disk $D^2 \subset \mathbb{R}^2$. The map

$$g = \eta \circ G_{S^2} \circ \eta^{-1} \tag{1.28}$$

is a C^∞ diffeomorphism of the disk D^2 that preserves area, has nonzero Lyapunov exponents almost everywhere, and is ergodic.

Our goal now is to show that the disk D^2 can be embedded into any surface and that the diffeomorphism g can be carried over to a diffeomorphism of the surface which has all the desired properties: it preserves area, is ergodic, and has nonzero Lyapunov exponents.

To this end let $\rho = \{\rho_n\}_{n \in \mathbb{N}}$ be a strictly decreasing sequence of numbers converging to zero. We say that a C^∞ diffeomorphism G of the disk is ρ_n-flat at the boundary if it is the identity on the boundary and there is a strictly increasing sequence of numbers $r_n \to 1$ such that

$$\|G - \mathrm{Id}\|_{C^n(D^2 \setminus D_n^2)} \leq \rho_n,$$

where D_n^2 is the closed disk in $\mathbb{R}^2$ centered at zero of radius r_n.

Proposition 1.34. *Given a compact surface M, there exists a continuous map $h \colon D^2 \to M$ and a strictly decreasing sequence of numbers $\rho = \{\rho_n\}_{n \in \mathbb{N}}$ converging to zero such that:*

(1) *the restriction $h|\operatorname{int} D^2$ is a diffeomorphic embedding;*

(2) $h(D^2) = M$;

(3) *the map h preserves area; more precisely, $h_* m = \mu$ where m is the area in D^2 and μ is the area in M; moreover, $\mu(M \setminus h(\operatorname{int} D^2)) = 0$;*

(4) *if G is a C^∞ diffeomorphism of the disk, which is ρ_n-flat at the boundary, then the map f_M given by*

$$f_M(x) = \begin{cases} h(G(h^{-1}(x))), & x \in h(\operatorname{int} D^2), \\ x, & \text{otherwise} \end{cases} \tag{1.29}$$

is a C^∞ diffeomorphism of M.

1.6. Area-preserving diffeomorphisms with nonzero exponents

Outline of the proof. A smooth compact connected oriented surface M of genus $\kappa > 0$ can be cut along closed geodesics γ_i, $i = 1, \ldots, 2\kappa$ (which represent distinct elements of the fundamental group $\pi(M)$) in such a way that the resulting surface with boundary $\tilde{M}$ is homeomorphic to a regular polygon P via a homeomorphism $\tau \colon P \to \tilde{M}$; this is a well-known topological construction. In fact, the map τ can be chosen in such a way that $\tau|\mathrm{int}\, P \to \mathrm{int}\, \tilde{M}$ is a C^∞ diffeomorphism.

Let $A_1, A_2, \ldots, A_p$, $p = 4\kappa$, be the vertices of the polygon and let O be its center. For $i = 1, \ldots, p$ denote by B_i the point on the line segment $A_i O$ for which $|A_i B_i| = \frac{1}{3}|A_i O|$. Consider the closed set

$$P^* := \left(\bigcup_{i=1}^p A_i A_{i+1}\right) \cup \left(\bigcup_{i=1}^p A_i B_i\right), \quad A_{p+1} = A_1,$$

whose complement U is an open simply connected domain in $\mathbb{R}^2$; see Figure 1.10.

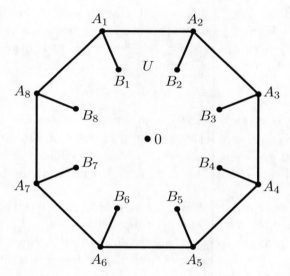

Figure 1.10. The regular polygon P, the set P^*, and the open simply connected domain U.

By the Riemann Mapping Theorem, there exists a C^∞ diffeomorphism $\tilde{h} \colon \mathrm{int}\, D^2 \to U$ which can be extended to a homeomorphism between the disk D^2 and the polygon P. One can now choose a strictly decreasing sequence of numbers $\{\rho_n\}_{n \in \mathbb{N}}$ which converges to zero so fast that for any C^∞ diffeomorphism G, which is ρ_n-flat at the boundary, the diffeomorphism f_P, given by

$$f_P(x) = \begin{cases} \tilde{h}(G(\tilde{h}^{-1}(x))), & x \in \tilde{h}(\mathrm{int}\, D^2), \\ x, & \text{otherwise} \end{cases}$$

is a C^∞ diffeomorphism of P. We now define the map

$$f_{\tilde{M}}(x) = \begin{cases} \tau(f_P(\tau^{-1}(x))), & x \in \tau(U), \\ x, & \text{otherwise.} \end{cases}$$

Since $f_P(U) = U$ and the restriction $f_P|P^*$ is the identity map, we obtain that the restriction $f_{\tilde{M}}|\tau(P^*)$ is also the identity map and, in view of the construction of the set P^*, the map $f_{\tilde{M}}$ is well-defined and is a C^∞ diffeomorphism of $\tilde{M}$. Denote

$$\mu = (\tilde{h}^{-1} \circ \tau^{-1})_* m_{\tilde{M}}.$$

This is a measure on the disk D^2. Since both m_{D^2} and $m_{\tilde{M}}$ are normalized Lebesgue measures, we have

$$\int_{D^2} dm_{D^2} = 1 = \int_M dm_{\tilde{M}} = \int_{D^2} d\mu.$$

This relation allows us to apply a version of Moser's theorem (see Lemma 1 in [50]) to two volume forms associated to the measures μ and μ_{D^2}, and to obtain a C^∞ diffeomorphism $\theta\colon D^2 \to D^2$ that can be continuously extended to ∂D^2 such that $\theta(\partial D^2) = \partial D^2$ and $\theta_*\mu = m_{D^2}$. Set

$$h = \tau \circ \tilde{h} \circ \theta \colon D^2 \to \tilde{M}.$$

Then the map f_M given by (1.29) is an area-preserving C^∞ diffeomorphism of $\tilde{M}$. Since f_M is the identity on the boundary of $\tilde{M}$, after "gluing" $\tilde{M}$ along the closed geodesics γ_i, $i = 1, \ldots, 2\kappa$, it becomes an area-preserving C^∞ diffeomorphism of M. This completes the proof of the proposition. $\square$

To complete our construction, note that the map g of the disk D^2 given by (1.28) is the identity on the boundary ∂D^2. Using (1.17) and (1.26)–(1.27), it is not difficult to show that given a strictly decreasing sequence of numbers $\{\rho_n\}_{n\in\mathbb{N}}$, which converges to zero, the slow-down function ψ can be chosen such that g is ρ_n-flat at the boundary. Proposition 1.34 now applies to the map $G = g$ so that the map f_M in (1.29) has the desired properties: it preserves area, has nonzero Lyapunov exponents almost everywhere, and is ergodic.

Thus we obtain the following result by Katok [61].

Theorem 1.35. *Given a compact surface M, there exists a C^∞ area-preserving ergodic diffeomorphism $f_M\colon M \to M$ with nonzero Lyapunov exponents almost everywhere.*

One can show using results in Chapter 9 that the map f_M is a Bernoulli diffeomorphism.

1.6.3. Further developments. Dolgopyat and Pesin [43] have extended Theorem 1.35 to any compact smooth manifold of dimension greater than 2.

Theorem 1.36. *Given a compact smooth Riemannian manifold M of dimension greater than 2, there exists a volume-preserving C^∞ diffeomorphism $f\colon M \to M$ with nonzero Lyapunov exponents almost everywhere which is a Bernoulli diffeomorphism.*

The mechanism of slowing down trajectories in a neighborhood of a hyperbolic fixed point is not robust in the C^1 topology as the following result by Bochi [20] demonstrates (see also [21]).

Theorem 1.37. *There exists an open set U in the space of area-preserving C^1 diffeomorphisms of $\mathbb{T}^2$ such that:*

(1) *the map $G_{\mathbb{T}^2}$ lies on the boundary of U;*
(2) *there is a G_δ subset $A \subset U$ such that every $f \in A$ is an area-preserving diffeomorphism whose Lyapunov exponents are all zero almost everywhere.*

1.7. A volume-preserving flow with nonzero Lyapunov exponents

We present an example of a dynamical system with continuous time which is nonuniformly hyperbolic. It was constructed in [87] by a "surgery" of an Anosov flow. Let φ_t be a volume-preserving ergodic Anosov flow on a compact three-dimensional manifold M and let X be the vector field of the flow.[8] Fix a point $p_0 \in M$ and introduce a coordinate system x, y, z in the ball $B(p_0, d)$ centered at p_0 of some radius $d > 0$ such that p_0 is the origin (i.e., $p_0 = 0$) and $X = \partial/\partial z$.

For each $\varepsilon > 0$, let $T_\varepsilon = S^1 \times D_\varepsilon \subset B(0, d)$ be the solid torus obtained by rotating the disk

$$D_\varepsilon = \left\{(x, y, z) \in B(0, d) : x = 0 \text{ and } (y - d/2)^2 + z^2 \leq (\varepsilon d)^2\right\}$$

around the z-axis. Every point on the solid torus can be represented as (θ, y, z) with $\theta \in S^1$ and $(y, z) \in D_\varepsilon$.

For every $0 \leq \alpha \leq 2\pi$, we consider the cross-section of the solid torus $\Pi_\alpha = \{(\theta, y, z) : \theta = \alpha\}$ and construct a new vector field $\tilde{X}$ on $M \setminus T_\varepsilon$ (see Figure 1.11).

---

[8]For example, one can take φ_t to be the geodesic flow on a compact surface of negative curvature.

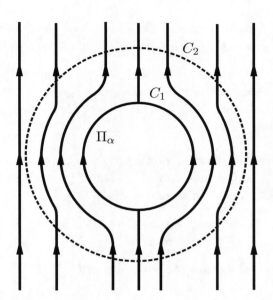

Figure 1.11. A cross-section Π_α and the flow $\tilde{\varphi}_t$. The circles C_1 and C_2 are the sets of points such that $y^2 + z^2 = \varepsilon^2$ and $y^2 + z^2 = 4\varepsilon^2$.

Lemma 1.38. *There exists a smooth vector field $\tilde{X}$ on $M \setminus T_\varepsilon$ such that the flow $\tilde{\varphi}_t$ generated by $\tilde{X}$ has the following properties:*

(1) $\tilde{X}|(M \setminus T_{2\varepsilon}) = X|(M \setminus T_{2\varepsilon})$;

(2) *for any $0 \leq \alpha, \beta \leq 2\pi$, the vector field $\tilde{X}|\Pi_\beta$ is the image of the vector field $\tilde{X}|\Pi_\alpha$ under the rotation around the z-axis that moves Π_α onto Π_β;*

(3) *for every $0 \leq \alpha \leq 2\pi$, the unique two fixed points of the flow $\tilde{\varphi}_t|\Pi_\alpha$ are those in the intersection of Π_α with the planes $z = \pm \varepsilon d$;*

(4) *for every $0 \leq \alpha \leq 2\pi$ and $(y, z) \in D_{2\varepsilon} \setminus \operatorname{int} D_\varepsilon$, the trajectory of the flow $\tilde{\varphi}_t|\Pi_\alpha$ passing through the point (y, z) is invariant under the symmetry $(\alpha, y, z) \mapsto (\alpha, y, -z)$;*

(5) *the flow $\tilde{\varphi}_t|\Pi_\alpha$ preserves the conditional measure induced by volume on the set Π_α.*

Proof. We shall only describe the construction of $\tilde{X}$ on the cross-section Π_0. The vector field $\tilde{X}|\Pi_\alpha$ on an arbitrary cross-section is the image of $\tilde{X}|\Pi_0$ under the rotation around the z-axis that moves Π_0 onto Π_α.

Consider the Hamiltonian function

$$H(y, z) = y(\varepsilon^2 - y^2 - z^2), \quad y, z \in \Pi_0.$$

1.7. A volume-preserving flow with nonzero Lyapunov exponents

The corresponding Hamiltonian vector field

$$\frac{\partial H}{\partial z} = -2yz, \quad -\frac{\partial H}{\partial y} = 3y^2 + z^2 - \varepsilon^2$$

has positive second component in the annulus $\varepsilon^2 < y^2 + z^2 < 4\varepsilon^2$. To make the first component zero when $y^2 + z^2 = 4\varepsilon^2$, we consider a C^∞ function $p\colon [\varepsilon, \infty) \to [0, 1]$ such that $p(t) = 1$ for $t \in [\varepsilon, 3\varepsilon/2]$, $p(t) = 0$ for $t \in [2\varepsilon, \infty)$, and p is strictly decreasing in $(3\varepsilon/2, 2\varepsilon)$. The flow generated by the system of differential equations

$$\begin{cases} y' = -2yzp(\sqrt{y^2 + z^2}), \\ z' = (3y^2 + z^2 - \varepsilon^2)p(\sqrt{y^2+z^2}) + 1 - p(\sqrt{y^2+z^2}) \end{cases}$$

behaves as shown in Figure 1.11. We denote it by $\overline{\varphi}_t$ and we denote by $\overline{X}$ the corresponding vector field.

We now show that it is possible to effect a time change of the flow $\overline{\varphi}_t$ in the annulus $\varepsilon^2 \leq y^2 + z^2 \leq 4\varepsilon^2$ so that the new flow $\tilde{\varphi}_t$ preserves volume. To achieve this, we shall construct a C^3 function $\tau\colon (T_{2\varepsilon} \setminus \operatorname{int} T_\varepsilon) \times \mathbb{R} \to \mathbb{R}$ such that

$$\tilde{\varphi}_t(x) = \overline{\varphi}_{\tau(x,t)}(x).$$

It is easy to see that the vector field $\tilde{X}(x)$ that generates the flow $\tilde{\varphi}_t(x)$ satisfies $\tilde{X}(x) = h(x)\overline{X}(x)$, where $h(x) = (\partial \tau / \partial t)(x, 0)$. The flow $\tilde{\varphi}_t$ would preserve volume if and only if

$$\operatorname{div}(h\overline{X}) = \nabla h \cdot \overline{X} + h \operatorname{div}(\overline{X}) = 0. \tag{1.30}$$

Since the two semicircles defined by $y^2 + z^2 = 4\varepsilon^2$, $y = \pm 2\varepsilon$ are not characteristics of the partial differential equation (1.30), there exists a unique solution h of (1.30) such that $h = 1$ outside $\Pi_0 \cap T_{2\varepsilon}$. This completes the construction of the desired vector field $\tilde{X}$. □

One can see that the orbits of the flows φ_t and $\tilde{\varphi}_t$ coincide on $M \setminus T_{2\varepsilon}$, that the flow $\tilde{\varphi}_t$ preserves volume, and that the only fixed points of this flow are those on the circles $\{(\theta, y, z) : z = -\varepsilon d\}$ and $\{(\theta, y, z) : z = \varepsilon d\}$.

On the set $T_{2\varepsilon} \setminus \operatorname{int} T_\varepsilon$ we introduce coordinates θ_1, θ_2, r with $0 \leq \theta_1, \theta_2 < 2\pi$ and $\varepsilon d \leq r \leq 2\varepsilon d$ such that the set of fixed points of $\tilde{\varphi}_t$ is composed of those for which $r = \varepsilon d$, and $\theta_1 = 0$ or $\theta_1 = \pi$.

Consider the flow on $T_{2\varepsilon} \setminus \operatorname{int} T_\varepsilon$ defined by

$$(\theta_1, \theta_2, r, t) \mapsto (\theta_1, \theta_2 + [2 - r/(\varepsilon d)]^4 t \cos \theta_1, r),$$

and let $\hat{X}$ be the corresponding vector field. Set
$$Y(x) = \begin{cases} X(x), & x \in M \setminus \operatorname{int} T_{2\varepsilon}, \\ \tilde{X}(x) + \hat{X}(x), & x \in \operatorname{int} T_{2\varepsilon} \setminus \operatorname{int} T_{\varepsilon}. \end{cases}$$

The vector field Y on $M \setminus \operatorname{int} T_\varepsilon$ generates the flow ψ_t on $M \setminus \operatorname{int} T_\varepsilon$.

Lemma 1.39. *The following properties hold:*

(1) *the flow ψ_t preserves volume and is ergodic;*

(2) *the flow ψ_t has no fixed points;*

(3) *for almost every $x \in M \setminus T_{2\varepsilon}$,*

$$\chi(x,v) < 0 \text{ for } v \in E^s(x) \quad \text{and} \quad \chi(x,v) > 0 \text{ for } v \in E^u(x),$$

where $E^u(x)$ and $E^s(x)$ are, respectively, the one-dimensional stable and unstable subspaces of the Anosov flow φ_t at the point x.

Proof. By construction, ψ_t preserves volume. Since the orbits of φ_t and ψ_t coincide in $M \setminus \operatorname{int} T_{2\varepsilon}$, the flow ψ_t is ergodic. The second statement follows from the construction of the flow ψ_t. In order to prove the third statement, consider the function
$$T(x,t) = \int_0^t I_{T_{2\varepsilon}}(\varphi_\tau x) \, d\tau,$$
where $I_{T_{2\varepsilon}}$ denotes the characteristic function of the set $T_{2\varepsilon}$. By the Birkhoff Ergodic Theorem, for almost every $x \in M$,
$$\lim_{t \to +\infty} \frac{T(x,t)}{t} = \mu(T_{2\varepsilon}).$$

Fix a point $x \in M \setminus T_{2\varepsilon}$. Consider the moment of time t_1 at which the trajectory $\psi_t(x)$ enters the set $T_{2\varepsilon}$ and the next moment of time t_2 at which this trajectory exits the set $T_{2\varepsilon}$. Given a vector $v \in E^u(x)$, denote by $\tilde{v}_i$ the orthogonal projection of the vector $d_x\psi_{t_i}v$ onto the (x,y)-plane for $i = 1, 2$. It follows from the construction of the flows $\tilde{\varphi}_t$ and ψ_t that $\|\tilde{v}_1\| \geq \|\tilde{v}_2\|$. Since the unstable subspaces $E^u(x)$ depend continuously on x, there exists $K \geq 1$ (independent of x, t_1, and t_2) such that
$$\|d_x\psi_t v\| \geq K \|d_x\varphi_{t-T(x,t)}v\|.$$
It follows that for almost every $x \in M \setminus T_{2\varepsilon}$ and $v \in E^u(x)$,
$$\chi(x,v) = \limsup_{t \to +\infty} \frac{1}{t} \log \|d_x\psi_t v\|$$
$$\geq (1 - \mu(T_{2\varepsilon})) \limsup_{t \to +\infty} \frac{1}{t} \log \|d_x\varphi_t v\| > 0$$

provided that ε is sufficiently small. Repeating the above argument with respect to the inverse flow ψ_{-t}, one can show that $\chi(x,v) < 0$ for almost every $x \in M \setminus T_{2\varepsilon}$ and $v \in E^s(x)$. □

We say that a smooth flow on a compact Riemannian manifold has nonzero Lyapunov exponents almost everywhere with respect to an invariant measure μ if it has nonzero Lyapunov exponents at μ-almost every point x and every nonzero tangent vector v which lies outside of the flow direction. In particular, we have shown that the flow ψ_t has nonzero Lyapunov exponents almost everywhere with respect to volume.

Set $M_1 = M \setminus T_\varepsilon$ and consider a copy $(\tilde{M}_1, \tilde{\psi}_t)$ of the flow (M_1, ψ_t). One can glue the manifolds M_1 and $\tilde{M}_1$ along their boundaries ∂T_ε and obtain a three-dimensional smooth Riemannian manifold D without boundary. We define a flow F_t on D by

$$F_t(x) = \begin{cases} \psi_t(x), & x \in M_1, \\ \tilde{\psi}_t(x), & x \in \tilde{M}_1. \end{cases}$$

It is clear that the flow F_t preserves volume and has nonzero Lyapunov exponents almost everywhere. It is easy to see that the flow $F_t|M_1$ is ergodic and so is the flow $F_t|M_2$.

Exercise 1.40. Modify the definition of the flow F_t to obtain a new flow that is ergodic on the whole manifold D.

One can show that any compact smooth Riemannian manifold M of dimension greater than 2 admits a volume-preserving C^∞ flow $\varphi_t \colon M \to M$ with nonzero Lyapunov exponents almost everywhere that is a Bernoulli flow; see [**57**].

1.8. A slow-down of the Smale–Williams solenoid

Let f be a $C^{1+\alpha}$ diffeomorphism of a compact smooth Riemannian manifold M possessing a trapping region $U \subset M$ (so that $\overline{f(U)} \subset U$) and the hyperbolic attractor $\Lambda_f = \bigcap_{n \in \mathbb{N}} f^n(U)$. We impose some additional restrictions on the action of f on Λ_f. Namely, we assume that (see [**114**]):

(1) $f|\Lambda_f$ is topologically transitive;
(2) for every $x \in \Lambda_f$ the unstable subspace $E^u(x)$ is one-dimensional;
(3) the map f has a fixed point $p \in \Lambda_f$ and there is a neighborhood V of p with local coordinates in which p is identified with 0 and f is the time-one map of the flow generated by a linear vector field $\dot{x} = Ax$ which acts conformally along the stable and unstable subspaces,
$$A = A_u \oplus A_s, \quad A_u = \gamma \operatorname{Id}_u, \quad A_s = -\beta \operatorname{Id}_s, \text{ and } \beta > \gamma > 0.$$

Exercise 1.41. Show that the Smale–Williams solenoid described in Section 1.4 with $\lambda_1 = \lambda_2$ (see (1.11)) is an example of an attractor which satisfies assumptions (1)–(3) above.

We now describe a version of the slow-down procedure applied to the map f (compare to Section 1.5.1). Fix $0 < r_0 < r_1$ so that $B(0, r_1) \subset V$. Choose a $C^{1+\alpha}$ function $\psi \colon V \to [0, 1]$ satisfying the following:

(1) $\psi(x) = \|x\|^\alpha$ for $\|x\| \leq r_0$;
(2) $\psi(x) = 1$ for $\|x\| \geq r_1$ and ψ is C^∞ away from 0.

Let φ_t be the flow generated by the system $\dot{x} = \psi(\|x\|)Ax$ on V. Define $g \colon U \to M$ by
$$g(x) = \begin{cases} \varphi_1(x), & x \in V, \\ f(x), & x \in U \setminus V. \end{cases}$$
It is not difficult to show that the map g is of class $C^{1+\alpha}$ and that $\overline{g(U)} \subset U$. Then the set
$$\Lambda_g = \bigcap_{n \in \mathbb{N}} g^n(U)$$
is an attractor for g. The map g is a *slow-down* of f.

Let m be the normalized volume in U. Consider the sequence μ_n of averages of the push-forward measures (1.12) (with f replaced by g). The following result is an analog of Theorem 1.19 and was proven in **[39]** (see also **[114]** and Section 11.1.3).

Theorem 1.42. *The sequence of measures μ_n converges in the weak*-topology to a measure μ which is supported on Λ_g and is invariant under g; it has nonzero Lyapunov exponents almost everywhere and is ergodic.*

Chapter 2

General Theory of Lyapunov Exponents

In this chapter we introduce the fundamental notion of Lyapunov exponent in a formal axiomatic setting and we will describe its basic properties. We then discuss the crucial concept of Lyapunov–Perron regularity and describe various criteria that guarantee that a given trajectory of the system is Lyapunov–Perron regular. Finally, we study Lyapunov exponents and Lyapunov–Perron regularity in some particular situations—linear differential equations and sequences of matrices.

2.1. Lyapunov exponents and their basic properties

2.1.1. The definition of the Lyapunov exponent. We follow the approach developed in [35]. Let V be a p-dimensional real vector space. A function $\chi\colon V \to \mathbb{R} \cup \{-\infty\}$ is called a *Lyapunov characteristic exponent* or simply a *Lyapunov exponent* on V if:

(1) $\chi(\alpha v) = \chi(v)$ for each $v \in V$ and $\alpha \in \mathbb{R} \setminus \{0\}$;

(2) $\chi(v + w) \leq \max\{\chi(v), \chi(w)\}$ for each $v, w \in V$;

(3) $\chi(0) = -\infty$ (*normalization property*).

We describe some basic properties of Lyapunov exponents.

Theorem 2.1. *The following statements hold:*

(1) *if $v, w \in V$ are such that $\chi(v) \neq \chi(w)$, then*
$$\chi(v + w) = \max\{\chi(v), \chi(w)\};$$

(2) *if $v_1, \ldots, v_m \in V$ and $\alpha_1, \ldots, \alpha_m \in \mathbb{R} \setminus \{0\}$, then*
$$\chi(\alpha_1 v_1 + \cdots + \alpha_m v_m) \leq \max\{\chi(v_i) : 1 \leq i \leq m\};$$
if, in addition, there exists i such that $\chi(v_i) > \chi(v_j)$ for all $j \neq i$, then
$$\chi(\alpha_1 v_1 + \cdots + \alpha_m v_m) = \chi(v_i);$$

(3) *if for some $v_1, \ldots, v_m \in V \setminus \{0\}$ the numbers $\chi(v_1), \ldots, \chi(v_m)$ are distinct, then the vectors $v_1, \ldots, v_m$ are linearly independent;*

(4) *the function χ attains at most p distinct finite values.*

Proof. Suppose that $\chi(v) < \chi(w)$. We have
$$\chi(v + w) \leq \chi(w) = \chi(v + w - v) \leq \max\{\chi(v + w), \chi(v)\}.$$
It follows that if $\chi(v + w) < \chi(v)$, then $\chi(w) \leq \chi(v)$, which contradicts our assumption. Hence, $\chi(v + w) \geq \chi(v)$, and thus, $\chi(v + w) = \chi(w)$. Statement (1) follows. Statement (2) is an immediate consequence of statement (1) and properties (1) and (2) in the definition of Lyapunov exponent.

In order to prove statement (3), assume to the contrary that the vectors $v_1, \ldots, v_m$ are linearly dependent, i.e., $\alpha_1 v_1 + \cdots + \alpha_m v_m = 0$ with not all α_i equal to zero, while $\chi(v_1), \ldots, \chi(v_m)$ are distinct. By statement (2) and property (3) in the definition of Lyapunov exponent, we obtain
$$-\infty = \chi(\alpha_1 v_1 + \cdots + \alpha_m v_m)$$
$$= \max\{\chi(v_i) : 1 \leq i \leq m \text{ and } \alpha_i \neq 0\} \neq -\infty.$$
This contradiction implies statement (3). Statement (4) follows from statement (3). $\square$

By Theorem 2.1, the Lyapunov exponent χ can take only finitely many distinct *values* on $V \setminus \{0\}$. We denote them by
$$\chi_1 < \cdots < \chi_s$$
for some $s \leq p$. In general, χ_1 may be $-\infty$.

2.1.2. The filtration associated with the Lyapunov exponent. For each $1 \leq i \leq s$, define
$$V_i = \{v \in V : \chi(v) \leq \chi_i\}. \tag{2.1}$$
Put $V_0 = \{0\}$. It follows from Theorem 2.1 that V_i is a linear subspace of V and
$$\{0\} = V_0 \subsetneq V_1 \subsetneq \cdots \subsetneq V_s = V. \tag{2.2}$$

2.1. Lyapunov exponents and their basic properties

Any collection $\mathcal{V} = \{V_i : i = 0, \ldots, s\}$ of linear subspaces of V satisfying (2.2) is called a *linear filtration* or simply a *filtration* of V.

The following result gives an equivalent characterization of Lyapunov exponents in terms of filtrations.

Theorem 2.2. *A function $\chi \colon V \to \mathbb{R} \cup \{-\infty\}$ is a Lyapunov exponent if and only if there exist numbers $\chi_1 < \cdots < \chi_s$ for some $1 \le s \le p$, and a filtration $\mathcal{V} = \{V_i : i = 0, \ldots, s\}$ of V such that:*

(1) $\chi(v) \le \chi_i$ *for every* $v \in V_i$;

(2) $\chi(v) = \chi_i$ *for every* $v \in V_i \setminus V_{i-1}$;

(3) $\chi(0) = -\infty$.

Proof. If χ is a Lyapunov exponent, then the filtration
$$\mathcal{V} = \{V_i : i = 0, \ldots, s\}$$
defined by (2.1) satisfies conditions (1) and (3) of the theorem. Moreover, for any $v \in V_i \setminus V_{i-1}$ we have $\chi_{i-1} < \chi(v) \le \chi_i$. Since χ attains no value strictly between χ_{i-1} and χ_i, we obtain $\chi(v) = \chi_i$ and condition (2) follows.

Now suppose that a function χ and filtration $\mathcal{V}$ satisfy the conditions of the theorem. Observe that $v \in V_i \setminus V_{i-1}$ if and only if $\alpha v \in V_i \setminus V_{i-1}$ for any $\alpha \in \mathbb{R} \setminus \{0\}$. Therefore, by condition (2), $\chi(\alpha v) = \chi(v)$. Now choose vectors $v_1, v_2 \in V$. Let $\chi(v_j) = \chi_{i_j}$ for $j = 1, 2$. It follows from conditions (1) and (2) that the subspace V_i can be characterized by (2.1). Therefore, $v_j \in V_{i_j}$ for $j = 1, 2$. Without loss of generality we may assume that $i_1 < i_2$. This implies that $v_1 + v_2 \in V_{i_1} \cup V_{i_2} = V_{i_2}$. Hence, by condition (1), we have
$$\chi(v_1 + v_2) \le \chi_{i_2} = \max\{\chi(v_1), \chi(v_2)\},$$
and thus, χ is a Lyapunov exponent. $\square$

We refer to the filtration $\mathcal{V} = \{V_i : i = 0, \ldots, s\}$ defined by (2.1) as the *filtration of V associated to χ* and denote it by $\mathcal{V}_\chi$. The number
$$k_i = \dim V_i - \dim V_{i-1}$$
is called the *multiplicity* of the value χ_i, and the collection of pairs
$$\mathrm{Sp}\,\chi = \{(\chi_i, k_i) : 1 \le i \le s\}$$
is called the *Lyapunov spectrum* of χ.

Given a filtration $\mathcal{V} = \{V_i : i = 0, \ldots, s\}$ of V and numbers χ_i, $1 \le i < s$, define a function $\chi \colon V \to \mathbb{R} \cup \{-\infty\}$ by $\chi(v) = \chi_i$ for every $v \in V_i \setminus V_{i-1}$, and $\chi(0) = -\infty$. It is easy to see that the function χ and the filtration $\mathcal{V}$ satisfy the conditions of Theorem 2.2 and thus determine a Lyapunov exponent on V.

2.1.3. Bases which are subordinate to filtrations. To every filtration $\mathcal{V} = \{V_i : i = 0, \ldots, s\}$ of V one can associate a special class of bases which are well adapted to the filtration. A basis $\mathbf{v} = (v_1, \ldots, v_p)$ of V is said to be *subordinate* to $\mathcal{V}$ if for every $1 \leq i \leq s$ there exists a basis of V_i composed of p_i vectors from $\{v_1, \ldots, v_p\}$.[1] A subordinate basis $\mathbf{v}$ is *ordered* if for every $1 \leq i \leq s$, the vectors $v_1, \ldots, v_{p_i}$ form a basis of V_i. Note that $p_i - p_{i-1} = k_i$.

Exercise 2.3. Show that for every filtration $\mathcal{V}$ there exists a basis that is subordinate to $\mathcal{V}$ and is ordered. Moreover, two filtrations coincide if and only if each basis that is subordinate to one of them is also subordinate to the other one.

Let $\mathcal{V}_\chi$ be the filtration associated to a Lyapunov exponent χ and let $\mathbf{v}$ be a basis that is subordinate to $\mathcal{V}_\chi$. We note that among the numbers $\chi(v_1), \ldots, \chi(v_p)$ the value χ_i occurs exactly k_i times, for each $i = 1, \ldots, s$. Hence,

$$\sum_{j=1}^{p} \chi(v_j) = \sum_{i=1}^{s} k_i \chi_i. \tag{2.3}$$

We use this observation to prove the following result.

Theorem 2.4. *A basis $\mathbf{v}$ is subordinate to a filtration $\mathcal{V}_\chi$ if and only if*

$$\inf\left\{\sum_{j=1}^{p} \chi(w_j) : \mathbf{w} = (w_1, \ldots, w_p) \text{ is a basis of } V\right\} = \sum_{j=1}^{p} \chi(v_j). \tag{2.4}$$

Proof. We begin with the following statement.

Lemma 2.5. *Let $\mathbf{v}$ be a basis that is subordinate to the filtration $\mathcal{V}_\chi$ that is ordered and let $\mathbf{w}$ be a basis of V for which $\chi(w_1) \leq \cdots \leq \chi(w_p)$. Then:*

(1) *$\chi(w_j) \geq \chi(v_j)$ for $1 \leq j \leq p$ and $\chi(w_j) = \chi(v_j)$ for $p - p_s + 1 \leq j \leq p$;*

(2) *$\sum_{j=1}^{p} \chi(w_j) \geq \sum_{j=1}^{p} \chi(v_j)$;*

(3) *$\mathbf{w}$ is subordinate to the filtration $\mathcal{V}_\chi$ if and only if $\chi(w_j) = \chi(v_j)$ for every $1 \leq j \leq p$;*

(4) *$\mathbf{w}$ is subordinate to the filtration $\mathcal{V}_\chi$ if and only if*

$$\sum_{j=1}^{p} \chi(w_j) = \sum_{j=1}^{p} \chi(v_j).$$

---

[1] In the literature, subordinate bases are also called *normal* (see [11, 13]). However, we use here the term "subordinate" since it reflects better the meaning of this notion.

2.1. Lyapunov exponents and their basic properties

Proof of the lemma. Since χ_1 is the minimal value of χ on $V \setminus \{0\}$, we have $\chi(w_j) \geq \chi(v_j) = \chi_1$ for all $j = 1, \ldots, p_1$. Assume that $\chi(w_{p_1+1}) = \chi_1$. Then $\chi(w_1) = \cdots = \chi(w_{p_1+1}) = \chi_1$ and
$$p_1 \geq \dim \operatorname{span}\{w_1, \ldots, w_{p_1+1}\} = p_1 + 1,$$
where $\operatorname{span} Z$ denotes the linear space generated by the set of vectors in Z. This contradiction implies that $\chi(w_{p_1+1}) \geq \chi_2$, and hence, $\chi(w_j) \geq \chi(v_j) = \chi_2$ for every $j = p_1 + 1, \ldots, p_2$.

Repeating the same argument finitely many times, we obtain $\chi(w_j) \geq \chi(v_j)$ for every $1 \leq j \leq p$. In particular, $\chi(w_p) = \chi(v_p)$ since $\chi(v_p)$ is the maximum value of χ. Statement (1) follows. Statement (2) is an immediate consequence of statement (1).

By (2.1), $\chi(w_j) = \chi(v_j)$ for every $1 \leq j \leq p$ if and only if $w_1, \ldots, w_{p_i}$ is a basis of V_i for every $1 \leq i \leq s$, and hence, if and only if the basis $\mathbf{w}$ is subordinate to the filtration $\mathcal{V}_\chi$. This implies statement (3). The last statement is a consequence of statements (2) and (3). □

We turn to the proof of the theorem. By the lemma, the infimum in (2.4) is equal to
$$\inf\left\{\sum_{j=1}^{p} \chi(w_j) : \mathbf{w} \text{ is subordinate to the filtration } \mathcal{V}_\chi\right\} = \sum_{i=1}^{s} k_i \chi_i.$$

In view of statement (4) of the lemma, the basis $\mathbf{v}$ is subordinate to the filtration $\mathcal{V}_\chi$ if and only if the relation (2.3) holds, and hence, if and only if the relation (2.4) holds. The theorem follows. □

Given a filtration $\mathcal{V}$, there is a useful construction of subordinate bases due to Lyapunov, which we now describe. Starting with a basis $\mathbf{v} = (v_1, \ldots, v_p)$, we define a sequence of bases $\mathbf{w}_n = (w_{n1}, \ldots, w_{np})$, $n \geq 0$, $\mathbf{w}_0 = \mathbf{v}$, which makes the sum $\sum_{j=1}^{p} \chi(w_{nj})$ decrease as n increases. Since χ takes on only finitely many values, this process ends after a finite number of steps. More precisely, assume that there is a linear combination $x = \sum_{j=1}^{p} \alpha_j v_j$, with not all α_j being zero, such that
$$\chi(x) < \max\{\chi(v_j) : 1 \leq j \leq p \text{ and } \alpha_j \neq 0\}.$$
Choose a vector v_k for which
$$\chi(v_k) = \max\{\chi(v_j) : 1 \leq j \leq p \text{ and } \alpha_j \neq 0\}$$
and replace it with the vector x. Using the fact that $\alpha_k \neq 0$, we obtain that the vectors $v_1, \ldots, v_{k-1}, x, v_{k+1}, \ldots, v_p$ are linearly independent and, hence, form a basis $\mathbf{w}_1$. Continuing in a similar fashion, we obtain a basis

$\mathbf{w} = (w_1, \ldots, w_p)$ with the property that for any linear combination $x = \sum_{j=1}^{p} \alpha_j w_j$, with not all α_j being zero,

$$\chi(x) = \max\{\chi(w_j) : 1 \leq j \leq p \text{ and } \alpha_j \neq 0\}. \tag{2.5}$$

We claim that the basis $\mathbf{w}$ is subordinate to the filtration $\mathcal{V}$. Otherwise, there exists i such that the number of vectors in the basis lying in the space V_i is strictly less than $\dim V_i$. Hence, these vectors fail to form a basis of V_i and there is a vector $x \in V_i$ which is linearly independent of these vectors and $\chi(x) \leq \chi_i$. On the other hand, we can write $x = \sum_{j=1}^{p} \alpha_j w_j$ with some $\alpha_j \neq 0$ such that the vector w_j lies outside of the space V_i. Since $\chi(w_j) > \chi_i$, this contradicts (2.5) and hence the basis $\mathbf{w}$ is subordinate to $\mathcal{V}$.

We now describe the behavior of subordinate bases under linear transformations. Let $A \colon V \to V$ be an invertible linear transformation of a vector space V. Moreover, let $\mathcal{V} = \{V_i : i = 0, \ldots, s\}$ be a filtration of V and let $\mathbf{v}$ be a basis that is subordinate to $\mathcal{V}$ and is ordered.

Exercise 2.6. Show that the following properties are equivalent:

(1) the basis $(Av_1, \ldots, Av_p)$ is subordinate to $\mathcal{V}$ and is ordered;

(2) the transformation A preserves the filtration $\mathcal{V}$, i.e., $AV_i = V_i$ for every $1 \leq i \leq s$;

(3) the transformation A, with respect to the basis $\mathbf{v}$, has the lower block-triangular form

$$\begin{pmatrix} A_1 & 0 & \cdots & 0 \\ & A_2 & \ddots & \vdots \\ & & \ddots & 0 \\ & & & A_s \end{pmatrix},$$

where each A_i is a $k_i \times k_i$ matrix with $\det A_i \neq 0$.

Let $\mathcal{V}$ and $\mathcal{W}$ be filtrations of V. As a consequence of the above statement one can show that *there exists a basis that is subordinate to both filtrations*. Indeed, let $\mathbf{v}$ be a basis that is subordinate to $\mathcal{V}$ and is ordered. There exists a lower-triangular $p \times p$ matrix A such that the basis $\mathbf{w} = (Av_1, \ldots, Av_p)$ is subordinate to $\mathcal{W}$. It follows from Exercise 2.6 that this basis is also subordinate to $\mathcal{V}$.

2.2. The Lyapunov and Perron irregularity coefficients

In this section we introduce the concept of regularity of the Lyapunov exponent which is crucial to our study of stability of solutions of differential equations (the continuous-time case) and of trajectories of diffeomorphisms (the discrete-time case). We begin by defining two positive quantities which originated in works of Lyapunov [78] and Perron [86] and were designed to measure the amount of "irregularity" of the exponent, so that it becomes "regular" if those quantities are equal to zero. For the purpose of stability this is the "best scenario" but as we will see in Chapters 6 and 7 a "small" amount of irregularity[2] may still be sufficient to guarantee stability.

2.2.1. The irregularity coefficients associated with the Lyapunov exponent. Let V be a vector space. Consider the dual vector space V^* to V which consists of the linear functionals on V. If $v \in V$ and $v^* \in V^*$, then $\langle v, v^* \rangle$ denotes the value of v^* on v. Let $\mathbf{v} = (v_1, \ldots, v_p)$ be a basis in V and let $\mathbf{v}^* = (v_1^*, \ldots, v_p^*)$ be a basis in V^*. We say that $\mathbf{v}$ is *dual* to $\mathbf{v}^*$ and write $\mathbf{v} \sim \mathbf{v}^*$ if

$$\langle v_i, v_j^* \rangle = v_j^*(v_i) = \delta_{ij} \text{ for each } i \text{ and } j.$$

Let χ be a Lyapunov exponent on V and let χ^* be a Lyapunov exponent on V^*. We say that the exponents χ and χ^* are *dual* and write $\chi \sim \chi^*$ if for any pair of dual bases $\mathbf{v}$ and $\mathbf{v}^*$ and every $1 \leq i \leq p$ we have

$$\chi(v_i) + \chi^*(v_i^*) \geq 0.$$

We denote by $\chi_1' \leq \cdots \leq \chi_p'$ the values of χ counted with their multiplicities. Similarly, we denote by $\chi_1^{*'} \geq \cdots \geq \chi_p^{*'}$ the values of χ^* counted with their multiplicities. We define:

(1) the *irregularity coefficient*

$$\gamma(\chi, \chi^*) = \min \max\{\chi(v_i) + \chi^*(v_i^*) : 1 \leq i \leq p\}, \qquad (2.6)$$

where the minimum is taken over all pairs of dual bases $\mathbf{v}$ and $\mathbf{v}^*$ of V and V^*;

(2) the *Perron coefficient* of χ and χ^*

$$\pi(\chi, \chi^*) = \max\{\chi_i' + \chi_i^{*'} : 1 \leq i \leq p\}.$$

Exercise 2.7. Show that there is always a pair of dual bases $\mathbf{v} = (v_i)$ and $\mathbf{v}^* = (v_i^*)$ for which the minimum is achieved in (2.6), i.e.,

$$\gamma(\chi, \chi^*) = \max\{\chi(v_i) + \chi^*(v_i^*) : 1 \leq i \leq p\}.$$

The following theorem establishes some relations between the two coefficients.

Theorem 2.8. *The following statements hold:*

(1) $\pi(\chi, \chi^*) \leq \gamma(\chi, \chi^*)$;
(2) *if* $\chi \sim \chi^*$, *then* $0 \leq \pi(\chi, \chi^*) \leq \gamma(\chi, \chi^*) \leq p \, \pi(\chi, \chi^*)$.

---

[2] This amount of irregularity depends on some global parameters of the system.

Proof. We begin with the following lemma.

Lemma 2.9. *Given numbers $\lambda_1 \leq \cdots \leq \lambda_p$ and $\mu_1 \geq \cdots \geq \mu_p$ and a permutation σ of $\{1, \ldots, p\}$, we have*

$$\min\{\lambda_j + \mu_{\sigma(j)} : 1 \leq j \leq p\} \leq \min\{\lambda_i + \mu_i : 1 \leq i \leq p\},$$

$$\max\{\lambda_i + \mu_{\sigma(i)} : 1 \leq i \leq p\} \geq \max\{\lambda_i + \mu_i : 1 \leq i \leq p\}.$$

Proof of the lemma. Notice that the second inequality follows from the first one in view of the following relations:

$$\begin{aligned}
\max\{\lambda_i + \mu_{\sigma(i)} : 1 \leq i \leq p\} &= -\min\{-\mu_{\sigma(i)} - \lambda_i : 1 \leq i \leq p\} \\
&= -\min\{-\mu_i - \lambda_{\sigma^{-1}(i)} : 1 \leq i \leq p\} \\
&\geq -\min\{-\mu_i - \lambda_i : 1 \leq i \leq p\}, \\
&= \max\{\lambda_i + \mu_i : 1 \leq i \leq p\}.
\end{aligned}$$

We now prove the first inequality. We may assume that σ is not the identity permutation (otherwise the result is trivial). Fix an integer i such that $1 \leq i \leq p$. If $i \leq \sigma(i)$, then $\mu_{\sigma(i)} \leq \mu_i$ and

$$\min\{\lambda_j + \mu_{\sigma(j)} : 1 \leq j \leq p\} \leq \lambda_i + \mu_{\sigma(i)} \leq \lambda_i + \mu_i.$$

If $i > \sigma(i)$, then there exists $k < i$ such that $i \leq \sigma(k)$. Otherwise, we would have $\sigma(1), \ldots, \sigma(i-1) \leq i-1$, and hence, $\sigma(i) \geq i$. It follows that

$$\min\{\lambda_i + \mu_{\sigma(i)} : 1 \leq i \leq p\} \leq \lambda_k + \mu_{\sigma(k)} \leq \lambda_i + \mu_i.$$

The desired result now follows. $\square$

We proceed with the proof of the theorem. Consider dual bases $\mathbf{v}$ and $\mathbf{v}^*$. Without loss of generality, we may assume that $\chi(v_1) \leq \cdots \leq \chi(v_p)$. Let σ be a permutation of $\{1, \ldots, p\}$ such that the numbers $\mu_{\sigma(i)} = \chi^*(v_i^*)$ satisfy $\mu_1 \geq \cdots \geq \mu_p$. We have $\chi(v_i) \geq \chi_i'$ and $\mu_i \geq \chi_i^{*\prime}$. By Lemma 2.9, we obtain

$$\begin{aligned}
\max\{\chi(v_i) + \chi^*(v_i^*) : 1 \leq i \leq p\} &\geq \max\{\chi(v_i) + \mu_i : 1 \leq i \leq p\} \\
&\geq \max\{\chi_i' + \chi_i^{*\prime} : 1 \leq i \leq p\} \\
&= \pi(\chi, \chi^*).
\end{aligned}$$

Therefore, $\gamma(\chi, \chi^*) \geq \pi(\chi, \chi^*)$ and statement (1) follows.

We assume now that $\chi \sim \chi^*$. Let $\mathcal{V}_\chi = \{V_i : i = 1, \ldots, s\}$ be the filtration associated to χ and let $\mathcal{V}_{\chi^*} = \{V_i^* : i = 1, \ldots, s^*\}$ be the filtration associated to χ^*. We have that

$$\{0\} = V_0 \subsetneq V_1 \subsetneq \cdots \subsetneq V_s = V, \quad V^* = V_1^* \supsetneq \cdots \supsetneq V_{s^*+1}^* = \{0\}.$$

One can choose dual bases $\mathbf{v}$ and $\mathbf{v}^*$ such that $\mathbf{v}$ is subordinate to the filtration $\mathcal{V}_\chi$ while $\mathbf{v}^*$ is subordinate to the filtration $\mathcal{V}_{\chi^*}$. Indeed, consider

2.2. The Lyapunov and Perron irregularity coefficients

the filtration $\mathcal{V}_{\chi^*}^\perp$ that is comprised of the orthogonal complements $V_i^{*\perp}$ to the subspaces forming the filtration $\mathcal{V}_{\chi^*}$, i.e.,

$$V_1^{*\perp} \subset \cdots \subset V_{s^*}^{*\perp}.$$

There is a basis $\mathbf{v}$ of V which is subordinate to both filtrations $\mathcal{V}_\chi$ and $\mathcal{V}_{\chi^*}^\perp$. Then the basis $\mathbf{v}^*$ of V^* that is dual to $\mathbf{v}$ is subordinate to $\mathcal{V}_{\chi^*}$.

We assume that the basis $\mathbf{v}$ is ordered. It follows that $\chi(v_i) = \chi_i'$ and $\mu_i = \chi_i^{*\prime}$ for each i. Thus,

$$\gamma(\chi, \chi^*) \leq \max\{\chi(v_i) + \chi^*(v_i^*) : 1 \leq i \leq p\}$$
$$\leq \sum_{i=1}^p (\chi(v_i) + \chi^*(v_i^*)) = \sum_{i=1}^p (\chi_i' + \chi_i^{*\prime})$$
$$\leq p \max\{\chi_i' + \chi_i^{*\prime} : 1 \leq i \leq p\} = p\,\pi(\chi, \chi^*).$$

Finally, since $\chi \sim \chi^*$, we have $\gamma(\chi, \chi^*) \geq 0$. This implies that $\pi(\chi, \chi^*) \geq 0$, and statement (2) follows. $\square$

2.2.2. Regularity of the Lyapunov exponent. We introduce the crucial concept of regularity of a pair of Lyapunov exponents χ and χ^* in dual vector spaces V and V^*. Roughly speaking, regularity means that the filtrations $\mathcal{V}_\chi$ and $\mathcal{V}_{\chi^*}$ are well adapted to each other (in particular, they are orthogonal; see Theorem 2.10 below). This yields some special properties of Lyapunov exponents which determine their role in the stability theory. At first glance the regularity requirements seem quite strong and even a bit artificial. However, they hold in "typical" situations.

The pair of Lyapunov exponents (χ, χ^*) is said to be *regular* if $\chi \sim \chi^*$ and $\gamma(\chi, \chi^*) = 0$.[3] By Theorem 2.8, this holds if and only if $\pi(\chi, \chi^*) = 0$ and also if and only if $\chi_i^{*\prime} = -\chi_i'$.

Theorem 2.10. *If the pair (χ, χ^*) is regular, then the filtrations $\mathcal{V}_\chi$ and $\mathcal{V}_{\chi^*}$ are orthogonal, that is, $s = s^*$, $\dim V_i + \dim V_{s-i}^* = p$, and $\langle v, v^* \rangle = 0$ for every $v \in V_i$ and $v^* \in V_{s-i}^*$.*

Proof. Set $m_i = p - \dim V_{s^*-i}^* + 1$. Then

$$V_{s-i}^* = \{v^* \in V^* : \chi^*(v^*) \leq \chi_{m_i}^{*\prime}\}$$

and $\chi_{m_i}^{*\prime} = -\chi_{m_i}'$ in view of Theorem 2.8. Let $\mathbf{v}$ be a basis of V that is subordinate to $\mathcal{V}_\chi$ and is ordered, and let $\mathbf{v}^*$ be the basis of V^* that is dual

---

[3] In traditional terminology (which we followed in the first edition of this book) the number $\gamma(\chi, \chi^*)$ is called the regularity coefficient. However, in view of the definition of regularity of the pair of Lyapunov exponents (χ, χ^*), this coefficient actually measures the level of irregularity of this pair thus justifying calling $\gamma(\chi, \chi^*)$ the irregularity coefficient.

to **v**. Since $\chi \sim \chi^*$, we obtain
$$\begin{aligned} V^*_{s-i} &= \{v^* \in V^* : \chi'_{m_i} + \chi^*(v^*) \leq 0\} \\ &= \{v^* \in V^* : \chi(v_j) + \chi^*(v^*) < 0 \text{ if and only if } j < m_i\} \\ &= \text{span}\{v^*_{m_i}, \ldots, v^*_p\}. \end{aligned}$$

This implies that the basis $\mathbf{v}^*$ is subordinate to $\mathcal{V}_{\chi^*}$ and is ordered. Fix i and consider any basis $\tilde{\mathbf{v}} = (\tilde{v}_1, \ldots, \tilde{v}_{p_i}, v_{p_i+1}, \ldots, v_p)$ of V that is subordinate to $\mathcal{V}_\chi$ and is ordered. Let $\tilde{\mathbf{v}}^*$ be the basis of V^* that is dual to $\tilde{\mathbf{v}}$. Then the last $p - p_i$ components of $\tilde{\mathbf{v}}^*$ coincide with those of $\mathbf{v}^*$. This implies that $s^* = s$ and
$$V^*_{s-i} = \text{span}\{v^*_{n_i+1}, \ldots, v^*_p\} = V_i^\perp.$$

The desired result now follows. □

2.3. Lyapunov exponents for linear differential equations

Consider a linear differential equation
$$\dot{v} = A(t)v, \tag{2.7}$$

where $v(t) \in \mathbb{C}^p$ and $A(t)$ is a $p \times p$ matrix with complex entries depending continuously on $t \in \mathbb{R}$. For every $v_0 \in \mathbb{C}^p$ there exists a unique solution $v(t) = v(t, v_0)$ of equation (2.7) that is defined for every $t \in \mathbb{R}$ and satisfies the initial condition $v(0, v_0) = v_0$. In what follows we will always assume that the matrix function $A(t)$ is bounded, i.e.,
$$\sup\{\|A(t)\| : t \in \mathbb{R}\} < \infty. \tag{2.8}$$

This requirement guarantees that for any $v_0 \in \mathbb{C}^p$ equation (2.7) has a unique solution which is well-defined for all $t \in \mathbb{R}$.

2.3.1. The definition of the Lyapunov exponents.
Consider the function $\chi^+ \colon \mathbb{C}^p \to \mathbb{R} \cup \{-\infty\}$ given by the formula
$$\chi^+(v_0) = \limsup_{t \to +\infty} \frac{1}{t} \log\|v(t)\|, \tag{2.9}$$

for each $v_0 \in \mathbb{C}^p$, where $v(t)$ is the unique solution of (2.7) satisfying the initial condition $v(0) = v_0$. Condition (2.8) ensures that the function χ^+ is well-defined and bounded on $\mathbb{R} \setminus \{0\}$.

Exercise 2.11. Show that $\chi^+(v_0)$ is a Lyapunov exponent in $\mathbb{C}^p$.

By Theorem 2.1, the function χ^+ attains only finitely many distinct values $\chi_1^+ < \cdots < \chi_{s^+}^+$ on $\mathbb{C}^p \setminus \{0\}$ where $s^+ \leq p$. Each number χ_i^+ is finite and occurs with some multiplicity k_i so that $\sum_{i=1}^{s^+} k_i = p$. We denote by $\mathcal{V}^+$ the filtration of $\mathbb{C}^p$ associated to χ^+:
$$\{0\} = V_0^+ \subsetneq V_1^+ \subsetneq \cdots \subsetneq V_{s^+}^+ = \mathbb{C}^p,$$

2.3. Lyapunov exponents for linear differential equations

where
$$V_i^+ = \{v \in \mathbb{C}^p : \chi^+(v) \leq \chi_i^+\}.$$
We also denote by
$$\operatorname{Sp}\chi^+ = \{(\chi_i^+, k_i^+) : 1 \leq i \leq s^+\}$$
the Lyapunov spectrum of χ^+.

Note that for every $\varepsilon > 0$ there exists $C_\varepsilon > 0$ such that for every solution $v(t)$ of (2.7) and any $t \geq 0$ we have
$$\|v(t)\| \leq C_\varepsilon e^{(\chi_{s^+}^+ + \varepsilon)t}\|v(0)\|.$$
In particular, if $\chi_{s^+}^+ < 0$, then the zero solution of equation (2.7) is exponentially stable.

2.3.2. The dual Lyapunov exponent. Consider the linear differential equation that is dual to (2.7),
$$\dot{w} = -A(t)^* w, \qquad (2.10)$$
where $w(t) \in \mathbb{C}^p$ and $A(t)^*$ denotes the complex-conjugate transpose of $A(t)$. Condition (2.8) ensures that for any $w_0 \in \mathbb{C}^p$ there exists a unique global solution $w(t)$ of this equation satisfying the initial condition $w(0) = w_0$. The function $\chi^{*+} \colon \mathbb{C}^p \to \mathbb{R} \cup \{-\infty\}$ given by
$$\chi^{*+}(w) = \limsup_{t \to +\infty} \frac{1}{t} \log\|w(t)\|$$
defines the Lyapunov exponent associated with equation (2.10). Note that the exponents χ^+ and χ^{*+} are dual. To see that, let $v(t)$ be a solution of equation (2.7) and let $v^*(t)$ be a solution of the dual equation (2.10). Observe that for every $t \in \mathbb{R}$,
$$\frac{d}{dt}\langle v(t), v^*(t)\rangle = \langle A(t)v(t), v^*(t)\rangle + \langle v(t), -A(t)^* v^*(t)\rangle$$
$$= \langle A(t)v(t), v^*(t)\rangle - \langle A(t)v(t), v^*(t)\rangle = 0,$$
where $\langle \cdot, \cdot \rangle$ denotes the standard inner product in $\mathbb{C}^p$. Hence,
$$\langle v(t), v^*(t)\rangle = \langle v(0), v^*(0)\rangle \qquad (2.11)$$
for any $t \in \mathbb{R}$. Now choose dual bases $(v_1, \ldots, v_p)$ and $(v_1^*, \ldots, v_p^*)$ of $\mathbb{C}^p$. Let $v_i(t)$ be the unique solution of (2.7) such that $v_i(0) = v_i$, and let $v_i^*(t)$ be the unique solution of (2.10) such that $v_i^*(0) = v_i^*$, for each i. We obtain for every $t \in \mathbb{R}$,
$$\|v_i(t)\| \cdot \|v_i^*(t)\| \geq \langle v_i(t), v_i(t)\rangle = \langle v_i(0), v_i(0)\rangle = 1.$$
Hence,
$$\chi^+(v_i) + \chi^{*+}(v_i^*) \geq 0$$
for every i. It follows that the exponents χ^+ and χ^{*+} are dual.

2.3.3. Regularity of the pair (χ^+, χ^{*+}) of Lyapunov exponents.

We discuss the regularity of the pair of Lyapunov exponents (χ^+, χ^{*+}). Let $\mathbf{v} = (v_1, \ldots, v_p)$ be a basis of $\mathbb{C}^p$. Denote by $\mathrm{Vol}_m^{\mathbf{v}}(t)$ the volume of the m-parallelepiped generated by the vectors $v_i(t)$, $i = 1, \ldots, m$, each of which is the solution of (2.7) satisfying the initial condition $v_i(0) = v_i$. Let $V_m(t)$ be the $m \times m$ matrix whose entries are $\langle v_i(t), v_j(t) \rangle$. Then

$$\mathrm{Vol}_m^{\mathbf{v}}(t) = |\det V_m(t)|^{1/2}.$$

In particular, $\mathrm{Vol}_1^{\mathbf{v}}(t) = |v_1(t)|$ and

$$\mathrm{Vol}_p^{\mathbf{v}}(t) = \mathrm{Vol}_p^{\mathbf{v}}(0) \exp\left(\int_0^t \mathrm{tr}\, A(\tau)\, d\tau\right). \tag{2.12}$$

The following theorem provides some crucial criteria for the pair (χ^+, χ^{*+}) to be regular.

Theorem 2.12. *The following statements are equivalent:*

(1) *the pair (χ^+, χ^{*+}) is regular;*

(2) *we have*

$$\lim_{t \to +\infty} \frac{1}{t} \int_0^t \mathrm{tr}\, A(\tau)\, d\tau = \sum_{i=1}^{s^+} k_i \chi_i^+; \tag{2.13}$$

(3) *for any basis $\mathbf{v}$ of $\mathbb{C}^p$ that is subordinate to $\mathcal{V}_{\chi^+}$ and is ordered and for any $1 \leq m \leq p$ the following limit exists:*

$$\lim_{t \to +\infty} \frac{1}{t} \log \mathrm{Vol}_m^{\mathbf{v}}(t). \tag{2.14}$$

In addition, if the pair (χ^+, χ^{*+}) is regular, then for any basis $\mathbf{v}$ of $\mathbb{C}^p$ that is subordinate to $\mathcal{V}_{\chi^+}$ and is ordered:

(a) *for any $1 \leq m \leq p$ we have*

$$\lim_{t \to +\infty} \frac{1}{t} \log \mathrm{Vol}_m^{\mathbf{v}}(t) = \sum_{i=1}^m \chi^+(v_i);$$

(b) *for any $1 \leq m < p$ the angle $\sigma(t)$ between $\mathrm{span}\{v_i(t) : i \leq m\}$ and $\mathrm{span}\{v_i(t) : i > m\}$ satisfies*

$$\lim_{t \to +\infty} \frac{1}{t} \log \sin \sigma(t) = 0.$$

Remark 2.13. As can be seen from Theorem 2.12, regularity of the pair (χ^+, χ^{*+}) of Lyapunov exponents can be ensured by requiring that either the limit in (2.14) exists for all $1 \leq m \leq p$ or this limit exists only for $m = p$ and is equal to the total sum of Lyapunov exponents; see Section 2.3.4 for more discussions on these requirements.

2.3. Lyapunov exponents for linear differential equations

Proof of the theorem. We adopt the following notation. Given a function $f\colon (0,\infty) \to \mathbb{R}$, we set

$$\overline{\chi}(f) = \limsup_{t\to+\infty} \frac{1}{t}\log|f(t)| \quad \text{and} \quad \underline{\chi}(f) = \liminf_{t\to+\infty} \frac{1}{t}\log|f(t)|. \qquad (2.15)$$

If, in addition, f is integrable, we shall also write

$$\overline{f} = \limsup_{t\to+\infty} \frac{1}{t}\int_0^t f(\tau)\,d\tau \quad \text{and} \quad \underline{f} = \liminf_{t\to+\infty} \frac{1}{t}\int_0^t f(\tau)\,d\tau.$$

We first show that statement (1) implies statement (2). We start with an auxiliary result. Let

$$\Delta(t) = \exp\left(\int_0^t \operatorname{tr} A(\tau)\,d\tau\right).$$

Lemma 2.14. *The following statements hold:*

(1) $\underline{\chi}(\Delta) = \operatorname{Re}\underline{\operatorname{tr} A}$ *and* $\overline{\chi}(\Delta) = \operatorname{Re}\overline{\operatorname{tr} A}$;

(2) *if* $(v_1,\ldots,v_p)$ *is a basis of* $\mathbb{C}^p$, *then*

$$-\sum_{i=1}^p \chi^{*+}(v_i) \le \underline{\chi}(\Delta) \le \overline{\chi}(\Delta) \le \sum_{i=1}^p \chi^+(v_i).$$

Proof of the lemma. We have

$$\underline{\chi}(\Delta) = \liminf_{t\to+\infty} \frac{1}{t}\operatorname{Re}\int_0^t \operatorname{tr} A(\tau)\,d\tau = \operatorname{Re}\underline{\operatorname{tr} A}$$

and

$$\overline{\chi}(\Delta) = \limsup_{t\to+\infty} \frac{1}{t}\operatorname{Re}\int_0^t \operatorname{tr} A(\tau)\,d\tau = \operatorname{Re}\overline{\operatorname{tr} A}.$$

This proves the first statement.

Observe that for a collection $\mathbf{v} = (v_1,\ldots,v_p)$ of linearly independent vectors the volume of the parallelepiped formed by the vectors $v_1(t),\ldots,v_p(t)$ is given by $\operatorname{Vol}_p^{\mathbf{v}}(t) = \operatorname{Vol}_p^{\mathbf{v}}(0)\Delta(t)$. Therefore,

$$\Delta(t) \le \prod_{i=1}^p \|v_i(t)\|,$$

and hence,

$$\overline{\chi}(\Delta) \le \sum_{i=1}^p \chi^+(v_i).$$

In a similar way,

$$-\underline{\chi}(\Delta) = -\operatorname{Re}\underline{\operatorname{tr} A} = \operatorname{Re}\overline{\operatorname{tr}(-A^*)} = \overline{\chi}(\Delta^*) \le \sum_{i=1}^p \chi^{*+}(v_i),$$

where
$$\Delta^*(t) = \exp\left(-\int_0^t \operatorname{tr}(A(\tau)^*)\, d\tau\right).$$
The lemma follows. □

We proceed with the proof of the theorem. Let χ'_i and $\chi^{*\prime}_i$ be the values of the Lyapunov exponents χ^+ and χ^{*+}, counted with their multiplicities. Choose a basis $(v_1, \ldots, v_p)$ of $\mathbb{C}^p$ that is subordinate to $\mathcal{V}_{\chi^+}$. It follows from Lemma 2.14 that
$$-\sum_{i=1}^p \chi^{*\prime}_i \leq \underline{\chi}(\Delta) \leq \overline{\chi}(\Delta) \leq \sum_{i=1}^p \chi'_i.$$
Therefore,
$$\overline{\chi}(\Delta) - \underline{\chi}(\Delta) \leq \sum_{i=1}^p (\chi'_i + \chi^{*\prime}_i) \leq p\,\pi(\chi^+, \chi^{*+}).$$
This shows that if the pair (χ^+, χ^{*+}) is regular, then
$$\underline{\chi}(\Delta) = \overline{\chi}(\Delta) = \sum_{i=1}^p \chi'_i = -\sum_{i=1}^p \chi^{*\prime}_i,$$
and (2.13) holds.

We now show that statement (3) implies statement (1). We split the proof into two steps, first showing how to reduce the case of a general matrix function $A(t)$ to the case when for every $t \in \mathbb{R}$ the matrix $A(t)$ is triangular. Then we prove the theorem in this latter case.

Step 1. For every $t \geq 0$ consider a linear coordinate change in $\mathbb{C}^p$ given by a differentiable matrix function $U(t)$. Setting $z(t) = U(t)^{-1}v(t)$, we obtain
$$\dot{v}(t) = \dot{U}(t)z(t) + U(t)\dot{z}(t)$$
$$= A(t)v(t) = A(t)U(t)z(t).$$
It follows that $\dot{z} = B(t)z$ where the matrix function $B(t) = (b_{ij}(t))$ is defined by
$$B(t) = U(t)^{-1}A(t)U(t) - U(t)^{-1}\dot{U}(t). \tag{2.16}$$
We need the following lemma of Perron. Its main manifestation is to show how to reduce the linear differential equation (2.7) with a general matrix function $A(t)$ to a linear differential equation with a triangular matrix function.

Lemma 2.15. *There exists a differentiable matrix function $U(t)$ such that:*
   (1) *$U(t)$ is unitary;*
   (2) *the matrix $B(t)$ is upper-triangular;*
   (3) *$\sup\{|b_{ij}(t)| : t \geq 0, i \neq j\} < \infty$;*

2.3. Lyapunov exponents for linear differential equations

(4) for $k = 1, \ldots, p$,
$$\operatorname{Re} b_{kk}(t) = \frac{d}{dt} \log \frac{\operatorname{Vol}_k^{\mathbf{v}}(t)}{\operatorname{Vol}_{k-1}^{\mathbf{v}}(t)}.$$

Proof of the lemma. Given a basis $\mathbf{v} = (v_1, \ldots, v_p)$, we construct the desired matrix function $U(t)$ by applying the Gram–Schmidt orthogonalization procedure to the basis $\mathbf{v}(t) = (v_1(t), \ldots, v_p(t))$ where $v_i(t)$ is the solution of (2.7) satisfying the initial condition $v_i(0) = v_i$. Thus, we obtain a collection of functions $u_1(t), \ldots, u_p(t)$ such that $\langle u_i(t), u_j(t) \rangle = \delta_{ij}$ where δ_{ij} is the Kronecker symbol. Let $V(t)$ and $U(t)$ be the matrices with columns $v_1(t), \ldots, v_p(t)$ and $u_1(t), \ldots, u_p(t)$, respectively. The matrix $U(t)$ is unitary. Moreover, the Gram–Schmidt procedure can be effected in such a way that each function $u_k(t)$ is a linear combination of functions $v_1(t), \ldots, v_k(t)$. It follows that the matrix $Z(t) = U(t)^{-1} V(t)$ is upper-triangular.

The columns $z_1(t) = U(t)^{-1} v_1(t), \ldots, z_p(t) = U(t)^{-1} v_p(t)$ of the matrix $Z(t)$ form a basis of the space of solutions of the linear differential equation $\dot{z} = B(t) z$. Writing this equation in matrix form, we obtain
$$\dot{Z}(t) = B(t) Z(t). \tag{2.17}$$

It follows that $B(t) = \dot{Z}(t) Z(t)^{-1}$ and as $Z(t)$ is upper-triangular, so is the matrix $B(t)$. Since $U(t)$ is unitary, using (2.16), we obtain

$$\begin{aligned} B(t) + B(t)^* &= U(t)^* (A(t) + A(t)^*) U(t) - (U(t)^* \dot{U}(t) + \dot{U}(t)^* U(t)) \\ &= U(t)^* (A(t) + A(t)^*) U(t) - \frac{d}{dt}(U(t)^* U(t)) \\ &= U(t)^* (A(t) + A(t)^*) U(t). \end{aligned}$$

Since $B(t)$ is triangular, we conclude that $|b_{ij}(t)| \le 2\|A(t)\| < \infty$ uniformly over $t \ge 0$ and $i \ne j$, proving the third statement.

In order to prove the last statement of the lemma, assume first that all entries of the matrix $Z(t) = (z_{ij}(t))$ are real. Then the entries of the matrix $B(t)$ are also real and
$$b_{kk}(t) = \frac{\dot{z}_{kk}(t)}{z_{kk}(t)} = \frac{d}{dt} \log z_{kk}(t).$$

Since $V(t) = U(t) Z(t)$ and $Z(t)$ is upper-triangular, we obtain
$$v_i(t) = \sum_{1 \le \ell \le i} u_\ell(t) z_{\ell i}(t).$$

Therefore,
$$\langle v_i(t), v_j(t)\rangle = \sum_{1\le \ell \le i, 1\le m\le j} \delta_{\ell m} z_{\ell i}(t)\overline{z_{mj}(t)}$$
$$= \sum_{1\le \ell \le \min\{i,j\}} z_{\ell i}(t)\overline{z_{\ell j}(t)} = \langle z_i(t), z_j(t)\rangle.$$

Set $\mathbf{z} = (z_1(0), \ldots, z_n(0))$. This implies that $\mathrm{Vol}^{\mathrm{v}}_k(t) = \mathrm{Vol}^{\mathrm{z}}_k(t)$ for each k, and thus,
$$\frac{\mathrm{Vol}^{\mathrm{v}}_k(t)}{\mathrm{Vol}^{\mathrm{v}}_{k-1}(t)} = \frac{\mathrm{Vol}^{\mathrm{z}}_k(t)}{\mathrm{Vol}^{\mathrm{z}}_{k-1}(t)} = z_{kk}(t).$$

In the general case (when the entries of $Z(t)$ are not necessarily real) we can write
$$\frac{\mathrm{Vol}^{\mathrm{v}}_k(t)}{\mathrm{Vol}^{\mathrm{v}}_{k-1}(t)} = |z_{kk}(t)|$$

and
$$\frac{d}{dt}\log|z_{kk}(t)| = \frac{1}{2}\frac{d}{dt}\log(\overline{z_{kk}(t)}z_{kk}(t))$$
$$= \frac{1}{2}\left(\frac{\overline{\dot z_{kk}(t)}}{\overline{z_{kk}(t)}} + \frac{\dot z_{kk}(t)}{z_{kk}(t)}\right)$$
$$= \frac{1}{2}(\overline{b_{kk}(t)} + b_{kk}(t)) = \mathrm{Re}\, b_{kk}(t).$$

This completes the proof of the lemma. $\square$

Consider a linear differential equation $\dot z = B(t)z$ with an upper-triangular matrix $B(t)$ and the corresponding matrix equation $\dot Z = B(t)Z$. We will give explicit formulae for the entries of the solution matrix $Z(t)$. To this end let $Z(t) = (z_{ij}(t))$ be a $p \times p$ matrix function defined as follows: given a collection of constants a_{ij}, $1 \le i < j \le p$, set

$$\begin{aligned}z_{ij}(t) &= 0 \quad \text{if } j < i,\\ z_{ij}(t) &= \exp\left(\int_0^t b_{ii}(\tau)\,d\tau\right) \quad \text{if } j = i,\\ z_{ij}(t) &= \int_{a_{ij}}^t \sum_{k=i+1}^j b_{ik}(s)z_{kj}(s)e^{\int_s^t b_{ii}(\tau)\,d\tau}ds \quad \text{if } j > i.\end{aligned} \qquad (2.18)$$

Lemma 2.16. *For any constants a_{ij}, $1 \le i < j \le p$, the columns of the matrix $Z(t)$ form a basis of solutions of the equation $\dot z = B(t)z$.*

2.3. Lyapunov exponents for linear differential equations

Proof of the lemma. For each i we have $\dot{z}_{ii}(t) = b_{ii}(t)z_{ii}(t)$ and for each $j > i$,

$$\dot{z}_{ij}(t) = \sum_{k=i+1}^{j} b_{ik}(t)z_{kj}(t) + b_{ii}(t)z_{ij}(t) = \sum_{k=i}^{j} b_{ik}(t)z_{kj}(t).$$

This shows that $\dot{Z}(t) = B(t)Z(t)$, and hence, the columns of $Z(t)$ (i.e., the vectors $\mathbf{z}_i(t) = (z_{1i}(t), \ldots, z_{pi}(t))$) are solutions of the equation $\dot{z} = B(t)z$. Since $Z(t)$ is upper-triangular, we have

$$\det Z(t) = \exp\left(\sum_{i=1}^{n} \int_0^t b_{ii}(\tau)\, d\tau\right) \neq 0,$$

and hence, the vectors $\mathbf{z}_i(t)$ form a basis. $\square$

Step 2. Assume that $\overline{\chi}(\mathrm{Vol}^{\mathbf{v}}_m) = \underline{\chi}(\mathrm{Vol}^{\mathbf{v}}_m)$ for any basis $\mathbf{v}$ that is subordinate to $\mathcal{V}_{\chi^+}$ and is ordered and for any $1 \leq m \leq p$. We show that the pair (χ^+, χ^{*+}) is regular. By Lemma 2.15, it suffices to consider the equation $\dot{z} = B(t)z$ where $B(t)$ is a $p \times p$ upper-triangular matrix for every t.

Lemma 2.17. *If $B_i := \overline{\mathrm{Re}\, b_{ii}} = \underline{\mathrm{Re}\, b_{ii}}$ for each $i = 1, \ldots, p$, then:*

(1) *the pair of Lyapunov exponents corresponding to the equations $\dot{z} = B(t)z$ and $\dot{w} = -B(t)^* w$ is regular;*

(2) *the numbers $B_1, \ldots, B_p$ are the values of the Lyapunov exponent χ^+;*

(3) *the numbers $-B_1, \ldots, -B_p$ are the values of the Lyapunov exponent χ^{*+}.*

Proof of the lemma. Consider the solutions of the equation $\dot{z} = B(t)z$ described in Lemma 2.16. We wish to show that for each vector

$$z_i(t) = (z_{1i}(t), \ldots, z_{pi}(t))$$

which is the ith column of the matrix $Z(t)$ we have for an appropriate choice of the constants a_{ij} in (2.18) that

$$\chi^+(z_i) = \limsup_{t \to +\infty} \frac{1}{t} \log \|z_i(t)\| = B_i. \tag{2.19}$$

Since $\overline{\mathrm{Re}\, b_{ii}} = B_i$, we clearly have $\chi^+(z_{ii}) = B_i$. Assume now that

$$\chi^+(z_{kj}) \leq B_j \quad \text{for each } i+1 \leq k \leq j. \tag{2.20}$$

We will show that $\chi^+(z_{ij}) \leq B_j$. By Lemma 2.15, $|b_{ij}(t)| \leq K$ for some $K > 0$ independent of i, j, and t. Hence, by (2.15), (2.20), and the fact that $\underline{\mathrm{Re}\, b_{ii}} = B_i$, for each $\varepsilon > 0$ there exists $K' > 0$ such that for all $s \geq 0$

$$\left| \sum_{k=i+1}^{j} b_{ik}(s)z_{kj}(s)e^{-\int_0^s b_{ii}(\tau)\, d\tau} \right| \leq K' e^{(B_j - B_i + \varepsilon)s}.$$

Since for $j > i$

$$z_{ij}(t) = e^{\int_0^t b_{ii}(\tau)\,d\tau} \int_{a_{ij}}^t \sum_{k=i+1}^j b_{ik}(s) z_{kj}(s) e^{-\int_0^s b_{ii}(\tau)\,d\tau}\,ds,$$

for each $\varepsilon > 0$ we have

$$\chi^+(z_{ij}) \leq \limsup_{t \to +\infty} \frac{1}{t}\left(\log\left|e^{\int_0^t b_{ii}(\tau)\,d\tau}\right|\right.$$
$$\left. + \log\left|\int_{a_{ij}}^t \sum_{k=i+1}^j b_{ik}(s) z_{kj}(s) e^{-\int_0^s b_{ii}(\tau)\,d\tau}\,ds\right|\right)$$
$$\leq B_i + \limsup_{t \to +\infty} \frac{1}{t} \log\left|\int_{a_{ij}}^t K' e^{(B_j - B_i + \varepsilon)s}\,ds\right|.$$

For each $j > i$ set $a_{ij} = 0$ if $B_j - B_i \geq 0$ and $a_{ij} = +\infty$ if $B_j - B_i < 0$. Then for every sufficiently small $\varepsilon > 0$ we have

$$\chi^+(z_{ij}) \leq B_i + \limsup_{t \to +\infty} \frac{1}{t} \log \frac{K'(e^{(B_j - B_i + \varepsilon)t} - 1)}{B_j - B_i + \varepsilon}$$

if $B_j - B_i \geq 0$ and

$$\chi^+(z_{ij}) \leq B_i + \limsup_{t \to +\infty} \frac{1}{t} \log \frac{K' e^{(B_j - B_i + \varepsilon)t}}{B_j - B_i + \varepsilon}$$

if $B_j - B_i < 0$. Therefore,

$$\chi^+(z_{ij}) \leq B_i + B_j - B_i + \varepsilon = B_j + \varepsilon.$$

Since ε is arbitrary, we obtain $\chi^+(z_{ij}) \leq B_j$. This shows that $\chi^+(z_i) = B_i$ for each $1 \leq i \leq p$.

Consider the dual linear differential equation $\dot{w}(t) = -B(t)^* w(t)$. In a way similar to the above, one can show that there exists a lower-triangular matrix $W(t)$ such that

$$\dot{W}(t) = -B(t)^* W(t)$$

(compare to (2.17)) and such that the entries of the matrix $W(t)$ are defined by (compare to (2.18))

$$w_{ij}(t) = 0 \quad \text{if } j > i,$$
$$w_{ij}(t) = \exp\left(-\int_0^t \overline{b_{jj}(\tau)}\,d\tau\right) \quad \text{if } j = i,$$
$$w_{ij}(t) = -\int_{a_{ji}}^t \sum_{k=j}^{i-1} \overline{b_{ki}(s)} w_{kj}(s) e^{-\int_s^t \overline{b_{ii}(\tau)}\,d\tau}\,ds \quad \text{if } j < i,$$

2.3. Lyapunov exponents for linear differential equations

where the constants a_{ji} are chosen as above. Since $\operatorname{Re} b_{ii} = B_i$, one can show in a similar manner that the columns $w_1(t), \ldots, w_p(t)$ of $W(t)$ satisfy

$$\chi^{*+}(w_i) = \limsup_{t \to +\infty} \frac{1}{t} \log\|w_i(t)\| = -B_i. \tag{2.21}$$

Note that $\chi^+(z_i) + \chi^{*+}(w_i) = 0$ for each i. In order to prove that the pair (χ^+, χ^{*+}) is regular, it remains to show that the bases $\mathbf{z}$ and $\mathbf{w}$ are dual. Clearly, $\langle z_i(0), w_j(0) \rangle = 0$ for every $i < j$. Moreover, $\langle z_i(0), w_i(0) \rangle = 1$ for each $1 \le i \le p$. We show that $\langle z_i(0), w_j(0) \rangle = 0$ for every $i > j$. For such an i and $t > 0$ we have that

$$\langle z_i(t), w_j(t) \rangle = \sum_{k=j}^{i} z_{ki}(t) \overline{w_{kj}(t)}. \tag{2.22}$$

By (2.19) and (2.21) we have $\chi^+(z_{ij}) \le B_j$ and $\chi^{*+}(w_{ij}) \le -B_j$. Hence, for every i, j, and $\varepsilon > 0$, we obtain (recall (2.15))

$$\overline{\chi}(\langle z_i, w_j \rangle) \le \max_{j+1 \le k \le i-1} \overline{\chi}(z_{ki} \overline{w_{kj}})$$

$$\le \max_{j+1 \le k \le i-1} \limsup_{t \to +\infty} \frac{1}{t} \left(\log \left| \int_{a_{ki}}^{t} K' e^{(B_i - B_k + \varepsilon)s} \, ds \right| \right.$$

$$\left. + \log \left| \int_{a_{jk}}^{t} K' e^{(-B_j + B_k + \varepsilon)s} \, ds \right| \right)$$

$$\le \max_{j+1 \le k \le i-1} (B_i - B_k - B_j + B_k + 2\varepsilon) = B_i - B_j + 2\varepsilon.$$

Since ε is arbitrary, if $B_i - B_j < 0$, we obtain

$$\overline{\chi}(\langle z_i, w_j \rangle) < 0$$

and using (2.11) we conclude that

$$\langle z_i(0), w_j(0) \rangle = \lim_{t \to +\infty} \langle z_i(t), w_j(t) \rangle = 0.$$

If $B_i - B_j \ge 0$, then $a_{ji} = 0$. Moreover, for each k we have $B_i - B_k \ge 0$ or $B_k - B_j \ge 0$, and hence, $a_{ki} = 0$ or $a_{jk} = 0$. Letting $t = 0$ in (2.22) yields

$$\langle z_i(0), w_j(0) \rangle = z_{ji}(0) \overline{w_{jj}(0)} + z_{ii}(0) \overline{w_{ij}(0)} + \sum_{k=j+1}^{i-1} z_{ki}(0) \overline{w_{kj}(0)}.$$

Since $i > j$ and $a_{ji} = 0$, we obtain $z_{ji}(0) = w_{ij}(0) = 0$. Moreover, for each k such that $j + 1 \le k \le i - 1$ we have $a_{ki} = 0$ or $a_{jk} = 0$, and hence, $z_{ki}(0) = 0$ or $w_{kj}(0) = 0$. Therefore, each term in the above sum is zero. Thus, $\langle z_i(0), w_j(0) \rangle = 0$ for every $i > j$, and the lemma follows. $\square$

By Lemma 2.15 and statement (3) of the theorem, we have

$$\frac{1}{t}\int_0^t \operatorname{Re} b_{ii}(\tau)\,d\tau = \frac{1}{t}\int_0^t \frac{d}{d\tau}\log\frac{\operatorname{Vol}_i^{\mathbf{v}}(t)}{\operatorname{Vol}_{i-1}^{\mathbf{v}}(t)}\,d\tau$$

$$= \frac{1}{t}\log\frac{\operatorname{Vol}_i^{\mathbf{v}}(t)/\operatorname{Vol}_{i-1}^{\mathbf{v}}(t)}{\operatorname{Vol}_i^{\mathbf{v}}(0)/\operatorname{Vol}_{i-1}^{\mathbf{v}}(0)} \to \overline{\chi}(\operatorname{Vol}_i^{\mathbf{v}}) - \overline{\chi}(\operatorname{Vol}_{i-1}^{\mathbf{v}})$$

as $t \to +\infty$. Since $\overline{\chi}(\operatorname{Vol}_i^{\mathbf{v}}) = \underline{\chi}(\operatorname{Vol}_i^{\mathbf{v}})$, this implies that

$$\overline{\operatorname{Re} b_{ii}} = \limsup_{t\to\infty} \frac{1}{t}\int_0^t \operatorname{Re} b_{ii}(\tau)\,d\tau = \overline{\chi}(\operatorname{Vol}_i^{\mathbf{v}}) - \overline{\chi}(\operatorname{Vol}_{i-1}^{\mathbf{v}})$$

$$= \underline{\chi}(\operatorname{Vol}_i^{\mathbf{v}}) - \underline{\chi}(\operatorname{Vol}_{i-1}^{\mathbf{v}})$$

$$= \liminf_{t\to\infty} \frac{1}{t}\int_0^t \operatorname{Re} b_{ii}(\tau)\,d\tau = \underline{\operatorname{Re} b_{ii}}.$$

We can, therefore, apply Lemma 2.17 and conclude that the Lyapunov exponent corresponding to the equation $\dot{z} = B(t)z$ is regular.

We now show that statement (2) implies statement (3). By Lemma 2.15, for every $1 \leq m \leq p$ we have

$$\operatorname{Vol}_m^{\mathbf{v}}(t) = \operatorname{Vol}_m^{\mathbf{z}}(t) = \prod_{k=1}^m z_{kk}(t). \tag{2.23}$$

By Lyapunov's construction of subordinate bases (see the description of the construction after Theorem 2.4), there exists a basis $(v_1(0),\ldots,v_p(0))$ of $\mathbb{C}^p$ that is subordinate to $\mathcal{V}_{\chi^+}$ such that $z_k(0) = e_k + f_k$ for some $f_k \in \operatorname{span}\{e_1,\ldots,e_{k-1}\}$ where $(e_1,\ldots,e_p)$ is the canonical basis of $\mathbb{C}^p$. Since the matrix solution $Z(t)$ is upper-triangular for each t, the vectors e_k and $Z(t)f_k$ are orthogonal. Therefore, $\|Z(t)f_k\| \leq |z_{kk}(t)|$, and

$$\chi^+(z_k) \geq \limsup_{t\to\infty} \frac{1}{t}\log|z_{kk}(t)| =: \lambda_k. \tag{2.24}$$

Without loss of generality we may use the norm in $\mathbb{C}^p$ given by

$$\|(w_1,\ldots,w_n)\| = |w_1| + \cdots + |w_p|.$$

By (2.24), we obtain

$$\lim_{t\to\infty} \frac{1}{t}\log\operatorname{Vol}_p^{\mathbf{v}}(t) = \sum_{i=1}^s k_i\chi_i^+ \geq \sum_{k=1}^p \chi^+(z_k) \geq \sum_{k=1}^p \lambda_k. \tag{2.25}$$

Furthermore, by (2.12) and Lemma 2.16, we have

$$\operatorname{Vol}_p^{\mathbf{z}}(t)/\operatorname{Vol}_p^{\mathbf{z}}(0) = \exp\left(\int_0^t \operatorname{tr} B(\tau)\,d\tau\right) = \prod_{k=1}^p z_{kk}(t),$$

2.3. Lyapunov exponents for linear differential equations

and hence, by (2.23),
$$\lim_{t\to\infty} \frac{1}{t} \log \mathrm{Vol}_p^v(t) \le \sum_{k=1}^p \lambda_k. \qquad (2.26)$$

It follows from (2.24), (2.25), and (2.26) that
$$\chi^+(v_k) = \chi^+(z_k) = \lim_{t\to+\infty} \frac{1}{t} \log\|z_k(t)\| = \lambda_k$$
for each $k = 1,\ldots,p$. Thus, again by (2.23), for each $1 \le m \le p$ we conclude that
$$\underline{\chi}(\mathrm{Vol}_m^v) = \overline{\chi}(\mathrm{Vol}_m^v) = \sum_{k=1}^m \chi^+(v_k).$$

This completes the proof of Theorem 2.12.

□

2.3.4. Examples of nonregular Lyapunov exponents. We stress that the relation (2.13) includes two requirements:

(1) the limit $\lim_{t\to+\infty} \frac{1}{t} \int_0^t \mathrm{tr}\, A(\tau)\, d\tau$ exists;
(2) this limit is equal to $\sum_{i=1}^{s^+} k_i \chi_i^+$.

The following example illustrates that the second requirement cannot be dropped. Consider the system of differential equations
$$\dot{v}_1 = -a(t)v_1, \quad \dot{v}_2 = a(t)v_2 \qquad (2.27)$$
for $t > 0$ where
$$a(t) = \cos\log t - \sin\log t - 1.$$

Exercise 2.18. Show that the general solution of the system can be written in the form
$$v_1(t) = C_1 b(t)^{-1}, \quad v_2(t) = C_2 b(t), \qquad (2.28)$$
for some constants C_1 and C_2, where
$$b(t) = \exp\left(\int_1^t a(\tau)\, d\tau\right) = \exp(t(\cos\log t - 1)). \qquad (2.29)$$

Observe that the matrix function $A(t)$ of the system of differential equations (2.27) is of the form
$$A(t) = \begin{pmatrix} -a(t) & 0 \\ 0 & a(t) \end{pmatrix},$$
and hence, $\int_0^t \mathrm{tr}\, A(\tau)\, d\tau = 0$ for every t. This shows that the first requirement holds.

Exercise 2.19. Show that

$$\liminf_{t\to+\infty} \frac{1}{t}\int_1^t a(\tau)\,d\tau = -2 \quad \text{and} \quad \limsup_{t\to+\infty} \frac{1}{t}\int_1^t a(\tau)\,d\tau = 0, \qquad (2.30)$$

and hence, the limit of 1-volumes does not exist.

The last exercise shows that the pair of Lyapunov exponents (χ^+, χ^{*+}) is not regular. For the function b in (2.29) we have

$$\limsup_{t\to+\infty} \frac{1}{t}\log(b(t)^{-1}) = 2 \quad \text{and} \quad \limsup_{t\to+\infty} \frac{1}{t}\log b(t) = 0.$$

Hence, it follows readily from (2.28) that for every $v = (v_1, v_2) \neq 0$ the Lyapunov exponent is $\chi^+(v) = 2$ if $v_1 \neq 0$ and $\chi^+(v) = 0$ otherwise. Hence,

$$0 = \lim_{t\to+\infty} \frac{1}{t}\int_0^t \operatorname{tr} A(\tau)\,d\tau < \chi_1^+ + \chi_2^+ = 2.$$

We now illustrate that the pair (χ^+, χ^{*+}) may not be regular if the limit in (2.13) does not exist. Consider the system of differential equations

$$\dot{v}_1 = v_2, \quad \dot{v}_2 = a(t)v_2 \qquad (2.31)$$

for $t > 0$ with the same function $a(t)$ as above.

Exercise 2.20. Show that the general solution of the system can be written in the form

$$v_1(t) = C_1 + C_2 \int_1^t b(\tau)\,d\tau, \quad v_2(t) = C_2 b(t)$$

for some constants C_1 and C_2. Show that for every vector $v = (v_1, v_2) \neq 0$,

$$\chi^+(v) = \limsup_{t\to+\infty} \frac{1}{t}\log\|v(t)\| = 0.$$

For equation (2.31) we have $\int_0^t \operatorname{tr} A(\tau)\,d\tau = \int_0^t a(\tau)\,d\tau$, and hence,

$$\limsup_{t\to+\infty} \frac{1}{t}\int_0^t \operatorname{tr} A(\tau)\,d\tau = 0 = \chi_1^+ + \chi_2^+.$$

On the other hand, it follows from (2.30) that the limit in (2.13) does not exist, and hence, the pair of Lyapunov exponents (χ^+, χ^{*+}) is not regular.

2.4. Forward and backward regularity. The Lyapunov–Perron regularity

2.4.1. Forward and backward Lyapunov exponents and their regularity. An important manifestation of Theorem 2.12 is that one can verify the regularity property of the pair of Lyapunov exponents (χ^+, χ^{*+}) dealing with the Lyapunov exponent χ^+ only. This justifies calling the Lyapunov exponent χ^+ *forward regular* (to stress that we only allow positive time) if the pair of Lyapunov exponents (χ^+, χ^{*+}) is regular.

In an analogous manner, reversing the time, we introduce the Lyapunov exponent $\chi^- \colon \mathbb{C}^p \to \mathbb{R} \cup \{-\infty\}$,

$$\chi^-(v) = \limsup_{t \to -\infty} \frac{1}{|t|} \log \|v(t)\|,$$

where $v(t)$ is the solution of (2.7) satisfying the initial condition $v(0) = v$. The function χ^- takes on only finitely many values $\chi_1^- > \cdots > \chi_{s^-}^-$ where $s^- \le p$. We denote by $\mathcal{V}^-$ the filtration of $\mathbb{C}^p$ associated to χ^-,

$$\mathbb{C}^p = V_0^- \supsetneq V_1^- \supsetneq \cdots \supsetneq V_{s^-}^- \supsetneq V_{s^-+1}^- = \{0\},$$

where

$$V_i^- = \{v \in \mathbb{C}^p : \chi^-(v) \le \chi_i^-\},$$

and by $k_i^- = \dim V_i^- - \dim V_{i+1}^-$ the multiplicity of the value χ_i^- such that $\sum_{i=1}^{s^-} k_i^- = p$. We also denote by

$$\operatorname{Sp} \chi^- = \{(\chi_i^-, k_i^-) : 1 \le i \le s^-\}$$

the Lyapunov spectrum of χ^-.

Consider the dual Lyapunov exponent $\chi^{*-} \colon \mathbb{C}^p \to \mathbb{R} \cup \{-\infty\}$ given by

$$\chi^{*-}(v^*) = \limsup_{t \to -\infty} \frac{1}{|t|} \log \|v^*(t)\|,$$

where $v^*(t)$ is the solution of the dual equation (2.10) satisfying the initial condition $v^*(0) = v^*$. Using an argument similar to the one in Section 2.3.2, one can show that the exponents χ^- and χ^{*-} are dual.

We say that the Lyapunov exponent χ^- is *backward regular* if the pair of Lyapunov exponents (χ^-, χ^{*-}) is regular. Reversing the time in Theorem 2.12, one can show that this pair has properties similar to the ones stated in this theorem. More precisely, we have the following result.

Theorem 2.21. *The following statements are equivalent:*

(1) *the pair (χ^-, χ^{*-}) is regular;*

(2)

$$\lim_{t \to -\infty} \frac{1}{t} \int_0^t \operatorname{tr} A(\tau) \, d\tau = \sum_{i=1}^{s^-} k_i^- \chi_i^-;$$

(3) *for any basis $\mathbf{v}$ of $\mathbb{C}^p$ that is subordinate to $\mathcal{V}_{\chi^-}$ and is ordered and for any $1 \le m \le p$ the following limit exists:*

$$\lim_{t \to -\infty} \frac{1}{t} \log \operatorname{Vol}_m^{\mathbf{v}}(t).$$

In addition, if the pair (χ^-, χ^{*-}) is regular, then for any basis $\mathbf{v}$ of $\mathbb{C}^p$ that is subordinate to $\mathcal{V}_{\chi^-}$ and is ordered:

(a) for any $1 \leq m \leq p$ we have
$$\lim_{t \to -\infty} \frac{1}{t} \log \mathrm{Vol}_m^{\mathbf{v}}(t) = \sum_{i=1}^m \chi^-(v_i);$$

(b) for any $1 \leq m < p$ the angle $\sigma(t)$ between $\mathrm{span}\{v_i(t) : i \leq m\}$ and $\mathrm{span}\{v_i(t) : i > m\}$ satisfies
$$\lim_{t \to -\infty} \frac{1}{t} \log \sin \sigma(t) = 0.$$

This means that one can verify the regularity of the pair (χ^-, χ^{*-}) of Lyapunov exponents dealing only with the Lyapunov exponent χ^-.

2.4.2. The Lyapunov–Perron regularity. We now introduce the crucial concept of Lyapunov–Perron regularity that substantially strengthens the notions of forward and backward regularity. Consider the forward χ^+ and backward χ^- Lyapunov exponents, the associated filtrations $\mathcal{V}^+$, $\mathcal{V}^-$, and the Lyapunov spectra
$$\mathrm{Sp}\,\chi^+ = \{(\chi_i^+, k_i^+) : 1 \leq i \leq s^+\},$$
$$\mathrm{Sp}\,\chi^- = \{(\chi_i^-, k_i^-) : 1 \leq i \leq s^-\},$$
respectively. We say that the filtrations $\mathcal{V}^+$ and $\mathcal{V}^-$ are *coherent* if the following properties hold:

(1) $s^+ = s^- =: s$;

(2) there exists a decomposition
$$\mathbb{C}^p = \bigoplus_{i=1}^s E_i \qquad (2.32)$$
into subspaces E_i such that
$$V_i^+ = \bigoplus_{j=1}^i E_j \quad \text{and} \quad V_i^- = \bigoplus_{j=i}^s E_j;$$

(3) $\chi_i^+ = -\chi_i^- =: \chi_i$;

(4) if $v \in E_i \setminus \{0\}$, then
$$\lim_{t \to \pm \infty} \frac{1}{t} \log \|v(t)\| = \chi_i$$
with uniform convergence on $\{v \in E_i : \|v\| = 1\}$ (recall that $v(t)$ is the solution of equation (2.7) with initial condition $v(0) = v$).

2.4. Forward and backward regularity

The decomposition (2.32) is called the *Oseledets decomposition* associated with the Lyapunov exponent χ^+ (or with the pair of Lyapunov exponents (χ^+, χ^-)).

We say that the Lyapunov exponent χ^+ is *Lyapunov–Perron regular* or simply *LP-regular* if the exponent χ^+ is forward regular, the exponent χ^- is backward regular, and the filtrations $\mathcal{V}^+$ and $\mathcal{V}^-$ are coherent.

Remark 2.22. We stress that simultaneous forward and backward regularity of the Lyapunov exponents does not imply the LP-regularity. Roughly speaking, whether the Lyapunov exponent is forward (respectively, backward) regular may not depend on the backward (respectively, forward) behavior of solutions of the system, i.e., the forward behavior of the system may "know nothing" about its backward behavior. To illustrate this, consider a flow φ_t of the sphere with the north pole a repelling fixed point with the rate of expansion $\lambda > 0$ and the south pole an attracting fixed point with the rate of contraction $\mu < 0$, so that every trajectory moves from the north pole to the south pole (see Figure 2.1). It is easy to see that every point of the flow is simultaneously forward and backward regular but if $\lambda \neq -\mu$, none of the points is LP-regular except for the north and south poles.

The LP-regularity requires some compatibility between the forward and backward behavior of solutions which is expressed in terms of the filtrations $\mathcal{V}^+$ and $\mathcal{V}^-$. Such compatibility can be expected if the trajectory of the flow is infinitely recurrent, i.e., it returns infinitely often to an arbitrarily small neighborhood of the initial point. This type of behavior occurs for trajectories that are typical with respect to an invariant measure for the flow due to the Poincaré Recurrence Theorem. While the recurrence property alone does not guarantee the LP-regularity, it turns out that given an

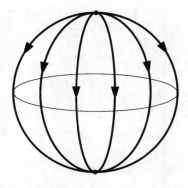

Figure 2.1. Flow φ_t.

invariant measure, almost every trajectory is LP-regular. This is due to the celebrated result by Oseledets known as the Multiplicative Ergodic Theorem (see Theorem 4.1).

2.4.3. Criteria for LP-regularity.

Theorem 2.23. *The Lyapunov exponent χ^+ is LP-regular if and only if there exist a decomposition*

$$\mathbb{C}^p = \bigoplus_{i=1}^{s} E_i$$

and numbers $\chi_1 < \cdots < \chi_s$ such that:

(1) *if $i = 1, \ldots, s$ and $v \in E_i \setminus \{0\}$, then*

$$\lim_{t \to \pm\infty} \frac{1}{t} \log \|v(t)\| = \chi_i;$$

(2)
$$\lim_{t \to \pm\infty} \frac{1}{t} \int_0^t \operatorname{tr} A(\tau) \, d\tau = \sum_{i=1}^{s} \chi_i \dim E_i. \qquad (2.33)$$

In addition, if the Lyapunov exponent χ^+ is LP-regular, then the following statements hold:

(a) *for any collection of vectors $\mathbf{v} = (v_1, \ldots, v_k)$ the limits*

$$\lim_{t \to \pm\infty} \frac{1}{t} \log \operatorname{Vol}_k^{\mathbf{v}}(t)$$

exist;

(b) $E_i = V_i^+ \cap V_i^-$ *and* $\dim E_i = k_i^+ = k_i^- =: k_i;$

(c) *if $\mathbf{v} = (v_1, \ldots, v_{k_i})$ is a basis of E_i, then*

$$\lim_{t \to \pm\infty} \frac{1}{t} \log \operatorname{Vol}_{k_i}^{\mathbf{v}}(t) = \chi_i k_i;$$

(d) *if $v_i(t)$ and $v_j(t)$ are solutions of equation (2.7) such that $v_i(0) \in E_i \setminus \{0\}$ and $v_j(0) \in E_j \setminus \{0\}$ with $i \neq j$, then*

$$\lim_{t \to \pm\infty} \frac{1}{t} \log \angle(v_i(t), v_j(t)) = 0.$$

Proof. First assume that the Lyapunov exponent χ^+ is LP-regular. Since the filtrations $\mathcal{V}^+$ and $\mathcal{V}^-$ are coherent, we have the decomposition (2.32) and numbers $\chi_1 < \cdots < \chi_s$ satisfying property (1). Moreover, since the exponent χ^+ is forward regular and the exponent χ^- is backward regular, property (2) follows readily from Theorems 2.12 and 2.21.

Now assume that we have the decomposition (2.32) and numbers $\chi_1 < \cdots < \chi_s$ satisfying properties (1) and (2). It follows from (1) that the first

2.4. Forward and backward regularity

three properties in the definition of coherent filtrations hold. It remains to verify that the convergence in (1) is uniform on the set
$$C_i = \{v \in E_i : \|v\| = 1\}.$$
We start with the following lemma.

Lemma 2.24. *If properties* (1) *and* (2) *hold, then given* $\mathbf{v} = (v_1, \ldots, v_p)$, *for each* $k = 1, \ldots, p-1$ *the following hold:*

(1) *we have*
$$\lim_{t \to \pm\infty} \frac{1}{t} \log \operatorname{Vol}_k^{\mathbf{v}}(t) = \sum_{i=1}^{k} \chi^+(v_i); \tag{2.34}$$

(2) *the angle* $\sigma(t)$ *between* $\operatorname{span}\{v_i(t) : i \leq k\}$ *and* $\operatorname{span}\{v_i(t) : i > k\}$ *satisfies*
$$\lim_{t \to \pm\infty} \frac{1}{t} \log \sin \sigma(t) = 0. \tag{2.35}$$

Proof of the lemma. Denote by $\operatorname{vol}(v_1, \ldots, v_k)$ the volume of the parallelepiped formed by the vectors $v_1, \ldots, v_k$. We have
$$\operatorname{Vol}_p^{\mathbf{v}}(t) \leq \operatorname{vol}(v_1(t), \ldots, v_k(t)) \operatorname{vol}(v_{k+1}(t), \ldots, v_p(t)) \sin \sigma(t).$$
Since
$$\lim_{t \to \pm\infty} \frac{1}{t} \log \operatorname{vol}(v_1(t), \ldots, v_k(t)) \leq \sum_{i=1}^{k} \chi^+(v_i), \tag{2.36}$$
$$\lim_{t \to \pm\infty} \frac{1}{t} \log \operatorname{vol}(v_{k+1}(t), \ldots, v_p(t)) \leq \sum_{i=k+1}^{p} \chi^+(v_i) \tag{2.37}$$
we obtain
$$\lim_{t \to \pm\infty} \frac{1}{t} \log \operatorname{Vol}_p^{\mathbf{v}}(t) \leq \sum_{i=1}^{p} \chi^+(v_i) + \liminf_{t \to \pm\infty} \frac{1}{t} \log \sin \sigma(t).$$
It follows from (2.33) that
$$\lim_{t \to \pm\infty} \frac{1}{t} \log \operatorname{Vol}_p^{\mathbf{v}}(t) = \sum_{i=1}^{p} \chi^+(v_i)$$
and so
$$\liminf_{t \to \pm\infty} \frac{1}{t} \log \sin \sigma(t) \geq 0.$$
Since $\sigma(t) \leq \pi/2$, this yields property (2.35). Therefore,
$$\sum_{i=1}^{p} \chi^+(v_i) \leq \lim_{t \to \pm\infty} \frac{1}{t} \log \operatorname{Vol}_k^{\mathbf{v}}(t) + \lim_{t \to \pm\infty} \frac{1}{t} \log \operatorname{vol}(v_{k+1}(t), \ldots, v_p(t))$$
and hence, (2.34) follows from (2.36) and (2.37). □

We proceed with the proof of the theorem. Let $\{e_1, \ldots, e_{k_i}\}$ be an orthonormal basis of E_i. We denote by $v(t) = v_j(t)$ the solution of equation (2.7) with an initial condition $v(0) = v_j(0)$. For every $t \in \mathbb{R}$ let $\rho_{t,j}$ and $\varphi_{t,j}$ be, respectively, the distance and the angle between $v_j(t)$ and $\text{span}\{v_k(t) : k \neq j\}$. Note that

$$\rho_{t,j} = \|v_j(t)\| \sin \varphi_{t,j}. \tag{2.38}$$

Now fix a number $t \in \mathbb{R}$ and choose a vector

$$v_t = \sum_{j=1}^{k_i} c_{t,j} e_j \in C_i$$

at which the map $C_i \ni v(0) \mapsto \|v(t)\|$ attains its minimum.[4] Denote by $\bar{v}_t$ the solution of equation (2.7) with the initial condition $\bar{v}_t(0) = v_t$ and choose a number $j(t)$ such that $|c_{t,j(t)}| = \max_j |c_{t,j}|$. Then $|c_{t,j(t)}| \geq 1/\sqrt{k_i}$. We have

$$\bar{v}_t(t) = c_{t,j(t)} \left(v_{j(t)}(t) + \sum_{j \neq j(t)} \frac{c_{t,j}}{c_{t,j(t)}} v_j(t) \right)$$

and hence, using (2.38), we find that

$$\|\bar{v}_t(t)\| \geq |c_{t,j(t)}| \rho_{t,j(t)} \geq \frac{1}{\sqrt{k_i}} \|v_{j(t)}(t)\| \sin \varphi_{t,j(t)}.$$

Since $j(t)$ can take only finitely many values, we obtain

$$\liminf_{t \to \pm\infty} \frac{1}{|t|} \log \|\bar{v}_t(t)\| \geq \liminf_{t \to \pm\infty} \frac{1}{|t|} \log \|v_{j(t)}(t)\| + \liminf_{t \to \pm\infty} \frac{1}{|t|} \log \sin \varphi_{t,j(t)}$$

$$\geq \min_j \liminf_{t \to \pm\infty} \frac{1}{|t|} \log \|v_j(t)\| + \min_j \liminf_{t \to \pm\infty} \frac{1}{|t|} \log \sin \varphi_{t,j}.$$

It follows from Lemma 2.24 that

$$\liminf_{t \to \pm\infty} \frac{1}{|t|} \log \|\bar{v}_t(t)\| \geq \chi_i.$$

Now for each $t \in \mathbb{R}$ take

$$w_t = \sum_{j=1}^{k_i} d_{t,j} e_j \in C_i$$

at which the map $C_i \ni v(0) \mapsto \|v(t)\|$ attains its maximum. Denote by $\bar{w}_t$ the solution with $\bar{w}_t(0) = w_t$. Then

$$\|\bar{w}_t(t)\| \leq \sum_{j=1}^{k_i} |d_{t,j}| \cdot \|v_j(t)\| \leq \sum_{j=1}^{k_i} \|v_j(t)\|$$

---

[4] Since this map is continuous and the set C_i is compact, such a vector always exists.

and
$$\limsup_{t \to \pm\infty} \frac{1}{|t|} \log\|\bar{w}_t(t)\| \leq \chi_i.$$

This establishes the uniform convergence on the set C_i. Finally, it follows from Theorems 2.12 and 2.21 that the exponent χ^+ is forward regular and the exponent χ^- is backward regular.

All the remaining statements follow easily from Lemma 2.24. □

We describe an example of a nonautonomous linear differential equation whose Lyapunov exponent is both forward and backward regular but is *not* LP-regular. Consider the system of equations
$$\dot{v}_1 = a(t)v_1, \quad \dot{v}_2 = a(-t)v_2$$
where $a \colon \mathbb{R} \to \mathbb{R}$ is a bounded continuous function such that $a(t) \to a_+$ as $t \to +\infty$ and $a(t) \to a_-$ as $t \to -\infty$, for some constants $a_+ \neq a_-$.

Exercise 2.25. Show that the values of the exponents χ^+ and χ^- coincide (up to the change of sign), but the filtrations $\mathcal{V}^+$ and $\mathcal{V}^-$ are not coherent.

2.5. Lyapunov exponents for sequences of matrices

In this section we study Lyapunov exponents associated with sequences of matrices, which can be viewed as the discrete-time version of the matrix functions we dealt with in the previous sections.

2.5.1. The forward Lyapunov exponent and forward regularity.
Consider a sequence of invertible $p \times p$ matrices $(A_m)_{m \in \mathbb{N}}$ with entries in $\mathbb{C}$ such that
$$\sup\{\|A_m^{\pm 1}\| : m \in \mathbb{N}\} < \infty. \tag{2.39}$$
Set
$$\mathcal{A}(m,n) = \begin{cases} A_{m-1} \cdots A_n, & m > n, \\ \mathrm{Id}, & m = n. \end{cases}$$
Consider the function $\chi^+ \colon \mathbb{C}^p \to \mathbb{R} \cup \{-\infty\}$ defined by
$$\chi^+(v) = \limsup_{m \to +\infty} \frac{1}{m} \log \|\mathcal{A}(m,1)v\|. \tag{2.40}$$

Condition (2.39) ensures that the function χ^+ is well-defined and bounded on $\mathbb{R} \setminus \{0\}$.

Exercise 2.26. Show that χ^+ is a Lyapunov exponent in $\mathbb{C}^p$.

By Theorem 2.1, the function χ^+ attains only finitely many distinct values $\chi_1^+ < \cdots < \chi_{s^+}^+$ on $\mathbb{C}^p \setminus \{0\}$ where $s^+ \leq p$. By (2.8), each number χ_i^+ is finite and occurs with some multiplicity k_i so that $\sum_{i=1}^{s^+} k_i = p$.

We denote by $\mathcal{V}^+$ the filtration of $\mathbb{C}^p$ associated to χ^+:
$$\{0\} = V_0^+ \subsetneqq V_1^+ \subsetneqq \cdots \subsetneqq V_{s^+}^+ = \mathbb{C}^p,$$
where
$$V_i^+ = \{v \in \mathbb{C}^p : \chi^+(v) \le \chi_i^+\}.$$
We also denote by
$$\operatorname{Sp}\chi^+ = \{(\chi_i^+, k_i^+) : 1 \le i \le s^+\}$$
the Lyapunov spectrum of χ^+.

Now we consider the sequence of dual matrices $B_m = (A_m^*)^{-1}$ and set
$$\mathcal{B}(m,n) = \begin{cases} (B_{m-1})^{-1} \cdots (B_n)^{-1}, & m > n, \\ \operatorname{Id}, & m = n. \end{cases}$$

The function $\chi^{*+} \colon \mathbb{C}^p \to \mathbb{R} \cup \{-\infty\}$ defined by
$$\chi^{*+}(v^*) = \limsup_{m \to +\infty} \frac{1}{m} \log \|\mathcal{B}(m,1)v^*\|$$
is a dual Lyapunov exponent. To see this, let us choose dual bases $(v_1, \ldots, v_p)$ and $(v_1^*, \ldots, v_p^*)$ of $\mathbb{C}^p$ and set $v_{im} = \mathcal{A}(m,1)v_i$ and $v_{im}^* = \mathcal{B}(m,1)v_i^*$. For every $m \in \mathbb{N}$ we have
$$\langle v_{im}, v_{im}^* \rangle = \langle \mathcal{A}(m,1)v_i, (\mathcal{A}(m,1)^*)^{-1} v_i^* \rangle = \langle v_i, v_i^* \rangle = 1.$$
Therefore, $1 \le \|\mathcal{A}(m,1)v_i\| \cdot \|\mathcal{B}(m,1)v_i^*\|$ and the exponents χ^+ and χ^{*+} are dual.

We will now discuss the regularity of the pair of Lyapunov exponents (χ^+, χ^{*+}). Let $\mathbf{v} = (v_1, \ldots, v_p)$ be a basis of $\mathbb{C}^p$. Denote by $\operatorname{Vol}_\ell^{\mathbf{v}}(m)$ the volume of the ℓ-parallelepiped generated by the vectors $v_{im} = \mathcal{A}(m,1)v_i$, $i = 1, \ldots, \ell$. The following result is a discrete-time version of Theorem 2.12.

Theorem 2.27. *The following statements are equivalent:*

(1) *the pair (χ^+, χ^{*+}) is regular;*

(2) *we have*

$$\lim_{m \to +\infty} \frac{1}{m} \log|\det \mathcal{A}(m,1)| = \sum_{i=1}^{s^+} k_i \chi_i^+; \qquad (2.41)$$

(3) *for any basis $\mathbf{v}$ of $\mathbb{C}^p$ that is subordinate to $\mathcal{V}_{\chi^+}$ and is ordered and for any $1 \le \ell \le p$ the following limit exists:*

$$\lim_{m \to +\infty} \frac{1}{m} \log \operatorname{Vol}_\ell^{\mathbf{v}}(m).$$

2.5. Lyapunov exponents for sequences of matrices

In addition, if the pair (χ^+, χ^{*+}) is regular, then for any basis $\mathbf{v}$ of $\mathbb{C}^p$ that is subordinate to $\mathcal{V}_{\chi^+}$ and is ordered:

(a) for any $1 \le \ell \le p$ we have

$$\lim_{m \to +\infty} \frac{1}{m} \log \operatorname{Vol}_\ell^{\mathbf{v}}(m) = \sum_{i=1}^\ell \chi^+(v_i);$$

(b) for any $1 \le \ell < p$ the angle σ_m between $\operatorname{span}\{\mathcal{A}(m,1)v_i : i \le \ell\}$ and $\operatorname{span}\{\mathcal{A}(m,1)v_i : i > \ell\}$ satisfies

$$\lim_{m \to +\infty} \frac{1}{m} \log \sin \sigma_m = 0.$$

We give only a brief outline of the proof of this theorem and refer the reader to [10] for full details.

The following exercise shows that if the pair of Lyapunov exponents (χ^+, χ^{*+}) is regular, then the relation (2.41) holds.

Exercise 2.28. Show that for any basis $(v_1, \ldots, v_p)$ of $\mathbb{C}^p$ we have

$$-\sum_{i=1}^p \chi^{*+}(v_i) \le \underline{\chi}(\det \mathcal{A}) \le \overline{\chi}(\det \mathcal{A}) \le \sum_{i=1}^p \chi^+(v_i),$$

where

$$\overline{\chi}(\det \mathcal{A}) = \limsup_{m \to +\infty} \frac{1}{m} \log |\det \mathcal{A}(m,1)|$$

and

$$\underline{\chi}(\det \mathcal{A}) = \liminf_{m \to +\infty} \frac{1}{m} \log |\det \mathcal{A}(m,1)|.$$

Use these inequalities to prove statement (2).

We now describe how statement (2) implies statement (3). We outline the argument assuming that the matrices A_m are upper-triangular and the canonical basis $e_1, \ldots, e_p$ is subordinate to the filtration associated to the Lyapunov exponent χ^+. The general case can always be reduced to this setting. Indeed, using the well-known Gram–Schmidt orthogonalization procedure, given an invertible matrix B, one can find an orthogonal matrix U such that the matrix $A = U^*BU$ is upper-triangular and the canonical basis is subordinate (and ordered). Now given a sequence of invertible matrices B_m and applying the above fact to the sequence of products of matrices $\tilde{B}_m = B_{m-1} \cdots B_1$, we find a sequence of orthogonal matrices U_m such that the matrices $A_m = U_{m+1}^* B_m U_m$ are upper-triangular and the canonical basis is subordinate (and ordered).

Let $\alpha_i(m)$ be the entries on the main diagonal of $\mathcal{A}(m,1)$. It follows from (2.41) that

$$\lim_{m \to +\infty} \frac{1}{m} \log|\det \mathcal{A}(m,1)| = \sum_{i=1}^{p} \chi^+(e_i) \geq \sum_{i=1}^{p} \limsup_{m \to +\infty} \frac{1}{m} \log|\alpha_i(m)|. \quad (2.42)$$

Here we use the fact that the basis $e_1, \ldots, e_p$ is subordinate to the filtration.

We also have

$$\lim_{m \to +\infty} \frac{1}{m} \log|\det \mathcal{A}(m,1)| = \lim_{m \to +\infty} \frac{1}{m} \sum_{i=1}^{p} \log|\alpha_i(m)|$$
$$\leq \sum_{i=1}^{p} \limsup_{m \to +\infty} \frac{1}{m} \log|\alpha_i(m)|. \quad (2.43)$$

It follows from (2.42) and (2.43) that

$$\lim_{m \to +\infty} \frac{1}{m} \sum_{i=1}^{p} \log|\alpha_i(m)| = \sum_{i=1}^{p} \limsup_{m \to +\infty} \frac{1}{m} \log|\alpha_i(m)|. \quad (2.44)$$

Let

$$\underline{\alpha}_j = \liminf_{m \to +\infty} \frac{1}{m} \log|\alpha_j(m)| \quad \text{and} \quad \overline{\alpha}_j = \limsup_{m \to +\infty} \frac{1}{m} \log|\alpha_j(m)|. \quad (2.45)$$

Assuming that $\underline{\alpha}_j < \overline{\alpha}_j$ for some j, choose a subsequence k_m such that

$$\lim_{m \to +\infty} \frac{1}{k_m} \log|\alpha_j(k_m)| = \underline{\alpha}_j.$$

Then

$$\lim_{m \to +\infty} \frac{1}{m} \sum_{i=1}^{p} \log|\alpha_i(m)| = \underline{\alpha}_j + \lim_{m \to +\infty} \frac{1}{k_m} \sum_{i \neq j} \log|\alpha_i(k_m)| < \sum_{i=1}^{p} \overline{\alpha}_i.$$

In view of (2.44) this implies that $\alpha_j := \underline{\alpha}_j = \overline{\alpha}_j$ for all j. For any $1 \leq \ell \leq p$ we have

$$\mathrm{Vol}_\ell^\mathbf{v}(m) = \prod_{i=1}^{\ell} |\alpha_i(m)|,$$

and hence for each $1 \leq \ell \leq p$,

$$\lim_{m \to +\infty} \mathrm{Vol}_\ell^\mathbf{v}(m) = \sum_{i=1}^{\ell} \alpha_i.$$

This establishes property (3) in Theorem 2.27.

Finally, considering again upper-triangular matrices A_m, one can show in a similar manner to that in the proof of Theorem 2.12 that if $\alpha_j := \underline{\alpha}_j = \overline{\alpha}_j$ for $j = 1, \ldots, p$ (see (2.45)), then the numbers $\alpha_1, \ldots, \alpha_p$ are the values of the Lyapunov exponent χ^+, the numbers $-\alpha_1, \ldots, -\alpha_p$ are the values of

2.5. Lyapunov exponents for sequences of matrices

the Lyapunov exponent χ^{*+}, and the pair of Lyapunov exponents (χ^+, χ^{*+}) is regular. This readily implies that if property (3) holds, then the pair (χ^+, χ^{*+}) is regular.

Now we outline the proofs of properties (a) and (b). Given a collection of vectors $v_1, \ldots, v_k$, denote
$$\chi^+(v_1, \ldots, v_k) = \limsup_{m \to +\infty} \frac{1}{m} \log \operatorname{Vol}_k^{\mathbf{v}}(m).$$

For $k = p$, applying the relation (2.41) we obtain that
$$\sum_{i=1}^{p} \chi^+(v_i) = \chi^+(v_1, \ldots, v_p). \tag{2.46}$$

Since $\operatorname{Vol}_k^{\mathbf{v}}(m) \le \prod_{i=1}^{k} \|\mathcal{A}(m, 1) v_i\|$, we obtain that
$$\chi^+(v_1, \ldots, v_k) \le \sum_{i=1}^{k} \chi^+(v_i).$$

This implies that
$$\chi^+(v_1, \ldots, v_\ell) \le \sum_{i=1}^{\ell} \chi^+(v_i), \quad \chi^+(v_{\ell+1}, \ldots, v_p) \le \sum_{i=\ell+1}^{p} \chi^+(v_i). \tag{2.47}$$

Now observe that
$$\begin{aligned}\operatorname{Vol}_p^{\mathbf{v}}(m) \le\ & \operatorname{vol}(\mathcal{A}(m,1)v_1, \ldots, \mathcal{A}(m,1)v_\ell) \\ & \times \operatorname{vol}(\mathcal{A}(m,1)v_{\ell+1}, \ldots, \mathcal{A}(m,1)v_p) \sin \sigma_m,\end{aligned} \tag{2.48}$$

where $\operatorname{vol}(v_1, \ldots, v_k)$ denotes the volume of the parallelepiped formed by the vectors $v_1, \ldots, v_k$. Hence,
$$\chi^+(v_1, \ldots, v_p) \le \sum_{i=1}^{p} \chi^+(v_i) + \liminf_{m \to +\infty} \frac{1}{m} \log \sin \sigma_m.$$

Therefore, in view of (2.46) we have that
$$\liminf_{m \to +\infty} \frac{1}{m} \log \sin \sigma_m \ge 0.$$

Since $\sin \sigma_m$ is bounded, we conclude that
$$\lim_{m \to +\infty} \frac{1}{m} \log \sin \sigma_m = 0.$$

Therefore, using (2.48) we find that
$$\sum_{i=1}^{p} \chi^+(v_i) = \chi^+(v_1, \ldots, v_p) \le \chi^+(v_1, \ldots, v_\ell) + \chi^+(v_{\ell+1}, \ldots, v_p).$$

It follows from (2.47) that

$$\chi^+(v_1,\ldots,v_\ell) = \sum_{i=1}^{\ell} \chi^+(v_i), \quad \chi^+(v_{\ell+1},\ldots,v_p) = \sum_{i=\ell+1}^{p} \chi^+(v_i).$$

This completes the proof of the theorem.

Theorem 2.27 shows that the regularity of the pair (χ^+, χ^{*+}) is completely determined by the Lyapunov exponent χ^+ alone. Therefore, similarly to the continuous-time case we call the Lyapunov exponent χ^+ *forward regular* if the pair of Lyapunov exponents (χ^+, χ^{*+}) is regular. In this case we also say that the sequence of matrices $\mathcal{A}^+ = (A_m)_{m \in \mathbb{N}}$ is *forward regular*.

2.5.2. The backward Lyapunov exponent and backward regularity.

Consider a sequence of invertible $p \times p$ matrices $(A_m)_{m \in \mathbb{Z}^-}$ with entries in $\mathbb{C}$ such that

$$\sup\{\|A_m^{\pm 1}\| : m \in \mathbb{Z}^-\} < \infty. \qquad (2.49)$$

Reversing the time, we introduce the Lyapunov exponent $\chi^- \colon \mathbb{C}^p \to \mathbb{R} \cup \{-\infty\}$ by

$$\chi^-(v) = \limsup_{m \to -\infty} \frac{1}{|m|} \log \|\mathcal{A}(m,1)v\|,$$

where

$$\mathcal{A}(m,n) = \begin{cases} A_m^{-1} \cdots A_{n-1}^{-1}, & m < n, \\ \mathrm{Id}, & m = n. \end{cases}$$

The function χ^- attains finitely many values $\chi_1^- > \cdots > \chi_{s^-}^-$ where $s^- \le p$. Let $\mathcal{V}^-$ be the filtration of $\mathbb{C}^p$ associated to χ^-,

$$\mathbb{C}^p = V_0^- \supsetneq V_1^- \supsetneq \cdots \supsetneq V_{s^-}^- \supsetneq V_{s^-+1}^- = \{0\},$$

where

$$V_i^- = \{v \in \mathbb{C}^p : \chi^-(v) \le \chi_i^-\},$$

and let $k_i^- = \dim V_i^- - \dim V_{i+1}^-$ be the *multiplicity* of the value χ_i^- such that $\sum_{i=1}^{s^-} k_i^- = p$. We also denote by

$$\mathrm{Sp}\,\chi^- = \{(\chi_i^-, k_i^-) : 1 \le i \le s^-\}$$

the Lyapunov spectrum of χ^-.

Consider the dual Lyapunov exponent $\chi^{*-} \colon \mathbb{C}^p \to \mathbb{R} \cup \{-\infty\}$ given by

$$\chi^{*-}(v^*) = \limsup_{m \to -\infty} \frac{1}{|m|} \log \|\mathcal{A}(m,1)^* v\|.$$

The regularity of the pair of Lyapunov exponents (χ^-, χ^{*-}) can be characterized by statements similar to the ones in Theorem 2.27 (in which the time should be reversed). In particular, the regularity of this pair is completely determined by the Lyapunov exponent χ^-. This justifies calling the

2.5. Lyapunov exponents for sequences of matrices

Lyapunov exponent χ^- *backward regular* if the pair of Lyapunov exponents (χ^-, χ^{*-}) is regular. In this case we also say that the sequence of matrices $\mathcal{A}^- = (A_m)_{m \in \mathbb{Z}^-}$ is *backward regular*.

2.5.3. The Lyapunov–Perron regularity. Consider a sequence of invertible $p \times p$ matrices $(A_m)_{m \in \mathbb{Z}}$ with entries in $\mathbb{C}$ satisfying (2.39) and (2.49).

We say that the Lyapunov exponent χ^+ is *Lyapunov–Perron regular* or simply *LP-regular* if the exponent χ^+ is forward regular, the exponent χ^- is backward regular, and the filtrations $\mathcal{V}^+$ and $\mathcal{V}^-$ are coherent (see Section 2.4.2).

We have a version of Theorem 2.23 in this setting.

Theorem 2.29. *The Lyapunov exponent χ^+ is LP-regular if and only if there exist a decomposition*

$$\mathbb{C}^p = \bigoplus_{i=1}^{s} E_i$$

and numbers $\chi_1 < \cdots < \chi_s$ such that:

(1) *if $i = 1, \ldots, s$ and $v \in E_i \setminus \{0\}$, then*

$$\lim_{m \to \pm\infty} \frac{1}{m} \log \|\mathcal{A}(m, 1)v\| = \chi_i;$$

(2)
$$\lim_{m \to \pm\infty} \frac{1}{m} \log |\det \mathcal{A}(m, 1)| = \sum_{i=1}^{s} \chi_i \dim E_i.$$

In addition, if the Lyapunov exponent χ^+ is LP-regular, then the following statements hold:

(a) *for any collection of vectors $\mathbf{v} = (v_1, \ldots, v_k)$ the limits*

$$\lim_{m \to \pm\infty} \frac{1}{m} \log \operatorname{Vol}_k^{\mathbf{v}}(m)$$

exist;

(b) $E_i = V_i^+ \cap V_i^-$ *and* $\dim E_i = k_i^+ = k_i^-$;

(c) *if $\mathbf{v} = (v_1, \ldots, v_{k_i})$ is a basis of E_i, then*

$$\lim_{m \to \pm\infty} \frac{1}{m} \log \operatorname{Vol}_{k_i}^{\mathbf{v}}(m) = \chi_i k_i;$$

(d) *if $v_i \in E_i \setminus \{0\}$ and $v_j \in E_j \setminus \{0\}$ with $i \neq j$, then*

$$\lim_{m \to \pm\infty} \frac{1}{m} \log \angle(\mathcal{A}(m, 1)v_i, \mathcal{A}(m, 1)v_j) = 0.$$

The proof follows closely the proof of Theorem 2.23.

We stress that the pure existence of the limit in (2.41) (without the requirement that it be equal to the sum of the Lyapunov exponents) does *not* guarantee that the pair (χ^+, χ^{*+}) is regular.

Example 2.30. Let $\mathcal{A}^+ = (A_m)_{m \in \mathbb{N}}$ be the sequence of matrices given by $A_1 = \begin{pmatrix} 1 & 0 \\ 2 & 4 \end{pmatrix}$ and $A_m = \begin{pmatrix} \frac{1}{2}m & 0 \\ -\frac{1}{2}m & 4 \end{pmatrix}$ for each $m > 1$ so that $\mathcal{A}(m,1) = \begin{pmatrix} 1 & 0 \\ 2^{m-1} & 4^{m-1} \end{pmatrix}$ for every $m > 1$. Given a vector $v = (a,b) \neq (0,0)$, we have $\chi^+(v) = \log 2$ if $b = 0$ and $\chi^+(v) = \log 4$ if $b \neq 0$. Let $v_1 = (1,0)$ and $v_2 = (0,1)$. Then $\chi^+(v_1) = \log 2$ and $\chi^+(v_2) = \log 4$. Since $\det \mathcal{A}(m,1) = 4^{m-1}$, there exists the limit
$$\chi^+(v_1, v_2) = \lim_{m \to +\infty} \frac{1}{m} \log|\det \mathcal{A}(m,1)| = \log 4.$$
On the other hand,
$$\log 4 < \log 2 + \log 4 = \chi^+(v_1) + \chi^+(v_2)$$
and the pair (χ^+, χ^{*+}) is not regular.

In the one-dimensional case the situation is different.

Exercise 2.31. Consider a sequence of real numbers (i.e., a sequence of one-dimensional matrices) $(A_m)_{m \in \mathbb{N}} \subset GL(1, \mathbb{R}) = \mathbb{R} \setminus \{0\}$. Show that the pair of Lyapunov exponents (χ^+, χ^{*+}) associated to this sequence is regular if and only if the limit in (2.41) exists.

Chapter 3

Cocycles over Dynamical Systems

Note that a sequence of matrices $(A_m)_{m\in\mathbb{Z}}$ in $\mathbb{R}^p$ generates a matrix function $\mathcal{A}\colon \mathbb{Z}\times\mathbb{Z} \to GL(p,\mathbb{R})$ given by

$$\mathcal{A}(m,\ell) = A_{m+\ell-1}\cdots A_m$$

which satisfies the particular property

$$\mathcal{A}(m,\ell+k) = \mathcal{A}(\sigma^k(m),\ell)\mathcal{A}(m,k),$$

where $\sigma\colon \mathbb{Z}\to\mathbb{Z}$ is the shift, i.e., $\sigma(m) = m+1$. This property identifies the function $\mathcal{A}$ as a (linear) cocycle over the shift σ. A general concept of cocycle is a far-reaching extension of this simple example in which the shift is replaced by an invertible map f acting on an abstract space X. One can then build the theory of Lyapunov exponents for cocycles extending the concepts of LP-regularity, nonuniform hyperbolicity, etc. This allows one to substantially broaden the applications of the theory. We stress that while LP-regularity and nonuniform hyperbolicity do not require the presence of any invariant measure, we shall see below that the study of cocycles over measurable transformations preserving finite measures provides many interesting new results.

Consider a diffeomorphism $f\colon M\to M$ of a compact smooth Riemannian manifold M of dimension p. Given a trajectory $\{f^m(x)\}_{m\in\mathbb{Z}}$, $x\in M$, we can identify each tangent space $T_{f^m(x)}M$ with $\mathbb{R}^p$ and thus obtain a sequence of matrices

$$(A_m(x))_{m\in\mathbb{Z}} = \{d_{f^m(x)}f\}_{m\in\mathbb{Z}}.$$

This generates a cocycle $\mathcal{A}_x$ over the transformation $f\colon X \to X$, where $X = \{f^m(x)\}_{m\in\mathbb{Z}}$. This cocycle can be viewed as the *individual derivative cocycle* associated with the trajectory of x. Dealing with such cocycles allows one to study LP-regularity of individual trajectories and, as we shall see below, construct stable and unstable local and global manifolds along such trajectories.

While individual derivative cocycles depend on the choice of the individual trajectory, one can build the "global" cocycle associated with the diffeomorphism f. To this end we represent the manifold M as a finite union $\bigcup_i \Delta_i$ of differentiable copies Δ_i of the p-simplex such that:

(1) in each Δ_i one can introduce local coordinates in such a way that $T(\Delta_i)$ can be identified with $\Delta_i \times \mathbb{R}^p$;

(2) all nonempty intersections $\Delta_i \cap \Delta_j$, for $i \neq j$, are $(p-1)$-dimensional submanifolds.

In each Δ_i the derivative of f can be interpreted as a linear cocycle. This implies that $df\colon M \to \mathbb{R}^p$ can be interpreted as a measurable cocycle $\mathcal{A}$ with $d_x f$ the matrix representation of df in local coordinates. We call $\mathcal{A}$ the *derivative cocycle* specified by the diffeomorphism f. One can show that it does not depend on the choice of the decomposition $\{\Delta_i\}$; more precisely, if $\{\Delta_i'\}$ is another decomposition of M into copies of the p-simplex satisfying conditions (1) and (2) above, then the corresponding cocycle $\mathcal{A}'$ is equivalent to the cocycle $\mathcal{A}$ (see Section 3.1.3 for the definition of equivalent cocycles).

The derivative cocycle allows one to apply the results about general cocycles to smooth dynamical systems. A remarkable manifestation of this fact is the Multiplicative Ergodic Theorem for smooth dynamical systems (see Theorem 5.3) that claims that a "typical" individual cocycle with respect to a finite invariant measure is LP-regular. This theorem is an immediate corollary of the corresponding result for cocycles in Chapter 4 (see Theorem 4.1).

3.1. Cocycles and linear extensions

3.1.1. Linear multiplicative cocycles. Consider an invertible measurable transformation $f\colon X \to X$ of a measure space X. We call the function $\mathcal{A}\colon X \times \mathbb{Z} \to GL(p, \mathbb{R})$ a *linear multiplicative cocycle over f* or simply a *cocycle* if it has the following properties:

(1) for each $x \in X$ we have $\mathcal{A}(x, 0) = \mathrm{Id}$, and given $m, k \in \mathbb{Z}$,
$$\mathcal{A}(x, m+k) = \mathcal{A}(f^k(x), m)\mathcal{A}(x, k);$$

(2) for every $m \in \mathbb{Z}$ the function $\mathcal{A}(\cdot, m)\colon X \to GL(p, \mathbb{R})$ is measurable.

3.1. Cocycles and linear extensions

Every cocycle is generated by a measurable function $A\colon X \to GL(p,\mathbb{R})$ called the *generator*. More precisely, every such function determines a cocycle by the formula
$$\mathcal{A}(x,m) = \begin{cases} A(f^{m-1}(x)) \cdots A(f(x))A(x) & \text{if } m > 0, \\ \text{Id} & \text{if } m = 0, \\ A(f^m(x))^{-1} \cdots A(f^{-2}(x))^{-1} A(f^{-1}(x))^{-1} & \text{if } m < 0. \end{cases}$$

On the other hand, a cocycle $\mathcal{A}$ is generated by the matrix function $A = \mathcal{A}(\cdot, 1)$.

A cocycle $\mathcal{A}$ over f induces a *linear extension* $F\colon X \times \mathbb{R}^p \to X \times \mathbb{R}^p$ of f to $X \times \mathbb{R}^p$ (also known as a *linear skew product*). It is given by the formula
$$F(x,v) = (f(x), A(x)v).$$

In other words, the action of F on the fiber over x to the fiber over $f(x)$ is given by the linear map $A(x)$. If $\pi\colon X \times \mathbb{R}^p \to X$ is the projection defined by $\pi(x,v) = x$, then the diagram

$$\begin{array}{ccc} X \times \mathbb{R}^p & \xrightarrow{F} & X \times \mathbb{R}^p \\ {\scriptstyle \pi}\downarrow & & \downarrow{\scriptstyle \pi} \\ X & \xrightarrow{f} & X \end{array}$$

is commutative and for each $m \in \mathbb{Z}$ we obtain
$$F^m(x,v) = (f^m(x), \mathcal{A}(x,m)v). \tag{3.1}$$

Linear extensions are particular cases of bundle maps of measurable vector bundles that we now consider. Let E and X be measure spaces and let $\pi\colon E \to X$ be a measurable transformation. E is called a *measurable vector bundle* over X if there exists a countable collection of measurable subsets $Y_i \subset X$, which cover X, and measurable maps $\pi_i\colon X \times \mathbb{R}^p \to X$ with $\pi_i^{-1}(Y_i) = Y_i \times \mathbb{R}^p$. A bundle map $F\colon E \to E$ over a measurable transformation $f\colon X \to X$ is a measurable map, which makes the following diagram commutative:

$$\begin{array}{ccc} E & \xrightarrow{F} & E \\ {\scriptstyle \pi}\downarrow & & \downarrow{\scriptstyle \pi} \\ X & \xrightarrow{f} & X. \end{array}$$

The following proposition shows that from the measure theory point of view every vector bundle over a compact metric space is trivial, and hence without loss of generality, one may always assume that $E = X \times \mathbb{R}^p$. In other words, every bundle map of E is essentially a linear extension provided that the base space X is a compact metric space.

Proposition 3.1. *If E is a measurable vector bundle over a compact metric space (X, ν), then there is a subset $Y \subset X$ such that $\nu(Y) = 1$ and $\pi^{-1}(Y)$ is (isomorphic to) a trivial vector bundle.*

Exercise 3.2. Prove Proposition 3.1. *Hint:* Using the compactness of the space X, construct a finite cover by balls $\{B(x_1, r_1), \ldots, B(x_k, r_k)\}$ such that $\nu(\partial B(x_i, r_i)) = 0$ and $\pi|B(x_i, r_i)$ is measurably isomorphic to $B(x_i, r_i) \times \mathbb{R}^p$ for each $i = 1, \ldots, k$.

3.1.2. Operations with cocycles. Starting from a given cocycle, one can build other cocycles using some basic constructions in ergodic theory and algebra (see [**13**] for more details).

Let $\mathcal{A}$ be a cocycle over a measurable transformation f of a Lebesgue space X. For each $m \geq 1$ consider the transformation $f^m \colon X \to X$ and the measurable cocycle $\mathcal{A}^m$ over f^m with the generator
$$A^m(x) := A(f^{m-1}(x)) \cdots A(x).$$
The cocycle $\mathcal{A}^m$ is called the *mth power cocycle* of $\mathcal{A}$.

Let $f \colon X \to X$ be a measure-preserving transformation of a Lebesgue space (X, ν) and let $Y \subset X$ be a measurable subset of positive ν-measure. By Poincaré's Recurrence Theorem the set $Z \subset Y$ of points $x \in Y$ such that $f^n(x) \in Y$ for infinitely many positive integers n has measure $\nu(Z) = \nu(Y)$. We define the transformation $f_Y \colon Y \to Y$ (mod 0) as follows: for each $x \in Z$, set
$$k_Y(x) = \min\{k \geq 1 : f^k(x) \in Y\} \quad \text{and} \quad f_Y(x) = f^{k_Y(x)}(x).$$
One can easily verify that the function k_Y and the map f_Y are measurable on Z. We call k_Y the *(first) return time* to Y and f_Y the *(first) return map* or *induced transformation* on Y.

Lemma 3.3 (See, for example, [**40**]).

(1) *We have that $\int_Y k_Y \, d\nu = \nu(\bigcup_{n \geq 0} f^n(Y))$ and $k_Y \in L^1(X, \nu)$.*

(2) *The measure ν is invariant under f_Y.*

Since $k_Y \in L^1(X, \nu)$, by Birkhoff's Ergodic Theorem, the function
$$\tau_Y(x) = \lim_{k \to +\infty} \frac{1}{k} \sum_{i=0}^{k-1} k_Y(f_Y^i(x))$$
is well-defined for ν-almost all $x \in Y$ and $\tau_Y \in L^1(X, \nu)$.

If $\mathcal{A}$ is a measurable linear cocycle over f with generator A, we define the *induced cocycle* $\mathcal{A}_Y$ over f_Y to be the cocycle with the generator
$$A_Y(x) = A^{k_Y(x)}(x).$$

3.1.3. Cohomology and tempered equivalence.
In this section we introduce the concept of two cocycles with the same base transformation being equivalent. Since cocycles act on fibers by linear transformations, we should first require that these linear actions be equivalent, i.e., that the corresponding matrices be conjugate by a linear coordinate change. We will then impose some requirements on how the coordinate change depends on the base point.

Following this line of thinking, consider two cocycles $\mathcal{A}$ and $\mathcal{B}$ over an invertible measurable transformation $f\colon X \to X$ and let $A, B\colon X \to GL(p,\mathbb{R})$ be their generators, respectively. The cocycles act on the fiber $\{x\} \times \mathbb{R}^p$ of $X \times \mathbb{R}^p$ by matrices $A(x)$ and $B(x)$, respectively. Now let $L\colon X \to GL(p,\mathbb{R})$ be a measurable matrix function such that for every $x \in X$,

$$A(x) = L(f(x))^{-1} B(x) L(x).$$

In other words, the actions of the cocycles $\mathcal{A}$ and $\mathcal{B}$ on the fibers are conjugate via the matrix function $L(x)$.

Exercise 3.4. Let $\mathcal{A}$ be a cocycle over an invertible measurable transformation $f\colon X \to X$ and let A be its generator. Also let $L\colon X \to GL(p,\mathbb{R})$ be a measurable matrix function. Given $x \in X$, let $v_{f(x)} = A(x)v_x$ and $u_x = L(x)v_x$. Show that the matrix function $B\colon X \to GL(p,\mathbb{R})$, for which $u_{f(x)} = B(x)u_x$ satisfies

$$A(x) = L(f(x))^{-1} B(x) L(x)$$

and generates a measurable cocycle $\mathcal{B}$ over f.

One can naturally think of the cocycles $\mathcal{A}$ and $\mathcal{B}$ as being equivalent. However, since the function L is in general only measurable, without any additional assumption on L the measure-theoretic properties of the cocycles $\mathcal{A}$ and $\mathcal{B}$ can be very different. We now introduce a sufficiently general class of coordinate changes, which have the important property of being tempered and which make the notion of equivalence productive.

Let $Y \subset X$ be an f-invariant nonempty measurable set. A measurable function $L\colon X \to GL(p,\mathbb{R})$ is said to be *tempered on Y with respect to f* or simply *tempered on Y* if for every $x \in Y$ we have

$$\lim_{m\to\pm\infty} \frac{1}{m} \log\|L(f^m(x))\| = \lim_{m\to\pm\infty} \frac{1}{m} \log\|L(f^m(x))^{-1}\| = 0. \quad (3.2)$$

A cocycle over f is said to be *tempered on Y* if its generator is tempered on Y.

If the real functions $x \mapsto \|L(x)\|, \|L(x)^{-1}\|$ are bounded or, more generally, have finite essential supremum, then the function L is tempered with respect to any invertible transformation $f\colon X \to X$ on any f-invariant nonempty measurable subset $Y \subset X$. The following statement provides a more general criterion for a function L to be tempered.

Proposition 3.5. *Let $f\colon X \to X$ be an invertible transformation preserving a probability measure ν, and let $L\colon X \to GL(p,\mathbb{R})$ be a measurable function. If*
$$\log\|L\|, \log\|L^{-1}\| \in L^1(X,\nu),$$
then L is tempered on some set of full ν-measure.

Proof. We need the following result.

Lemma 3.6. *Let $f\colon X \to X$ be a measurable transformation preserving a probability measure ν. Show that if the function $\varphi\colon X \to \mathbb{R}$ is in $L^1(X,\nu)$, then*
$$\lim_{m\to+\infty} \frac{\varphi(f^m(x))}{m} = 0$$
for ν-almost every $x \in X$.

Proof of the lemma. We have
$$\frac{1}{m+1}\sum_{k=0}^{m} \varphi(f^k(x)) = \frac{m}{m+1}\frac{1}{m}\sum_{k=0}^{m-1} \varphi(f^k(x)) + \frac{m}{m+1}\frac{\varphi(f^m(x))}{m}. \quad (3.3)$$

Since $\varphi \in L^1(X,\nu)$, it follows from Birkhoff's Ergodic Theorem and the assumption of the lemma that
$$\tilde\varphi(x) := \lim_{m\to\infty} \frac{1}{m+1}\sum_{k=0}^{m} \varphi(f^k(x)) = \lim_{m\to\infty} \frac{1}{m}\sum_{k=0}^{m-1} \varphi(f^k(x))$$
for ν-almost every $x \in X$. The result follows letting $m \to \infty$ in (3.3). □

The proposition follows by applying Lemma 3.6 to the functions $x \mapsto \log\|L(x)\|$ and $x \mapsto \log\|L(x)^{-1}\|$. □

We now proceed with the notion of equivalence for cocycles. Let $\mathcal{A}$ and $\mathcal{B}$ be two cocycles over an invertible measurable transformation f and let A, B be their generators, respectively. Given a measurable subset $Y \subset X$, we say that the cocycles are *equivalent on Y* or *cohomologous on Y* if there exists a measurable function $L\colon X \to GL(p,\mathbb{R})$, which is tempered on Y, such that for every $x \in Y$ we have
$$A(x) = L(f(x))^{-1} B(x) L(x). \quad (3.4)$$

This is clearly an equivalence relation and if two cocycles $\mathcal{A}$ and $\mathcal{B}$ are equivalent, we write $\mathcal{A} \sim_Y \mathcal{B}$. Equation (3.4) is called the *cohomological equation*.

It follows from (3.4) that for any $x \in Y$ and $m \in \mathbb{Z}$,

$$\mathcal{A}(x,m) = L(f^m(x))^{-1}\mathcal{B}(x,m)L(x). \tag{3.5}$$

Proposition 3.5 immediately implies the following result.

Corollary 3.7. *If* $L\colon X \to GL(p,\mathbb{R})$ *is a measurable function such that* $\log\|L\|$, $\log\|L^{-1}\| \in L^1(X,\nu)$, *then any two cocycles* $\mathcal{A}$ *and* $\mathcal{B}$ *satisfying* (3.5) *are equivalent cocycles.*

Notice that if two cocycles $\mathcal{A}$ and $\mathcal{B}$ are equivalent and $x = f^m(x)$ is a periodic point, then, by (3.5), the matrices $\mathcal{A}(x,m)$ and $\mathcal{B}(x,m)$ are conjugate. In a similar manner to that in Livshitz's theorem (see, for example, [**64**]), one can ask whether the converse holds. This is indeed true for some classes of cocycles with hyperbolic behavior. We refer the reader to [**66**] for a detailed discussion and to the survey [**103**] for more recent results on this topic.

3.2. Lyapunov exponents and Lyapunov–Perron regularity for cocycles

Let $\mathcal{A}$ be a cocycle over an invertible measure-preserving transformation f of a Lebesgue space (X,ν). For every $x \in X$, the cocycle $\mathcal{A}$ generates the sequence of matrices $\{A_m\}_{m\in\mathbb{Z}} = \{A(f^m(x))\}_{m\in\mathbb{Z}}$. Using the notion of the Lyapunov exponent for these sequences of matrices (see Section 2.5), we will introduce the notion of the Lyapunov exponent for the cocycle. In doing so, one should carefully examine the dependence of the Lyapunov exponent when moving from one such sequence of matrices to another one, i.e., moving from one trajectory of f to another one. This is a crucial new element in studying Lyapunov exponents for cocycles over dynamical systems versus studying Lyapunov exponents for sequences of matrices.

Another important element is that since the cocycle $\mathcal{A}$ is assumed to be only measurable, it may not be bounded, and hence, condition (2.39) may fail. We will replace this requirement with the following more general *integrability condition* on the generator A of the cocycle $\mathcal{A}$:

$$\log^+\|A\|,\ \log^+\|A^{-1}\| \in L^1(X,\nu), \tag{3.6}$$

where $\log^+ a = \max\{\log a, 0\}$.

Theorem 3.8. *Assume that cocycle $\mathcal{A}$ satisfies condition (3.6). Then there is a set $Y \subset X$ of full ν-measure such that for all $x \in Y$ the Lyapunov exponent for the sequence of matrices $\{A(f^m(x))\}_{m \in \mathbb{Z}}$ is well-defined, that is, for ν-almost all $x \in X$ the upper limit*

$$\limsup_{m \to +\infty} \frac{1}{m} \log \|\mathcal{A}(x,m)v\| \tag{3.7}$$

is finite for all $v \in \mathbb{R}^p \setminus \{0\}$.

Proof. Note that for every $v \in \mathbb{R}^p \setminus \{0\}$ and $m \geq 0$

$$\|\mathcal{A}(x,m)v\| \leq \|\mathcal{A}(x,m)\| \leq \prod_{i=0}^{m} \|A(f^i(x))\|$$

and

$$\|\mathcal{A}(x,m)v\| \geq \|\mathcal{A}(x,m)^{-1}\|^{-1} \geq \prod_{i=0}^{m} \|A(f^i(x))^{-1}\|^{-1}.$$

It follows that

$$-\frac{1}{m}\sum_{i=0}^{m-1} \log^+ \|A(f^i(x))^{-1}\| \leq \frac{1}{m}\log\|\mathcal{A}(x,m)v\| \leq \frac{1}{m}\sum_{i=0}^{m-1}\log^+ \|A(f^i(x))\|.$$

Applying now condition (3.6) and Birkhoff's Ergodic Theorem, we obtain that for ν-almost all $x \in X$ the limit in (3.7) is finite for all $v \in \mathbb{R}^p \setminus \{0\}$. □

We assume from now on that the integrability condition (3.6) holds. Given a point $(x,v) \in Y \times \mathbb{R}^p$, define the *forward Lyapunov exponent at (x,v) (with respect to the cocycle $\mathcal{A}$)* by

$$\chi^+(x,v) = \chi^+(x,v,\mathcal{A}) = \limsup_{m \to +\infty} \frac{1}{m}\log\|\mathcal{A}(x,m)v\|.$$

With the convention that $\log 0 = -\infty$, we obtain $\chi^+(x,0) = -\infty$ for $x \in X$.

Exercise 3.9. Show that for each $x \in X$ the function $\chi^+(x,\cdot)$ is a Lyapunov exponent in $\mathbb{R}^p$ and that it does not depend on the choice of the inner product on $\mathbb{R}^p$.

Fix $x \in Y$. By Theorem 2.1, there exist a positive integer $s^+(x) \leq n$, a collection of numbers

$$\chi_1^+(x) < \chi_2^+(x) < \cdots < \chi_{s^+(x)}^+(x),$$

and a filtration $\mathcal{V}_x^+$ of linear subspaces

$$\{0\} = V_0^+(x) \subsetneq V_1^+(x) \subsetneq \cdots \subsetneq V_{s^+(x)}^+(x) = \mathbb{R}^p,$$

such that:

(1) $V_i^+(x) = \{v \in \mathbb{R}^p : \chi^+(x,v) \le \chi_i^+(x)\}$;
(2) $\chi^+(x,v) = \chi_i^+(x)$ for $v \in V_i^+(x) \setminus V_{i-1}^+(x)$.

The numbers $\chi_i^+(x)$ are the values of the Lyapunov exponent χ^+ at x; the collection of linear spaces $V_i^+(x, \mathcal{A}) = V_i^+(x)$ is the filtration $\mathcal{V}_x^+$ of $\mathbb{R}^p$ associated to χ^+ at x; the number $k_i^+(x) = \dim V_i^+(x) - \dim V_{i-1}^+(x)$ is the multiplicity of the value $\chi_i^+(x)$. Finally, the Lyapunov spectrum of χ^+ at x is the collection of pairs

$$\mathrm{Sp}_x^+ \mathcal{A} = \{(\chi_i^+(x), k_i^+(x)) : i = 1, \ldots, s^+(x)\}.$$

Exercise 3.10. Show that the following properties hold:

(1) the functions χ^+ and s^+ are Borel measurable;
(2) $\chi^+ \circ F = \chi^+$ where F is the linear extension given by (3.1);
(3) for every $x \in Y$ and $1 \le i \le s^+(x)$ we have

$$A(x) V_i^+(x) = V_i^+(f(x)), \quad \chi_i^+(f(x)) = \chi_i^+(x),$$

$$k_i^+(f(x)) = k_i^+(x), \quad s^+(f(x)) = s^+(x).$$

In particular, $\mathrm{Sp}_{f(x)}^+ \mathcal{A} = \mathrm{Sp}_x^+ \mathcal{A}$ for any $x \in X$.

Similarly, reversing the time, we define the *backward Lyapunov exponent* at (x,v) *(with respect to the cocycle $\mathcal{A}$)* by

$$\chi^-(x,v) = \chi^-(x,v,\mathcal{A}) = \limsup_{m \to -\infty} \frac{1}{m} \log \|A(x,m)v\|.$$

We have then the associated Lyapunov spectrum of χ^- at x as

$$\mathrm{Sp}_x^- \mathcal{A} = \{(\chi_i^-(x), k_i^-(x)) : i = 1, \ldots, s^-(x)\}$$

and the associated filtration $\mathcal{V}_x^-$ of linear subspaces as

$$\mathbb{R}^p = V_1^-(x) \supsetneq V_2^-(x) \supsetneq \cdots \supsetneq V_{s^-(x)+1}^-(x) = \{0\}.$$

The backward Lyapunov exponent has properties similar to the ones described in Exercise 3.10 (with "+" replaced by "−").

The forward and backward Lyapunov exponents, their spectra, and associated filtrations are invariant under tempered coordinate changes.

Proposition 3.11. *Let $\mathcal{A}$ and $\mathcal{B}$ be equivalent cocycles over a measurable transformation $f\colon X \to X$ and let $L\colon X \to GL(p,\mathbb{R})$ be a measurable function that is tempered (see (3.4)). For any $x \in X$:*

(1) *the forward and backward Lyapunov spectra at x coincide, i.e.,*
$$\operatorname{Sp}_x^+ \mathcal{A} = \operatorname{Sp}_x^+ \mathcal{B} \quad \text{and} \quad \operatorname{Sp}_x^- \mathcal{A} = \operatorname{Sp}_x^- \mathcal{B};$$

(2) *$L(x)$ preserves the forward and backward filtrations of $\mathcal{A}$ and $\mathcal{B}$, i.e., for each $i = 1,\ldots, s^+(x)$,*
$$L(x) V_i^+(x, \mathcal{A}) = V_i^+(x, \mathcal{B}),$$
and for each $i = 1\ldots, s^-(x)$,
$$L(x) V_i^-(x, \mathcal{A}) = V_i^-(x, \mathcal{B}).$$

Proof. Since the function L satisfies (3.5), for each $v \in \mathbb{R}^p$ we obtain
$$\limsup_{m \to \pm\infty} \frac{1}{|m|} \log \|\mathcal{A}(x,m)v\| \leq \limsup_{m \to \pm\infty} \frac{1}{|m|} \log \|L(f^m(x))^{-1}\|$$
$$+ \limsup_{m \to \pm\infty} \frac{1}{|m|} \|\mathcal{B}(x,m)L(x)v\|.$$

Since L is tempered, this implies that
$$\chi^+(x, v, \mathcal{A}) \leq \chi^+(x, L(x)v, \mathcal{B}) \quad \text{and} \quad \chi^-(x, v, \mathcal{A}) \leq \chi^-(x, L(x)v, \mathcal{B}).$$

Writing $\mathcal{B}(x,m)L(x) = L(f^m(x))\mathcal{A}(x,m)$, we conclude in a similar way that
$$\chi^+(x, L(x)v, \mathcal{B}) \leq \chi^+(x, v, \mathcal{A}) \quad \text{and} \quad \chi^-(x, L(x)v, \mathcal{B}) \leq \chi^-(x, v, \mathcal{A}).$$

The desired result now follows. □

We say that a point $x \in Y$ is *forward regular* (respectively, *backward regular*) for $\mathcal{A}$ if the sequence of matrices $(A(f^m(x)))_{m \in \mathbb{Z}}$ is forward (respectively, backward) regular.

One can easily check that if x is a forward (respectively, backward) regular point for $\mathcal{A}$, then so is the point $f^m(x)$ for every $m \in \mathbb{Z}$. Moreover, for any forward (respectively, backward) regular point $x \in Y$ the conclusions of Theorem 2.27 (and of its analog for the backward Lyapunov exponent) hold when the uniform boundedness assumption (2.39) (respectively, (2.49)) is replaced with the integrability condition (3.6) with little modifications in the proofs.

It is easy to see that if $\mathcal{A}$ and $\mathcal{B}$ are equivalent cocycles on Y, then the point $x \in Y$ is forward (respectively, backward) regular for $\mathcal{A}$ if and only if it is forward (respectively, backward) regular for $\mathcal{B}$.

3.2. Lyapunov exponents, Lyapunov–Perron regularity for cocycles

Similarly to the above, we say that a point $x \in Y$ is *Lyapunov–Perron regular* or simply *LP-regular* for $\mathcal{A}$ if the sequence of matrices $(A(f^m(x)))_{m \in \mathbb{Z}}$ is Lyapunov–Perron regular. We shall discuss this crucial notion in more detail and introduce some relevant notations and terminology which will be used later in the book.

Consider the families of filtrations $\mathcal{V}^+ = \{V_x^+\}_{x \in Y}$ and $\mathcal{V}^- = \{V_x^-\}_{x \in Y}$ associated with the Lyapunov exponents χ^+ and χ^- of the cocycle $\mathcal{A}$. We say that these families of filtrations are *coherent* if the following properties hold:

(1) $s^+(x) = s^-(x) =: s(x)$;

(2) there exists a decomposition
$$\mathbb{R}^p = \bigoplus_{j=1}^{s(x)} E_j(x) \tag{3.8}$$
into subspaces $E_j(x)$ such that $A(x)E_j(x) = E_j(f(x))$ and
$$V_i^+(x) = \bigoplus_{j=1}^{i} E_j(x) \quad \text{and} \quad V_i^-(x) = \bigoplus_{j=i}^{s(x)} E_j(x);$$

(3) for $v \in E_i(x) \setminus \{0\}$, $i = 1, \ldots, s(x)$,
$$\lim_{m \to \pm \infty} \frac{1}{m} \log \|\mathcal{A}(x, m)v\| = \chi_i^+(x) = -\chi_i^-(x) =: \chi_i(x), \tag{3.9}$$
with uniform convergence on $\{v \in E_i(x) : \|v\| = 1\}$.

We call the decomposition (3.8) the *Oseledets decomposition*, and we call the subspaces $E_i(x)$ the *Oseledets subspaces* at x.

We remark that property (2) is equivalent to the following property: for $i = 1, \ldots, s(x)$ the spaces
$$E_i(x) = V_i^+(x) \cap V_i^-(x) \tag{3.10}$$
satisfy (3.8).

A point $x \in Y$ is *Lyapunov–Perron regular* or simply *LP-regular* for $\mathcal{A}$ if the following conditions hold:

(1) x is simultaneously forward and backward regular for $\mathcal{A}$;

(2) the families of filtrations $\mathcal{V}^+$ and $\mathcal{V}^-$ are coherent at x.

Condition (2) requires some degree of compatibility between forward and backward regularity.

We denote by $\mathcal{R} = \mathcal{R}_\mathcal{A}$ the set of LP-regular points. It is easy to see that if x is an LP-regular point for $\mathcal{A}$, then so is the point $f^m(x)$ for every $m \in \mathbb{Z}$. It follows that this set $\mathcal{R}$ is f-invariant. We remark that if $\mathcal{A}$ and $\mathcal{B}$

are equivalent cocycles on Y, then the point $y \in Y$ is LP-regular for $\mathcal{A}$ if and only if it is LP-regular for $\mathcal{B}$, i.e., $\mathcal{R}_\mathcal{A} = \mathcal{R}_\mathcal{B}$. We will show in Theorem 4.1 that under fairly general assumptions the set of LP-regular points has full measure with respect to any invariant measure.

For a given collection of linearly independent vectors $\mathbf{v} = (v_1, \ldots, v_k)$, $1 \le k \le n$, the number $\operatorname{Vol}_k^{\mathbf{v}}(x, m)$ denotes the volume of the parallelepiped formed by the vectors $\mathcal{A}(x, m)v_i$, $i = 1, \ldots, k$.

Theorem 3.12. *A point $x \in X$ is an LP-regular point for $\mathcal{A}$ if and only if there exist a decomposition*

$$\mathbb{R}^p = \bigoplus_{i=1}^{s(x)} E_i(x)$$

and numbers $\chi_1(x) < \cdots < \chi_{s(x)}(x)$ such that:

(1) *if $i = 1, \ldots, s(x)$ and $v \in E_i(x) \setminus \{0\}$, then*

$$\lim_{m \to \pm\infty} \frac{1}{m} \log \|\mathcal{A}(x, m) v\| = \chi_i(x);$$

(2)

$$\lim_{m \to \pm\infty} \frac{1}{m} \log |\det \mathcal{A}(x, m)| = \sum_{i=1}^{s(x)} \chi_i(x) \dim E_i(x).$$

In addition, if $x \in X$ is an LP-regular point for $\mathcal{A}$, then the following statements hold:

(a) *for any collection of vectors $\mathbf{v} = (v_1, \ldots, v_k)$ the limits*

$$\lim_{m \to \pm\infty} \frac{1}{m} \log \operatorname{Vol}_k^{\mathbf{v}}(x, m)$$

exist;

(b) *$E_i(x) = V_i^+(x) \cap V_i^-(x)$ and $\dim E_i(x) = k_i^+(x) = k_i^-(x) =: k_i(x)$;*

(c) *if $\mathbf{v} = (v_1, \ldots, v_{k_i(x)})$ is a basis of $E_i(x)$, then*

$$\lim_{m \to \pm\infty} \frac{1}{m} \log \operatorname{Vol}_{k_i}^{\mathbf{v}}(x, m) = \chi_i(x) k_i(x);$$

(d) *if $v_i \in E_i(x) \setminus \{0\}$ and $v_j \in E_j(x) \setminus \{0\}$ with $i \ne j$, then*

$$\lim_{m \to \pm\infty} \frac{1}{m} \log \angle(\mathcal{A}(x, m) v_i, \mathcal{A}(x, m) v_j) = 0.$$

3.3. Examples of measurable cocycles over dynamical systems

Measurable cocycles over dynamical systems appear naturally in many areas of mathematics. In this section we describe two rather simple situations.

3.3.1. Reducible cocycles. Let $\mathcal{A}$ be a cocycle over a measurable transformation $f\colon X \to X$. We say that $\mathcal{A}$ is *reducible* if it is equivalent to a cocycle whose generator is constant along trajectories. More precisely, if $A\colon X \to GL(p,\mathbb{R})$ is the generator of $\mathcal{A}$, then there exist measurable functions $B, L\colon X \to GL(p,\mathbb{R})$ such that:

(1) B is f-invariant and L is tempered on X;
(2) $A(x) = L(f(x))^{-1} B(x) L(x)$ for every $x \in X$.

In particular,
$$\mathcal{A}(x,m) = L(f^m(x))^{-1} B(x)^m L(x).$$

We note that if $\mathcal{A}$ is a reducible cocycle, then every point $x \in X$ is LP-regular and the values of the Lyapunov exponent at a point $x \in X$ are equal to $\log|\lambda_i(x)|$, where $\lambda_i(x)$ are the eigenvalues of the matrix $B(x)$.

Reducible cocycles are natural objects in the classical Floquet theory where the dynamical system in the base is a periodic flow. Before describing this example we briefly discuss cocycles over flows.

Let $\varphi\colon \mathbb{R} \times X \to X$ be a measurable flow. The function $\mathcal{A}\colon X \times \mathbb{Z} \to GL(p,\mathbb{R})$ is a *linear multiplicative cocycle over* φ or simply a *cocycle* if:

(1) for each $x \in X$ we have $\mathcal{A}(x,0) = \mathrm{Id}$, and given $s, t \in \mathbb{R}$,
$$\mathcal{A}(x, s+t) = \mathcal{A}(f^t(x), s) \mathcal{A}(x, t);$$
(2) for every $t \in \mathbb{R}$ the function $\mathcal{A}(\cdot, t)\colon X \to GL(p,\mathbb{R})$ is measurable.

Let $Y \subset X$ be a nonempty measurable set which is φ_t-invariant for all $t \in \mathbb{R}$, where $\varphi_t = \varphi(t, \cdot)$. A measurable function $L\colon X \to GL(p,\mathbb{R})$ is said to be *tempered on Y with respect to φ* or simply *tempered on Y* if for every $x \in Y$ we have
$$\lim_{t \to \pm\infty} \frac{1}{t} \log\|L(\varphi_t(x))\| = \lim_{t \to \pm\infty} \frac{1}{t} \log\|L(\varphi_t(x))^{-1}\| = 0.$$

We say that a cocycle $\mathcal{A}$ over φ is *reducible* if there exist measurable functions $D\colon X \to M(p,\mathbb{R})$ (the set of $p \times p$ matrices with real entries) and $L\colon X \to GL(p,\mathbb{R})$ such that:

(1) D is φ_t-invariant for all $t \in \mathbb{R}$ and L is tempered on X;
(2) $\mathcal{A}(x,t) = L(\varphi_t(x))^{-1} e^{D(x)t} L(x)$ for every $x \in X$ and $t \in \mathbb{R}$.

We note that if $\mathcal{A}$ is a reducible cocycle, then every point $x \in X$ is LP-regular and the values of the Lyapunov exponent at a point $x \in X$ are equal to $\operatorname{Re}\lambda_i(x)$, where $\lambda_i(x)$ are the eigenvalues of the matrix $D(x)$.

We now proceed with the example. Consider an autonomous differential equation $x' = F(x)$, and let $\varphi_t(x)$ be a T-periodic solution. The corresponding system of variational equations is given by

$$z' = C_x(t)z, \qquad (3.11)$$

where $C_x(t) = d_{\varphi_t(x)}F$ is a T-periodic matrix function. By Floquet theory, any fundamental solution of equation (3.11) is of the form $X(t) = P(t)e^{Dt}$ for some T-periodic matrix function $P(t)$ and a constant matrix D. Setting $L(\varphi_t(x)) = P(t)^{-1}$, the Cauchy matrix can be written in the form

$$X(t)X(0)^{-1} = L(\varphi_t(x))^{-1}e^{Dt}L(x).$$

Since $P(t)$ is T-periodic and thus bounded, the function L is tempered. It follows that the cocycle $X(t)X(0)^{-1}$ over the flow restricted to the orbit of x is reducible. More generally, if the flow defined by the equation $x' = F(x)$ has only constant or periodic trajectories (possibly with different periods), then the corresponding cocycle defined by the systems of variational equations is reducible.

We describe another example of reducible cocycles associated with translations on the torus. Given a translation vector $\omega \in \mathbb{R}^p$, let $f \colon \mathbb{T}^p \to \mathbb{T}^p$ be the translation $f(x) = x+\omega \bmod 1$. Consider the cocycle $\mathcal{A}(\omega)$ over f generated by the time-one map obtained from the solutions of a linear differential equation

$$y' = B(x + t\omega)y,$$

where $B \colon \mathbb{T}^p \to GL(p, \mathbb{R})$ is a given function. It was shown by Floquet that if $\omega/\pi \in \mathbb{Q}^p$, then the cocycle $\mathcal{A}(\omega)$ is reducible.

The case $\omega/\pi \notin \mathbb{Q}^p$ is more subtle and whether the cocycle $\mathcal{A}(\omega)$ is reducible depends on ω. One can show (see [**59**]) that there are translation vectors ω for which some points $x \in \mathbb{T}^p$ are not LP-regular for the cocycle $\mathcal{A}$. This implies that for these ω the cocycle is not reducible.

We shall describe a condition on ω that (along with some other requirements on the cocycle) guarantees that $\mathcal{A}(\omega)$ is reducible. If $\omega/\pi \notin \mathbb{Q}^p$, then the vector $\tilde{\omega} = (\omega, 2\pi)$ is rationally independent, that is, $\langle k, \tilde{\omega} \rangle \neq 0$ whenever $k \in \mathbb{Z}^{n+1} \setminus \{0\}$. We say that ω satisfies a *Diophantine condition* if the vector $\tilde{\omega}$ is Diophantine, that is,

$$|\langle k, \tilde{\omega} \rangle| \geq \frac{c}{\|k\|^\tau}$$

for every $k \in \mathbb{Z}^{n+1} \setminus \{0\}$ and some $c, \tau > 0$. One can show that the set of Diophantine vectors has full Lebesgue measure. Johnson and Sell [60] obtained the following criterion for reducibility.

Theorem 3.13. *If ω satisfies a Diophantine condition and all Lyapunov exponents are exact and simple everywhere, then the cocycle $\mathcal{A}(\omega)$ is reducible.*

We refer the reader to [47, 72] for further results on the reducibility of cocycles when ω satisfies a Diophantine condition.

3.3.2. Cocycles associated with Schrödinger operators. The Schrödinger equation is a linear partial differential equation that is designed to describe how the quantum state of a physical system changes in time. Its discrete-time version is an equation of the form

$$-(u_{m+1} + u_{m-1}) + V(f^m(x))u_m = Eu_m, \qquad (3.12)$$

where $f(x) = x + \alpha \pmod 1$ is an irrational rotation of the circle S^1, $E \in \mathbb{R}$ is the total energy of the system, and $V \colon S^1 \to \mathbb{R}$ is the potential energy.

Exercise 3.14. Show that equation (3.12) is equivalent to

$$\begin{pmatrix} u_{m+1} \\ u_m \end{pmatrix} = A(f^m(x)) \begin{pmatrix} u_m \\ u_{m-1} \end{pmatrix},$$

where $A \colon S^1 \to SL(2, \mathbb{R})$ is given by

$$A(x) = \begin{pmatrix} V(x) - E & -1 \\ 1 & 0 \end{pmatrix}.$$

The Schrödinger cocycle is the cocycle $\mathcal{A}$ over the map f generated by the matrix function A.

Let $\lambda_1(V, x)$ and $\lambda_2(V, x)$ be the Lyapunov exponents of the cocycle generated by A at a point $x \in S^1$, and set

$$\Lambda_i(V) = \int_{S^1} \lambda_i(V, x) \, dx, \quad i = 1, 2.$$

One can show (see [21]) that the map

$$C^0(S^1, \mathbb{R}) \ni V \mapsto (\Lambda_1(V), \Lambda_2(V))$$

is discontinuous at a potential V if and only if $\Lambda_1(V)$ and $\Lambda_2(V)$ are nonzero and E lies in the spectrum of the associated Schrödinger operator

$$(H_x \Psi)(m) = -(\Psi(m+1) + \Psi(m-1)) + V(f^m(x))\Psi(m)$$

in the space $\ell^2(\mathbb{Z})$ (we recall that the spectrum of a linear operator B is the complement of the set of numbers $E \in \mathbb{C}$ such that the inverse $(B - E)^{-1}$ is well-defined and bounded). This is due to the fact that E lies in the complement of the spectrum if and only if the cocycle is uniformly hyperbolic; see also Ruelle [100], Bourgain [25], and Bourgain and Jitomirskaya [26].

We present a result by Avila and Krikorian [9] on Lyapunov exponents for Schrödinger cocycles. We say that a number α satisfies the *recurrent Diophantine condition* if there are infinitely many $m > 0$ for which the mth image of α under the Gauss map satisfies the Diophantine condition with fixed constant c and power τ.

Theorem 3.15. *Assume that $V \in C^r(S^1, \mathbb{R})$ with $r = \omega$ or ∞ and that α satisfies the recurrent Diophantine condition. Then for almost every E the Schrödinger cocycle over the rotation by α either has nonzero Lyapunov exponents or is C^r-equivalent to a constant cocycle.*

Chapter 4

The Multiplicative Ergodic Theorem

The goal of this chapter is to present one of the principal results of this book, known as Oseledets' Multiplicative Ergodic Theorem.[1] It was first proved by Oseledets in [85] and since then other versions of this theorem have been obtained by Kingman, by Raghunathan, by Kaĭmanovich, and by Karlsson and Margulis; see [13].

Consider an invertible measure-preserving transformation f of a Lebesgue space (X, ν). We are interested in studying a cocycle $\mathcal{A}$ over f for which there exists a subset $Y \subset X$ of positive measure such that (1) the Lyapunov exponents on Y are nonzero and (2) every point $x \in Y$ is LP-regular. There are certain methods that allow one to estimate Lyapunov exponents and, in particular, to show that they are not equal to zero. However, establishing LP-regularity of a given trajectory is a daunting task. This is where the Multiplicative Ergodic Theorem comes into play: it provides a condition on the cocycle $\mathcal{A}$ and its invariant measure ν that guarantees that almost every trajectory of f is LP-regular.

Theorem 4.1. *Assume that the generator A of the cocycle $\mathcal{A}$ satisfies condition (3.6). Then the set of LP-regular points for $\mathcal{A}$ has full ν-measure.*

We emphasize that while Theorem 2.29 presents some criteria for LP-regularity of the Lyapunov exponent for a sequence of matrices $\{A_m\}_{m \in \mathbb{Z}}$ in terms of some properties of the exponent itself, Theorem 4.1 shows that for a cocycle $\mathcal{A}$ over a measurable transformation f of a Lebesgue space X

---

[1] The term "Multiplicative Ergodic Theorem" was coined by Oseledets.

and almost every point $x \in X$ with respect to an f-invariant measure the sequence of matrices $\{A(f^m(x))\}_{m \in \mathbb{Z}}$ is LP-regular. However, to determine if a particular point $x \in X$ is LP-regular, one needs to verify the conditions of Theorem 2.29 along the trajectory of x.

In the next two sections we present a proof of this theorem following the original Oseledets approach (see [85]). Its idea is to first reduce the general case to the case of triangular cocycles and then to prove the theorem for such cocycles.

4.1. Lyapunov–Perron regularity for sequences of triangular matrices

We begin by considering again sequences of matrices, and in addition to Theorem 2.27, we present a useful criterion for forward (and backward) regularity in the special case when the matrices in the sequence are lower-triangular. We will use this criterion in the proof of Theorem 4.1.

Let $\mathcal{A}^+ = (A_m)_{m \in \mathbb{Z}} \subset GL(p, \mathbb{R})$ be a sequence of lower-triangular matrices, that is, $A_m = (a_{ij}^m)$ where the entries $a_{ij}^m = 0$ if $i < j$. Let also χ^+ be the Lyapunov exponent associated with this sequence of matrices; see (2.40).

Theorem 4.2. *Assume that the following conditions hold:*

(1) *the limit*

$$\lim_{m \to +\infty} \frac{1}{m} \sum_{k=0}^{m} \log|a_{ii}^k| =: \lambda_i, \quad i = 1, \ldots, p,$$

exists and is finite;

(2) *we have*

$$\limsup_{m \to +\infty} \frac{1}{m} \log^+ |a_{ij}^m| = 0, \quad i, j = 1, \ldots, p.$$

Then the sequence of matrices $\mathcal{A}^+$ is forward regular, and the numbers λ_i are the values of the Lyapunov exponent χ^+ (counted with their multiplicities but possibly not ordered) associated to the sequence of matrices $\mathcal{A}^+$.

Proof. Before going into the detailed proof, let us explain the main point. For the sake of this discussion let us count each exponent according to its multiplicity; thus we have exactly p exponents. To verify the forward regularity, we will produce a basis $\{v_1, \ldots, v_p\}$, which is subordinate to the standard filtration (i.e., related to the standard basis by an upper-triangular coordinate change) and such that $\chi^+(v_i) = \lambda_i$.

If the exponents are ordered so that $\lambda_1 \geq \lambda_2 \geq \cdots \geq \lambda_p$, then the standard basis is in fact forward regular. To see this, notice that while multiplying lower-triangular matrices one obtains a matrix whose off-diagonal entries contain a polynomially growing number of terms each of which can be estimated by the growth of the product of diagonal terms above.

However, if the exponents are not ordered that way, then an element e_i of the standard basis will grow according to the maximal of the exponents λ_j for $j \geq i$. In order to produce the right growth, one has to compensate for the growth caused by off-diagonal terms by subtracting from the vector e_i a certain linear combination of vectors e_j for which $\lambda_j > \lambda_i$. This can be done in a unique fashion. The proof proceeds by induction in p.

For $p = 1$ the result follows immediately from condition (1). Given $p > 1$, we assume that the sequence of lower-triangular matrices $A_m \in GL(p+1, \mathbb{R})$ satisfies conditions (1) and (2). For each $m \geq 1$ let A'_m be the triangular $p \times p$ matrix obtained from A_m by deleting its first row and first column. The sequence of matrices $\mathcal{A}' = (A'_m)_{m \in \mathbb{Z}}$ satisfies conditions (1) and (2). Consider the Lyapunov exponent χ' associated with the sequence of matrices $\mathcal{A}'$ (see (2.40)) and the filtration

$$\{0\} = V_0 \subsetneq V_1 \subsetneq \cdots \subsetneq V_s = \mathbb{R}^p$$

associated to the Lyapunov exponent χ'. Write $\mathbb{R}^p = \bigoplus_{i=1}^{s} E_i$ where $E_i = V_{i-1}^{\perp} \cap V_i$ and let $\lambda'_2 \leq \cdots \leq \lambda'_{p+1}$ be the numbers $\lambda_2, \ldots, \lambda_{p+1}$ in condition (1) written in nondecreasing order. By the induction hypothesis, the sequence of matrices $\mathcal{A}'$ is forward regular, and hence, conditions (1) and (2) guarantee that for $i \geq 2$ and $v \in E_i \setminus \{0\}$,

$$\lim_{m \to +\infty} \frac{1}{m} \log \|\mathcal{A}'_m v\| = \lambda'_i, \qquad (4.1)$$

where $\mathcal{A}'_m = A'_{m-1} \cdots A'_0$.

It remains only to show that λ_1 is a value of the Lyapunov exponent χ for some vector $v \in \mathbb{R}^{p+1} \setminus \bigoplus_{i=1}^{s} E_i$. Indeed, if this were true, since the sequence of matrices $\mathcal{A}'$ is forward regular, we would obtain that

$$\lim_{m \to +\infty} \frac{1}{m} \log |\det \mathcal{A}_m| = \lim_{m \to +\infty} \frac{1}{m} \sum_{k=0}^{m} \log |a_{11}^k|$$
$$+ \lim_{m \to +\infty} \frac{1}{m} \log |\det \mathcal{A}'_m|$$
$$= \sum_{i=1}^{p+1} \lambda_i$$

and thus, the sequence of matrices $\mathcal{A}^+$ would be forward regular.

By Theorem 2.27, we have

$$\lim_{m \to +\infty} \frac{1}{m} \log \sin \angle(E_i^m, \widehat{E}_i^m) = 0, \qquad (4.2)$$

where
$$E_i^m = \mathcal{A}_m' E_i, \quad \widehat{E}_i^m = \bigoplus_{j \neq i} \mathcal{A}_m' E_i.$$

Let $\{v_1^m, \ldots, v_p^m\}$ be a basis of $\mathbb{R}^p$ such that $\{v_{n_{i-1}+1}^m, \ldots, v_{n_i}^m\}$ is an orthonormal basis of E_i^m, where $n_i = \dim V_i$, $i = 1, \ldots, s$. We denote by C_m the coordinate change from the standard basis $\{e_0, \ldots, e_p\}$ of $\mathbb{R}^{p+1}$ to the basis $\{e_0, v_1^m, \ldots, v_p^m\}$. It follows from (4.2) that

$$\lim_{m \to +\infty} \frac{1}{m} \log \|C_m\| = \lim_{m \to +\infty} \frac{1}{m} \log \|C_m^{-1}\| = 0. \tag{4.3}$$

Consider the sequence of matrices
$$B_m = C_{m+1}^{-1} A_m C_m$$
and note that for each i,
$$B_m \operatorname{span}\{e_{n_{i-1}+1}, \ldots, e_{n_i}\} = \operatorname{span}\{e_{n_{i-1}+1}, \ldots, e_{n_i}\}.$$

Hence, the matrix B_m has the form
$$B_m = \begin{pmatrix} a_{11}^m & & & & \\ g_1^m & B_1^m & & & \\ g_2^m & 0 & B_2^m & & \\ \vdots & \vdots & & \ddots & \\ g_s^m & 0 & \cdots & 0 & B_s^m \end{pmatrix},$$

where each B_i^m is a $k_i \times k_i$ matrix and each $g_i^m \in \mathbb{R}^{k_i}$ is a column vector. Set $\mathcal{B}_m = B_{m-1} \cdots B_0$ and observe that

$$\mathcal{B}_m = C_m^{-1} \mathcal{A}_m C_0. \tag{4.4}$$

The following lemma establishes a crucial property of the column vector g_i^m.

Lemma 4.3. *For each $i = 1, \ldots, s$ we have*
$$\limsup_{m \to +\infty} \frac{1}{m} \log^+ \|g_i^m\| = 0.$$

Proof of the lemma. Let $g^m \in \mathbb{R}^p$ be the column vector composed of the components of $g_1^m, \ldots, g_s^m$. We have
$$\|g^m\| \leq \|C_{m+1}^{-1}\| \cdot \|a_{21}^m e_1 + \cdots + a_{p+1,1}^m e_p\| \cdot \|C_m\|.$$

Therefore, using (4.3) and condition (2), we obtain
$$\limsup_{m \to +\infty} \frac{1}{m} \log^+ \|g^m\| \leq \limsup_{m \to +\infty} \frac{1}{m} \log^+ \max\{|a_{i1}^m| : 2 \leq i \leq p+1\} \tag{4.5}$$
$$= 0.$$

Moreover, since g_i^m is the projection of g^m onto E_i^{m+1} along $\widehat{E}_i^{m+1}$, we have
$$\|g_i^m\| \leq \|g^m\| / \sin \angle(E_i^{m+1}, \widehat{E}_i^{m+1}),$$

4.1. Lyapunov–Perron regularity for sequences of triangular matrices

and hence,
$$\frac{1}{m}\log\|g_i^m\| \leq \frac{1}{m}\log\|g^m\| - \frac{1}{m}\log\sin\angle(E_i^{m+1}, \widehat{E}_i^{m+1}).$$

By (4.2) and (4.5),
$$0 \leq \limsup_{m\to+\infty}\frac{1}{m}\log^+\|g_i^m\| \leq \limsup_{m\to+\infty}\frac{1}{m}\log^+\|g^m\| = 0$$

and the lemma follows. □

We proceed with the proof of the theorem.

Case 1: $\lambda_1 \geq \lambda'_{p+1}$ or $\lambda_1 \geq \lambda_j$ for all $2 \leq j \leq p+1$. This is a rather easy situation, since, as we will show, the off-diagonal elements of the matrix $\mathcal{B}_m$ grow sufficiently slowly compared to the growth rate of the first element on the diagonal of $\mathcal{B}_m$. Observe that

$$\mathcal{B}_m e_0 = \prod_{i=0}^{m-1} a_{11}^i e_0 + \sum_{j=1}^{s}\sum_{i=1}^{m-1} a_{11}^{i-1}\cdots a_{11}^0 B_j^{m-1}\cdots B_j^{i+1} g_j^i. \qquad (4.6)$$

Moreover, $\mathcal{B}_m v = B_j^{m-1}\cdots B_j^0 v$ for each $v \in \{0\}\times E_j$. Given j, $1 \leq j \leq s$, and $v \in C_0^{-1}(\{0\}\times E_j)\setminus\{0\}$, set

$$c_{0j}(v) = \frac{\|B_j^0 v\|}{\|v\|} \quad\text{and}\quad c_{mj}(v) = \frac{\|B_j^m\cdots B_j^0 v\|}{\|B_j^{m-1}\cdots B_j^0 v\|} \quad\text{for } m > 0.$$

Since the first component of $C_0 v$ vanishes, by (4.1) and (4.4) we have
$$\lambda'_j = \lim_{m\to+\infty}\frac{1}{m}\log\|\mathcal{A}_m C_0 v\| = \lim_{m\to+\infty}\frac{1}{m}\log\|C_m \mathcal{B}_m v\|.$$

Now observe that
$$\|C_m^{-1}\|^{-1}\cdot\|\mathcal{B}_m v\| \leq \|C_m \mathcal{B}_m v\| \leq \|C_m\|\cdot\|\mathcal{B}_m v\|.$$

Hence, it follows from (4.3) that
$$\lambda'_j = \lim_{m\to+\infty}\frac{1}{m}\log\|C_m \mathcal{B}_m v\| = \lim_{m\to+\infty}\frac{1}{m}\log\|\mathcal{B}_m v\|$$
$$= \lim_{m\to+\infty}\frac{1}{m}\log\prod_{l=0}^{m-1}\frac{\|\mathcal{B}_{l+1} v\|}{\|\mathcal{B}_l v\|} = \lim_{m\to+\infty}\frac{1}{m}\log\prod_{l=0}^{m-1} c_{lj}(v) \qquad (4.7)$$
$$= \lim_{m\to+\infty}\frac{1}{m}\sum_{l=0}^{m-1}\log c_{lj}(v).$$

Fix i, $0 \leq i < m$, and set
$$K_{mij} = \log\|a_{11}^{i-1}\cdots a_{11}^0 B_j^{m-1}\cdots B_j^{i+1} g_j^i\|.$$

Note that

$$K_{mij} = \sum_{\ell=0}^{i-1} \log|a_{11}^\ell| + \log\|B_j^{m-1}\cdots B_j^{i+1} g_j^i\|$$

$$= \sum_{\ell=0}^{i-1} \log|a_{11}^\ell| + \sum_{r=i+1}^{m-1} \log c_{rj}(v) + \log\|g_j^i\|,$$

where $v = (B_j^i \cdots B_j^0)^{-1} g_j^i$. By Lemma 4.3, for each $\varepsilon > 0$ there exists i_0 independent of m such that $\log^+\|g_j^i\| < \varepsilon i < \varepsilon m$ for every $i_0 \le i < m$. By condition (1) and (4.7), we may assume (choosing a larger i_0 if necessary) that

$$(\lambda_1 - \varepsilon)i < \sum_{\ell=0}^{i-1} \log|a_{11}^\ell| < (\lambda_1 + \varepsilon)i$$

and

$$(\lambda_j' - \varepsilon)(m-i) < \sum_{r=i+1}^{m-1} \log c_{rj}(v) < (\lambda_j' + \varepsilon)(m-i).$$

Since $\lambda_1 \ge \lambda_j'$, it follows that

$$K_{mij} < (\lambda_1 + \varepsilon)i + (\lambda_j' + \varepsilon)(m-i) + \varepsilon m \le (\lambda_1 + 2\varepsilon)m.$$

By (4.6), we conclude that

$$\exp[(\lambda_1 - \varepsilon)m] \le \|\mathcal{B}_m e_0\|$$

$$\le \exp[(\lambda_1 + \varepsilon)m] + \sum_{j=1}^{s}\sum_{i=0}^{m-1} \exp K_{mij}$$

$$\le \exp[(\lambda_1 + \varepsilon)m] + sm \exp[(\lambda_1 + 2\varepsilon)m]$$

$$\le (s+1)\exp[(\lambda_1 + 3\varepsilon)m]$$

for all sufficiently large m for which $m \le \exp(\varepsilon m)$. It follows from (4.3) and (4.4) that

$$\chi'(e_0) = \lim_{m\to+\infty} \frac{1}{m} \log\|\mathcal{B}_m e_0\| = \lambda_1.$$

This completes the proof of the theorem in the case $\lambda_1 \ge \lambda_{p+1}'$.

Case 2: $\lambda_1 < \lambda_j'$ for some j such that $2 \le j \le p+1$. Using again Lemma 4.3, given $\varepsilon > 0$, we find that $\|g_j^m\| < \exp(\varepsilon m)$. Furthermore,

$$\|(B_j^m \cdots B_j^0)^{-1} g_j^m\| \le \exp[(-\lambda_j' + \varepsilon)m]$$

4.1. Lyapunov–Perron regularity for sequences of triangular matrices

for all sufficiently large m. Therefore,

$$\sum_{m=1}^{\infty} \|a_{11}^{m-1} \cdots a_{11}^{0}(B_j^m \cdots B_j^0)^{-1} g_j^m\| \leq \sum_{m=1}^{\infty} \exp[(\lambda_1 + \varepsilon)m + (-\lambda_j' + \varepsilon)m]$$

$$\leq \sum_{m=1}^{\infty} \exp[(\lambda_1 - \lambda_j' + 2\varepsilon)m] < +\infty$$

for all $m \geq 1$ and all sufficiently small $\varepsilon > 0$. This shows that the formula

$$h_j = -\sum_{m=1}^{\infty} a_{11}^{m-1} \cdots a_{11}^{0}(B_j^m \cdots B_j^0)^{-1} g_j^m$$

defines a vector in E_j. Set

$$v = e_0 + \sum_{j:\lambda_1 < \lambda_j'} h_j$$

and denote by $\operatorname{proj}_{E_j^m}$ the projection onto E_j^m along $\widehat{E}_j^m$. We have

$$\operatorname{proj}_{E_j^m} \mathcal{B}_m v = \operatorname{proj}_{E_j^m} \mathcal{B}_m(e_0 + h_j)$$

$$= \sum_{i=1}^{m-1} a_{11}^{i-1} \cdots a_{11}^{0} B_j^{m-1} \cdots B_j^{i+1} g_j^i + \mathcal{B}_m h_j$$

$$= \sum_{i=1}^{m-1} a_{11}^{i-1} \cdots a_{11}^{0} B_j^{m-1} \cdots B_j^{i+1} g_j^i + B_j^{m-1} \cdots B_j^0 h_j$$

$$= -\sum_{i=m}^{\infty} a_{11}^{i-1} \cdots a_{11}^{0}(B_j^i \cdots B_j^m)^{-1} g_j^i.$$

Proceeding as before, we obtain

$$\|\operatorname{proj}_{E_j^m} \mathcal{B}_m v\| \leq \sum_{i=m}^{\infty} \|a_{11}^{i-1} \cdots a_{11}^{0}(B_j^i \cdots B_j^m)^{-1} g_j^i\|$$

$$\leq \sum_{i=m}^{\infty} \exp[(\lambda_1 + \varepsilon)i + (-\lambda_j' + \varepsilon)(i - m) + \varepsilon i]$$

$$\leq \exp[(\lambda_1 + \varepsilon)m] \sum_{i=m}^{\infty} \exp[(\lambda_1 - \lambda_j' + 2\varepsilon)(i - m) + \varepsilon i]$$

$$= D(\varepsilon) \exp[(\lambda_1 + 2\varepsilon)m]$$

(4.8)

for all sufficiently large $m \geq 1$. Here

$$D(\varepsilon) = \sum_{i=1}^{\infty} \exp[(\lambda_1 - \lambda'_j + 3\varepsilon)i] < +\infty,$$

provided $\varepsilon > 0$ is chosen such that $\lambda_1 - \lambda'_j + 3\varepsilon < 0$ (that is always possible since $\lambda_1 < \lambda'_j$).

Denote by proj_{e_0} the projection on e_0 along $\mathbb{R}^p = \bigoplus_{i=1}^{s} E_i$. Exploiting the special form of the matrix $\mathcal{B}_m$ described above, we obtain

$$\|\text{proj}_{e_0} \mathcal{B}_m v\| = \prod_{i=0}^{m-1} a_{11}^i.$$

By Theorem 2.27, for each j we have

$$\lim_{m \to +\infty} \frac{1}{m} \log \sin \angle(E_j^m, \widehat{E}_j^m) = 0.$$

Since e_0 is orthogonal to $\bigoplus_{i=1}^{s} E_i$, there exist $c_1 > 0$ and $c_2 > 0$ such that

$$c_1 \|\mathcal{B}_m v\| \leq \prod_{i=0}^{m-1} a_{11}^i + \sum_{j:\lambda_j > \lambda_1} \|\text{proj}_{E_j^m} \mathcal{B}_m v\| \leq c_2 \|\mathcal{B}_m v\|$$

for every $m \geq 1$. By (4.3), (4.8), and condition (1), we have

$$\chi'(C_0^{-1} v) = \lim_{m \to +\infty} \frac{1}{m} \log \|\mathcal{B}_m v\| = \lambda_1.$$

The above discussion implies that the Lyapunov exponent χ' is exact with respect to the vector

$$v_{p+1} = e_0 + C_0^{-1} \sum_{j:\lambda_1 < \lambda'_j} h_j$$

and that $\chi'(v_{p+1}) = \lambda_1$. This completes the proof of Theorem 4.2. $\square$

An analogous criterion for forward regularity holds for sequences of upper-triangular matrices.

Using the correspondence between forward and backward sequences of matrices, we immediately obtain the corresponding criterion for backward regularity.

Theorem 4.4. *Let $\mathcal{A}^- = (a_{ij}^m)_{m \in \mathbb{Z}} \subset GL(p, \mathbb{R})$ be a sequence of lower-triangular matrices such that:*

(1) *the limit*

$$\lim_{m \to -\infty} \frac{1}{|m|} \sum_{k=1}^{m} \log|a_{ii}^k| =: \lambda_i^-, \quad i = 1, \ldots, p,$$

exists and is finite;

(2) *we have*

$$\limsup_{m\to-\infty} \frac{1}{|m|} \log^+ |a_{ij}^m| = 0, \quad i,j = 1,\ldots,p.$$

Then the sequence of matrices $\mathcal{A}^-$ is backward regular, and the numbers λ_i^- are the values of the Lyapunov exponent χ^- (counted with multiplicities but possibly not ordered) associated to the sequence of matrices $\mathcal{A}^-$.

4.2. Proof of the Multiplicative Ergodic Theorem

We split the proof of the theorem into two steps. First we show how to reduce the general case to the case of triangular cocycles. Then we present the proof of the latter case.

Step 1. Reduction to triangular cocycles. We follow the original Oseledets approach and construct an extension of the transformation f,

$$F \colon X \times SO(p,\mathbb{R}) \to X \times SO(p,\mathbb{R}),$$

where $SO(p,\mathbb{R})$ is the group of orthogonal $p \times p$ matrices. Given $(x,U) \in X \times SO(p,\mathbb{R})$, one can apply the Gram–Schmidt orthogonalization procedure to the columns of the matrix $A(x)U$ and write

$$A(x)U = R(x,U)T(x,U), \tag{4.9}$$

where $R(x,U)$ is orthogonal and $T(x,U)$ is lower-triangular (with positive entries on the diagonal). The two matrices $R(x,U)$ and $T(x,U)$ are uniquely defined, and their entries are linear combinations of the entries of U. Also, they depend measurably on x and continuously on U. Set

$$F(x,U) = (f(x), R(x,U)).$$

The map F depends measurably on x and continuously on U and f is a factor-map of F with respect to the partition whose elements are $\{x\} \times SO(p,\mathbb{R})$.

Consider the projection $\pi\colon (x,U) \mapsto U$. By (4.9), we obtain

$$T(x,U) = ((\pi \circ F)(x,U))^{-1} A(x) \pi(x,U). \tag{4.10}$$

Let $\tilde{\mathcal{A}}$ and $\mathcal{T}$ be two cocycles over F with generators given, respectively, by $\tilde{A}(x,U) = A(x)$ and $\mathcal{T}(x,U) = T(x,U)$. Since $\|U\| = 1$ for every $U \in SO(p,\mathbb{R})$, it follows from (4.10) that the cocycles $\tilde{\mathcal{A}}$ and $\mathcal{T}$ are equivalent on $X \times SO(p,\mathbb{R})$ via the conjugacy matrix function $L(x,U) = \pi(x,U)$ which is bounded and, hence, tempered. Therefore, a point $(x,U) \in X \times SO(p,\mathbb{R})$ is LP-regular for $\tilde{\mathcal{A}}$ if and only if it is LP-regular for $\mathcal{T}$. Moreover, the Lyapunov spectra of the cocycles $\tilde{\mathcal{A}}$ and $\mathcal{T}$ coincide.

Now if $x \in X$ is an LP-regular point for the cocycle $\mathcal{A}$, then statements (1) and (2) of Theorem 3.12 hold. It follows that for any $U \in SO(p,\mathbb{R})$

these statements hold for the cocycle $\tilde{\mathcal{A}}$ at the point (x, U), and hence, this point is LP-regular for $\tilde{\mathcal{A}}$. Moreover, the Lyapunov spectrum of the cocycle $\mathcal{A}$ at any LP-regular point $x \in X$ coincides with the Lyapunov spectrum of the cocycle $\tilde{\mathcal{A}}$ at (x, U) for any $U \in SO(p, \mathbb{R})$.

Without loss of generality, since X is a Lebesgue space, we may assume that it is a compact metric space. Let $\mathcal{M}$ be the set of all Borel probability measures $\tilde{\nu}$ on $X \times SO(p, \mathbb{R})$ that satisfy

$$\tilde{\nu}(B \times SO(p, \mathbb{R})) = \nu(B) \tag{4.11}$$

for all measurable sets $B \subset X$. Since $SO(p, \mathbb{R})$ is compact, $\mathcal{M}$ is a compact convex subset of a locally convex topological vector space of signed measures with bounded variation. Define a map $F_*\colon \mathcal{M} \to \mathcal{M}$ by setting for any measurable $B \subset X$

$$(F_*\tilde{\nu})(B \times SO(p, \mathbb{R})) = \tilde{\nu}(F^{-1}(B \times SO(p, \mathbb{R}))).$$

F_* is a bounded linear (and, hence, continuous) operator. By the Tychonoff Fixed Point Theorem, there exists a fixed point $\tilde{\nu}_0 \in \mathcal{M}$ for the operator F_*, i.e., a measure $\tilde{\nu}_0$ such that for every measurable set $B \subset X$

$$\tilde{\nu}_0(F^{-1}(B \times SO(p, \mathbb{R}))) = \tilde{\nu}_0(B \times SO(p, \mathbb{R})).$$

By (4.11) and the above discussion about LP-regular points and Lyapunov exponents for the cocycles $\mathcal{A}$, $\tilde{\mathcal{A}}$, and $\mathcal{T}$, we conclude that the set of LP-regular points for $\mathcal{A}$ has full ν-measure if and only if the set of LP-regular points for $\tilde{\mathcal{A}}$ has full $\tilde{\nu}_0$-measure and, hence, if and only if the set of LP-regular points for $\mathcal{T}$ has full $\tilde{\nu}_0$-measure.

Step 2. Proof for triangular cocycles. From now on we assume that $A(x) = (a_{ij}(x))$ is a lower-triangular matrix (recall that this means $a_{ij}(x) = 0$ if $i < j$). Write $A(x)^{-1} = (b_{ij}(x))$ and note that $b_{ii}(x) = 1/a_{ii}(x)$ for each i. By (3.6), we obtain $\log^+|a_{ij}|, \log^+|b_{ij}| \in L^1(X, \nu)$. It follows from Lemma 3.6 that

$$\lim_{m \to +\infty} \frac{1}{m} \log^+|a_{ij}(f^m(x))| = \lim_{m \to -\infty} \frac{1}{m} \log^+|b_{ij}(f^m(x))| = 0 \tag{4.12}$$

for every $1 \le i, j \le p$ and ν-almost every $x \in X$. Note that

$$|\log|a_{ii}|| = \log^+|a_{ii}| - \log^-|a_{ii}| = \log^+|a_{ii}| - \log^+|b_{ii}|. \tag{4.13}$$

By (3.6) and (4.13), we have $\log|a_{ii}| = -\log|b_{ii}| \in L^1(X, \nu)$. Birkhoff's Ergodic Theorem guarantees the existence of measurable functions $\lambda_i \in L^1(X, \nu)$ for $i = 1, \ldots, p$, such that

$$\lim_{m \to +\infty} \frac{1}{m} \sum_{k=0}^{m-1} \log|a_{ii}(f^k(x))| = \lim_{m \to -\infty} \frac{1}{m} \sum_{k=m}^{-1} \log|b_{ii}(f^k(x))| = \lambda_i(x) \tag{4.14}$$

for each $i = 1, \ldots, p$ and ν-almost every $x \in X$. Let $Y \subset X$ be the set of points $x \in X$ for which (4.12) and (4.14) hold. Notice that Y is a set of full ν-measure. We will show that Y consists of LP-regular points for $\mathcal{A}$, and thus the set of LP-regular points for $\mathcal{A}$ has full ν-measure.

By Theorems 4.2 and 4.4, the sequence $A(f^m(x))_{m \in \mathbb{Z}}$ is simultaneously forward and backward regular for every $x \in Y$. Moreover, the numbers $\lambda_i(x)$ are the forward Lyapunov exponents counted with their multiplicities (but possibly not ordered) and are symmetric to the backward Lyapunov exponents counted with their multiplicities (but possibly not ordered). We conclude that $s^+(x) = s^-(x) =: s(x)$ and $\chi_i^-(x) = -\chi_i^+(x)$ for $i = 1, \ldots, s(x)$. We now show that the spaces $E_1(x), \ldots, E_{s(x)}(x)$ defined by (3.10) satisfy (3.8).

Lemma 4.5. *If $x \in Y$ and $v \in \mathbb{R}^p \setminus \{0\}$, then $\chi^+(x, v) + \chi^-(x, v) \geq 0$.*

Proof of the lemma. For every $v \in \mathbb{R}^p \setminus \{0\}$, there exists i such that $v = \alpha_i e_i + \cdots + \alpha_p e_p$ with $\alpha_i \neq 0$, where $e_1, \ldots, e_p$ is the standard basis of $\mathbb{R}^p$. For any $m > 0$ the projection of $\mathcal{A}(x, m)v$ over $\text{span}\{e_i\}$ along its orthogonal complement is

$$\prod_{k=0}^{m-1} a_{ii}(f^k(x)) \alpha_i e_i.$$

Thus,

$$\log \|\mathcal{A}(x, m)v\| \geq \sum_{k=0}^{m-1} \log |a_{ii}(f^k(x))| + \log |\alpha_i|,$$

and $\chi^+(x, v) \geq \lambda_i(x)$. In a similar way, for every $m < 0$,

$$\log \|\mathcal{A}(x, m)v\| \geq \sum_{k=-m}^{-1} \log |b_{ii}(f^k(x))| + \log |\alpha_i|,$$

and $\chi^-(x, v) \geq -\lambda_i(x)$. Hence, $\chi^+(x, v) + \chi^-(x, v) \geq 0$. □

Lemma 4.6. *If $x \in Y$, then $\bigoplus_{i=1}^{s(x)} E_i(x) = \mathbb{R}^p$.*

Proof of the lemma. Since

$$A(x) V_i^+(x) = V_i^+(f(x)) \quad \text{and} \quad A(x) V_i^-(x) = V_i^-(f(x)),$$

we have $A(x) E_i(x) = E_i(f(x))$ for each i. Take $v \in V_i^+(x) \cap V_{i+1}^-(x)$. Then

$$\chi^+(x, v) \leq \chi_i^+(x) \quad \text{and} \quad \chi^-(x, v) \leq -\chi_{i+1}^+(x).$$

Hence,

$$\chi^+(x, v) + \chi^-(x, v) \leq \chi_i^+(x) - \chi_{i+1}^+(x) < 0,$$

and by Lemma 4.5 we obtain $v = 0$. Therefore, $V_i^+(x) \cap V_{i+1}^-(x) = \{0\}$. Moreover, for each $i < j$ we have

$$E_i(x) \cap E_j(x) \subset V_i^+(x) \cap V_j^-(x) \subset V_i^+(x) \cap V_{i+1}^-(x) = \{0\}.$$

Observe that

$$\dim E_i(x) \geq \dim V_i^+(x) + \dim V_i^-(x) - p$$
$$= (k_1(x) + \cdots + k_i(x)) + (k_i(x) + \cdots + k_{s(x)}(x)) - p = k_i(x).$$

Therefore,

$$\dim \bigoplus_{i=1}^{s(x)} E_i(x) \geq k_1(x) + \cdots + k_{s(x)}(x) = p$$

and hence $\bigoplus_{i=1}^{s(x)} E_i(x) = \mathbb{R}^p$. This completes the proof of the lemma. $\square$

It remains to show that the convergence in (3.9) is uniform on $C_i(x) = \{v \in E_i(x) : \|v\| = 1\}$.

Lemma 4.7. *If $x \in Y$, then*

$$\lim_{m \to \pm\infty} \frac{1}{|m|} \log \inf_v \|\mathcal{A}(x,m)v\| = \lim_{m \to \pm\infty} \frac{1}{|m|} \log \sup_v \|\mathcal{A}(x,m)v\| = \chi_i(x),$$

with the infimum and supremum taken over all $v \in C_i(x)$.

Proof of the lemma. Take $x \in Y$ and let $\{e_1, \ldots, e_{k_i(x)}\}$ be an orthonormal basis of $E_i(x)$. Let now

$$u_m = \sum_{j=1}^{k_i(x)} c_{m,j} e_j$$

be a vector in $C_i(x)$ at which $v \mapsto \|\mathcal{A}(x,m)v\|$ attains its minimum. Choose an integer $j(m)$ such that $|c_{m,j(m)}| = \max_j |c_{m,j}|$. Since

$$\sum_{j=1}^{k_i(x)} c_{m,j}^2 = 1,$$

we have $|c_{m,j(m)}| \geq 1/\sqrt{k_i(x)}$. We denote by $\rho_{m,j}$ and $\varphi_{m,j}$, respectively, the distance and the angle between $\mathcal{A}(x,m)e_j$ and $\mathcal{A}(x,m)\operatorname{span}\{e_i : i \neq j\}$. Note that

$$\rho_{m,j} = \|\mathcal{A}(x,m)e_j\| \sin \varphi_{m,j}.$$

We have

$$\mathcal{A}(x,m)u_m = c_{m,j(m)}\left(\mathcal{A}(x,m)e_{j(m)} + \sum_{j \neq j(m)} \frac{c_{m,j}}{c_{m,j(m)}} \mathcal{A}(x,m)e_j\right),$$

4.2. Proof of the Multiplicative Ergodic Theorem

and thus,
$$\|\mathcal{A}(x,m)u_m\| \geq |c_{m,j(m)}|\rho_{m,j(m)}$$
$$\geq \frac{1}{\sqrt{k_i(x)}}\|\mathcal{A}(x,m)e_{j(m)}\|\sin\varphi_{m,j(m)}.$$

Since x is forward and backward regular, it follows from Theorem 2.27 (and the analogous statement in the case of backward regularity) that
$$\lim_{m\to\pm\infty}\frac{1}{m}\log\sin\varphi_{m,j}=0.$$

Since $j(m)$ can only take on a finite number of values, we obtain

$$\liminf_{m\to\pm\infty}\frac{1}{|m|}\log\|\mathcal{A}(x,m)u_m\|$$
$$\geq \liminf_{m\to\pm\infty}\frac{1}{|m|}\log\|\mathcal{A}(x,m)e_{j(m)}\| + \liminf_{m\to\pm\infty}\frac{1}{|m|}\log\sin\varphi_{m,j(m)}$$
$$\geq \min_{j}\liminf_{m\to\pm\infty}\frac{1}{|m|}\log\|\mathcal{A}(x,m)e_j\|$$
$$+ \min_{j}\liminf_{m\to\pm\infty}\frac{1}{|m|}\log\sin\varphi_{m,j} = \chi_i(x),$$
(4.15)

where the minima are taken over all $j \in \{1,\ldots,k_i(x)\}$. Choose a vector
$$v_m = \sum_{j=1}^{k_i(x)} d_{m,j}e_j$$

in $C_i(x)$ for which the function $v \mapsto \|\mathcal{A}(x,m)v\|$ attains its maximum. We have
$$\|\mathcal{A}(x,m)v_m\| \leq \sum_{j=1}^{k_i(x)}|d_{m,j}|\cdot\|\mathcal{A}(x,m)e_j\| \leq \sum_{j=1}^{k_i(x)}\|\mathcal{A}(x,m)e_j\|,$$

and hence,
$$\limsup_{m\to\pm\infty}\frac{1}{|m|}\log\|\mathcal{A}(x,m)v_m\| \leq \chi_i(x). \qquad (4.16)$$

The desired result now follows from (4.15) and (4.16). □

Lemmas 4.6 and 4.7 show that the filtrations $\mathcal{V}^+$ and $\mathcal{V}^-$ are coherent at every point $x \in Y$, and hence the set Y consists of LP-regular points for $\mathcal{A}$. This implies that the set of LP-regular points for $\mathcal{A}$ has full ν-measure.

4.3. Normal forms of cocycles

The goal of this section is to describe a "normal form" of a cocycle over a measurable transformation associated with its Lyapunov exponent. A construction of such normal forms is a manifestation of the Multiplicative Ergodic Theorem (see Theorem 4.1). In the simple case of a rigid cocycle $\mathcal{A}$ whose generator is a constant map (i.e., $\mathcal{A}(x,m) = A^m$ for some matrix $A \in GL(p,\mathbb{R})$) it is easy to verify that the cocycle $\mathcal{A}$ is equivalent to the rigid cocycle $\mathcal{B}$ whose generator is the Jordan block form of the matrix A. We consider $\mathcal{B}$ as the "normal form" of $\mathcal{A}$ and say that $\mathcal{A}$ is reduced to $\mathcal{B}$.

A cocycle $\mathcal{A}$ satisfying the integrability condition (3.6) is, so to speak, "weakly" rigid; in other words, it can be reduced to a constant cocycle up to arbitrarily small error (see Theorem 4.10 below). We consider this constant cocycle as a "normal form" of $\mathcal{A}$.

4.3.1. The strong Lyapunov inner product.
Consider a cocycle $\mathcal{A}$ over a measurable transformation $f\colon X \to X$ and a family of inner products $\langle \cdot, \cdot \rangle = \langle \cdot, \cdot \rangle_x$ on $\mathbb{R}^p$ for $x \in X$. For each LP-regular point $x \in X$ we will replace this inner product with a special one that is generated by the action of the cocycle along the trajectory $\{f^m(x)\}_{m \geq 0}$. It provides an important technical tool in studying hyperbolic properties of cocycles over dynamical systems (see the discussion in Section 5.3.3). We start with the following auxiliary result.

Exercise 4.8. For each $\varepsilon > 0$, each LP-regular point $x \in X$, and each $i = 1, \ldots, s(x)$, show that the formula

$$\langle u, v \rangle'_{x,i} = \sum_{m \in \mathbb{Z}} \langle \mathcal{A}(x,m)u, \mathcal{A}(x,m)v \rangle e^{-2\chi_i(x)m - 2\varepsilon|m|} \tag{4.17}$$

determines a scalar product in $E_i(x)$.

Let $x \in X$ be an LP-regular point. We fix $\varepsilon > 0$ and introduce a new inner product in $\mathbb{R}^p$ (which will thus depend on ε) by the formula

$$\langle u, v \rangle'_x = \sum_{i=1}^{s(x)} \langle u_i, v_i \rangle'_{x,i}, \tag{4.18}$$

where u_i and v_i are the projections of u and v on $E_i(x)$ along $\bigoplus_{j \neq i} E_j(x)$. We call $\langle \cdot, \cdot \rangle'_x$ the *strong Lyapunov inner product at x*, and we call the corresponding norm

$$\|v\|'_x = \left(\sum_{i=1}^{s(x)} \sum_{m \in \mathbb{Z}} \|\mathcal{A}(x,m)v_i\|^2 e^{-2\chi_i(x)m - 2\varepsilon|m|} \right)^{1/2} \tag{4.19}$$

the *strong Lyapunov norm at* x. For each $i = 1, \ldots, s(x)$ the sequence of weights $\{e^{-2\chi_i(x)m - 2\varepsilon|m|}\}_{m \in \mathbb{Z}}$ in (4.17) is called the *Pesin tempering kernel* (this is a particular case of the tempering kernels that we will introduce in Section 4.3.3).

Exercise 4.9. Show that (compare to Theorem 1.14):

(1) $\langle \cdot, \cdot \rangle'_x$ determines an inner product in $T_x M$ that depends Borel measurably on x on the set $\mathcal{R}_A$ of LP-regular points in X;

(2) for every LP-regular point $x \in X$ and $i \neq j$, the spaces $E_i(x)$ and $E_j(x)$ are orthogonal with respect to the strong Lyapunov inner product.

We stress that, in general, the strong Lyapunov inner product does not determine a Riemannian metric on M. It will only be used to study the action of the cocycle along its LP-regular trajectories.

A coordinate change $C_\varepsilon \colon X \to GL(p, \mathbb{R})$ is called a *Lyapunov change of coordinates* if for each LP-regular point $x \in X$ and $u, v \in \mathbb{R}^p$ it satisfies

$$\langle u, v \rangle = \langle C_\varepsilon(x) u, C_\varepsilon(x) v \rangle'_x. \tag{4.20}$$

We note that the formula (4.20) does not determine the function $C_\varepsilon(x)$ uniquely.

4.3.2. The Oseledets–Pesin Reduction Theorem.
The following principal result describes "normal forms" of cocycles (see [**13**] for more details).

Theorem 4.10 (Oseledets–Pesin Reduction Theorem). *Let* $f \colon X \to X$ *be an invertible measure-preserving transformation of the Lebesgue space* (X, ν), *and let* $\mathcal{A}$ *be a cocycle over* f. *Given* $\varepsilon > 0$, *if* x *is an LP-regular point for* $\mathcal{A}$ *then:*

(1) *any Lyapunov change of coordinates* C_ε *sends the orthogonal decomposition* $\bigoplus_{i=1}^{s(x)} \mathbb{R}^{k_i(x)}$ *to the decomposition* $\bigoplus_{i=1}^{s(x)} E_i(x)$ *of* $\mathbb{R}^p$;

(2) *the cocycle* $A_\varepsilon(x) = C_\varepsilon(f(x))^{-1} A(x) C_\varepsilon(x)$ *has the block form*

$$A_\varepsilon(x) = \begin{pmatrix} A_\varepsilon^1(x) & & \\ & \ddots & \\ & & A_\varepsilon^{s(x)}(x) \end{pmatrix}, \tag{4.21}$$

where each block $A_\varepsilon^i(x)$ *is a* $k_i(x) \times k_i(x)$ *matrix and the entries are zero above and below the matrices* $A_\varepsilon^i(x)$;

(3) *each block* $A_\varepsilon^i(x)$ *satisfies*

$$e^{\chi_i(x) - \varepsilon} \leq \|A_\varepsilon^i(x)^{-1}\|^{-1} \leq \|A_\varepsilon^i(x)\| \leq e^{\chi_i(x) + \varepsilon}; \tag{4.22}$$

(4) *if the integrability condition (3.6) holds, then the map C_ε is tempered ν-almost everywhere (see (3.2)), and the Lyapunov spectra of $\mathcal{A}$ and $\mathcal{A}_\varepsilon$ coincide ν-almost everywhere.*

Proof. Statement (1) follows immediately from (4.20). For each i we have
$$A_\varepsilon(x)\mathbb{R}^{k_i(x)} = C_\varepsilon(f(x))^{-1}A(x)E_i(x)$$
$$= C_\varepsilon(f(x))^{-1}E_i(f(x)) = \mathbb{R}^{k_i(f(x))} = \mathbb{R}^{k_i(x)}.$$

Thus, $A_\varepsilon(x)$ has the block form (4.21).

Since $\mathcal{A}(x, m+1) = \mathcal{A}(f(x), m)A(x)$, for every $v \in E_i(x)$ we have that (see (4.19))
$$(\|A(x)v\|'_{f(x)})^2 = \sum_{m \in \mathbb{Z}} \|\mathcal{A}(f(x), m)A(x)v\|^2 e^{-2m\chi_i(x) - 2\varepsilon|m|}$$
$$= \sum_{m \in \mathbb{Z}} \|\mathcal{A}(x, m+1)v\|^2 e^{-2m\chi_i(x) - 2\varepsilon|m|}$$
$$= \sum_{k \in \mathbb{Z}} \|\mathcal{A}(x, k)v\|^2 e^{-2k\chi_i(x) - 2\varepsilon|k| + \eta(x, \varepsilon)},$$
where $\eta(x, \varepsilon) = 2\chi_i(x) + 2\varepsilon(|k| - |k-1|)$. Therefore,
$$e^{\chi_i(x) - \varepsilon}\|v\|'_x \leq \|A(x)v\|'_{f(x)} \leq e^{\chi_i(x) + \varepsilon}\|v\|'_x \qquad (4.23)$$
for every $v \in E_i(x)$. By (4.20), for each $w \in \mathbb{R}^p$ we obtain $\|w\| = \|C_\varepsilon(x)w\|'_x$ and
$$\|A_\varepsilon(x)w\| = \|C_\varepsilon(f(x))^{-1}A(x)C_\varepsilon(x)w\| = \|A(x)C_\varepsilon(x)w\|'_{f(x)}.$$
Hence if $v = C_\varepsilon(x)w \in E_i(x)$, then
$$\|w\| = \|v\|'_x \quad \text{and} \quad \|A_\varepsilon(x)w\| = \|A(x)v\|'_{f(x)}.$$
Therefore, by (4.23), for $w \in \mathbb{R}^{k_i(x)}$,
$$e^{\chi_i(x) - \varepsilon}\|w\| \leq \|A_\varepsilon(x)w\| \leq e^{\chi_i(x) + \varepsilon}\|w\|.$$
This implies the inequalities in (4.22).

We now proceed with the proof of statement (4). Since for every $w \in \mathbb{R}^p$
$$\|w\| = \|C_\varepsilon(x)w\|'_x \geq \|C_\varepsilon(x)w\|,$$
we have
$$\|C_\varepsilon(x)\| \leq 1 \quad \text{and} \quad \|C_\varepsilon(x)^{-1}\| \geq 1 \qquad (4.24)$$
for every LP-regular point $x \in X$. Now observe that
$$\mathcal{A}_\varepsilon(x, m) = C_\varepsilon(f^m(x))^{-1}\mathcal{A}(x, m)C_\varepsilon(x).$$
For each $N > 0$, consider the set
$$X_N = \{x \in X : x \text{ is LP-regular}, \|C_\varepsilon(x)\| > N^{-1}, \text{ and } \|C_\varepsilon(x)^{-1}\| < N\}.$$

4.3. Normal forms of cocycles

By the Multiplicative Ergodic Theorem (see Theorem 4.1), $\nu(X_N) \to 1$ as $N \to \infty$. Let us fix $N > 0$ such that $\nu(X_N) > 0$. It follows from Poincaré's Recurrence Theorem that there exists a set $Y_N \subset X_N$ such that $\nu(X_N \setminus Y_N) = 0$ and the forward and backward orbits of each point $y \in Y_N$ return both infinitely many times to Y_N. Given $x \in Y_N$, let $m_k = m_k(x)$ be a sequence of positive integers such that $f^{m_k}(y) \in Y_N$ for all $k \geq 1$. For $i = 1, \ldots, s(x)$ we have

$$\|\mathcal{A}^i_\varepsilon(x, m_k)\| \leq \|C_\varepsilon(f^{m_k}(x))^{-1}\| \cdot \|\mathcal{A}(x, m_k)|E_i(x)\| \cdot \|C_\varepsilon(x)\|$$
$$\leq N\|\mathcal{A}(x, m_k)|E_i(x)\|,$$

where $\mathcal{A}^i_\varepsilon(x, m)$ is the cocycle generated by the block $A^i_\varepsilon(x)$. Hence,

$$\limsup_{m \to +\infty} \frac{1}{m} \log\|\mathcal{A}^i_\varepsilon(x, m)\| \leq \chi_i(x). \tag{4.25}$$

In a similar way, one can show that

$$\|\mathcal{A}(x, m_k)|E_i(x)\| \leq \|C_\varepsilon(f^{m_k}(x))\| \cdot \|\mathcal{A}^i_\varepsilon(x, m_k)\| \cdot \|C_\varepsilon(x)^{-1}\|$$
$$\leq N\|\mathcal{A}^i_\varepsilon(x, m_k)\|,$$

and hence,

$$\chi_i(x) \leq \liminf_{m \to +\infty} \frac{1}{m} \log\|\mathcal{A}^i_\varepsilon(x, m)\|. \tag{4.26}$$

We conclude from (4.25) and (4.26) that for each $x \in Y_N$ and each $v \in \mathbb{R}^{k_i(x)} = C_\varepsilon(x)^{-1}E_i(x)$,

$$\lim_{m \to +\infty} \frac{1}{m} \log\|\mathcal{A}_\varepsilon(x, m)v\| = \chi_i(x).$$

In a similar way, one can prove that

$$\lim_{m \to -\infty} \frac{1}{|m|} \log\|\mathcal{A}_\varepsilon(x, m)v\| = \chi_i(x).$$

Since $\nu(Y_N) \to 1$ as $N \to \infty$, the Lyapunov spectra of $\mathcal{A}$ and $\mathcal{A}_\varepsilon$ coincide ν-almost everywhere.

We now show that the function $C_\varepsilon \colon X \to GL(p, \mathbb{R})$ is tempered. Observe that

$$\|C_\varepsilon(f^m(x))^{-1}\| = \sup_{v \neq 0} \frac{\|\mathcal{A}_\varepsilon(x, m)C_\varepsilon(x)^{-1}\mathcal{A}(x, m)^{-1}v\|}{\|v\|}$$
$$= \max_i \sup_{w \in E_i(x)} \frac{\|\mathcal{A}_\varepsilon(x, m)C_\varepsilon(x)^{-1}w\|}{\|\mathcal{A}(x, m)w\|}.$$

Since $\|C_\varepsilon(x)^{-1}\| \geq 1$, using (4.25), we obtain that for ν-almost all LP-regular points $x \in X$,

$$0 \leq \liminf_{m \to \pm\infty} \frac{1}{|m|} \log\|C_\varepsilon(f^m(x))^{-1}\|$$
$$\leq \limsup_{m \to \pm\infty} \frac{1}{|m|} \log\|C_\varepsilon(f^m(x))^{-1}\|$$
$$= \max_i \sup_{w \in E_i(x)} \lim_{m \to \pm\infty} \frac{1}{|m|} \log \frac{\|\mathcal{A}_\varepsilon(x,m)C_\varepsilon(x)^{-1}w\|}{\|\mathcal{A}(x,m)w\|} = 0.$$

This implies that for ν-almost every $x \in X$,

$$\lim_{m \to \pm\infty} \frac{1}{m} \log\|C_\varepsilon(f^m(x))^{-1}\| = 0.$$

In a similar way one can show that for ν-almost every $x \in X$,

$$\lim_{m \to \pm\infty} \frac{1}{m} \log\|C_\varepsilon(f^m(x))\| = 0.$$

This implies that C_ε is a tempered function ν-almost everywhere and completes the proof of Theorem 4.10. $\square$

4.3.3. A tempering kernel. Let $f\colon X \to X$ be a measurable transformation of a measurable space X and let $K\colon X \to \mathbb{R}$ be a measurable function. We say that K is *tempered* at the point $x \in X$ if

$$\lim_{m \to \pm\infty} \frac{1}{m} \log K(f^m(x)) = 0. \tag{4.27}$$

We need a technical but crucial statement known as the Tempering Kernel Lemma. It is an immediate corollary of a more general Lemma 5.6 applied to the function $A(x, \varepsilon) = K(x)$ which satisfies a more general ε-tempered property.

Lemma 4.11. *If $K\colon X \to \mathbb{R}$ is a positive measurable function tempered on some subset $Z \subset X$, then for any $\varepsilon > 0$ there exists a positive measurable function $K_\varepsilon \colon Z \to \mathbb{R}$ such that $K(x) \leq K_\varepsilon(x)$ and for $x \in Z$,*

$$e^{-\varepsilon} \leq \frac{K_\varepsilon(f(x))}{K_\varepsilon(x)} \leq e^\varepsilon. \tag{4.28}$$

The function K_ε satisfying (4.28) is called a *tempering kernel*. Note that if f preserves a Lebesgue measure ν on the space X, then by Lemma 3.6, any positive function $K\colon X \to \mathbb{R}$ with $\log K \in L^1(X, \nu)$ satisfies (4.27).

4.4. Regular neighborhoods

In this section we consider the particular case of a derivative cocycle df associated with a diffeomorphism $f\colon M \to M$ of a compact smooth p-dimensional Riemannian manifold M (see the introduction to Chapter 3). Applying the Reduction Theorem (see Theorem 4.10) to the derivative cocycle df, given $\varepsilon > 0$ and an LP-regular point $x \in M$, there exists a linear transformation $C_\varepsilon(x)\colon \mathbb{R}^p \to T_x M$ such that:

(1) the matrix
$$A_\varepsilon(x) = C_\varepsilon(f(x))^{-1} d_x f \, C_\varepsilon(x)$$
has the Lyapunov block form (4.21) (see Theorem 4.10);

(2) $\{C_\varepsilon(f^m(x))\}_{m \in \mathbb{Z}}$ is a tempered sequence of linear transformations.

We wish to construct, for every LP-regular point $x \in M$, a neighborhood $N(x)$ of x such that f acts in $N(x)$ very much like the linear map $A_\varepsilon(x)$ in a neighborhood of the origin.

First, given a measurable vector field $X \ni x \mapsto v(x) \in \mathbb{R}^p \setminus \{0\}$, we shall show that the function $x \mapsto \|v(x)\|'_x / \|v(x)\|$ is tempered on the set of LP-regular points.

Proposition 4.12. *Given $\varepsilon > 0$, there is a positive measurable function $K_\varepsilon\colon X \to \mathbb{R}$ such that if $x \in X$ is an LP-regular point, then:*

(1) $K_\varepsilon(x) e^{-\varepsilon|m|} \le K_\varepsilon(f^m(x)) \le K_\varepsilon(x) e^{\varepsilon|m|}$ *for every $m \in \mathbb{Z}$;*

(2) $p^{-1/2} \|v\| \le \|v\|'_x \le K_\varepsilon(x) \|v\|$ *for every $v \in \mathbb{R}^p$;*

(3) *for every measurable vector field $X \ni x \mapsto v(x) \in \mathbb{R}^p \setminus \{0\}$, the function $x \mapsto \|v(x)\|'_x / \|v(x)\|$ is tempered on the set of LP-regular points.*

Proof. The first inequality in statement (2) follows from the inequality
$$p \sum_{i=1}^p a_i^2 \ge \left(\sum_{i=1}^p a_i\right)^2,$$
with equality if and only if $a_1 = \cdots = a_p$ (this inequality follows from the Cauchy–Schwarz inequality for the standard inner product applied to the vectors $(a_1, \ldots, a_p)$ and $(1, \ldots, 1)$). Given an LP-regular point $x \in X$ and $\varepsilon > 0$, let $C_\varepsilon(x)$ be the matrix defined by (4.20). Then $\|v\|'_x \le \|C_\varepsilon(x)^{-1}\| \cdot \|v\|$ for each $v \in \mathbb{R}^p$. By Theorem 4.10, the function C_ε is tempered. Thus, we can apply the Tempering Kernel Lemma (see Lemma 4.11) to the positive function $K(x) = \|C_\varepsilon(x)^{-1}\|$ and find a function K_ε satisfying the desired properties. $\square$

Consider the set $\mathcal{R}$ of LP-regular points for f and denote by $B(0, r)$ the standard Euclidean ball in $\mathbb{R}^p$ centered at the origin of radius r. The following result describes a particular choice of local coordinate charts around LP-regular points that provide a very useful tool in studying the dynamics of the diffeomorphism f because the action of the restriction of f to the chart is uniformly hyperbolic (see [**13**] for more details).

Theorem 4.13. *For every $\varepsilon > 0$ the following properties hold:*

(1) *there exists a tempered function $q: \mathcal{R} \to (0, 1]$ and a collection of embeddings $\Psi_x: B(0, q(x)) \to M$ for each $x \in \mathcal{R}$ such that $\Psi_x(0) = x$ and $e^{-\varepsilon} < q(f(x))/q(x) < e^{\varepsilon}$; these embeddings satisfy $\Psi_x = \exp_x \circ C_\varepsilon(x)$, where $C_\varepsilon(x)$ is the Lyapunov change of coordinates;*

(2) *if $f_x := \Psi_{f(x)}^{-1} \circ f \circ \Psi_x: B(0, q(x)) \to \mathbb{R}^p$, then $d_0 f_x$ has the Lyapunov block form (4.21);*

(3) *the C^1 distance $d_{C^1}(f_x, d_0 f_x) < \varepsilon$ in $B(0, q(x))$;*

(4) *there exist a constant $K > 0$ and a measurable function $A: \mathcal{R} \to \mathbb{R}$ such that for every $y, z \in B(0, q(x))$,*

$$K^{-1} \rho(\Psi_x y, \Psi_x z) \leq \|y - z\| \leq A(x) \rho(\Psi_x y, \Psi_x z)$$

with $e^{-\varepsilon} < A(f(x))/A(x) < e^{\varepsilon}$.

Proof. For each $x \in X$ consider the Lyapunov change of coordinates $C_\varepsilon(x)$ from $T_x M$ onto $\mathbb{R}^p$ (see (4.20)). Thus for each $x \in \mathcal{R}$, the matrix

$$A_\varepsilon(x) = C_\varepsilon(f(x))^{-1} d_x f C_\varepsilon(x) \qquad (4.29)$$

has the Lyapunov block form.

For each $x \in M$ and $r > 0$ set

$$T_x M(r) = \{w \in T_x M : \|w\| < r\}.$$

Now choose $r_0 > 0$ such that for every $x \in M$ the map $\exp_x: T_x M(r_0) \to M$ is an embedding, $\|d_w \exp_x\|, \|d_{\exp_x w} \exp_x^{-1}\| \leq 2$ for every $w \in T_x M(r_0)$, and $\exp_{f(x)}$ is injective on

$$\exp_{f(x)}^{-1} \circ f \circ \exp_x(T_x M(r_0)).$$

Define the map

$$f_x = C_\varepsilon(f(x))^{-1} \circ \exp_{f(x)}^{-1} \circ f \circ \exp_x \circ C_\varepsilon(x) \qquad (4.30)$$

on $C_\varepsilon(x) T_x M(r_0) \subset \mathbb{R}^p$. Observe that the Euclidean ball

$$\{w \in T_x M : \|w\| < r_0 \|C_\varepsilon(x)^{-1}\|^{-1}\}$$

is contained in $C_\varepsilon(x) T_x M(r_0)$.

4.4. Regular neighborhoods

We wish to compare the actions of f_x and of $A_\varepsilon(x)$ restricted to a small neighborhood of x. In order to do that, we first introduce the maps

$$r_x(u) = f_x(u) - A_\varepsilon(x)u,$$
$$g_x(w) = (\exp_{f(x)}^{-1} \circ f \circ \exp_x)w - d_x f w.$$

Since f is of class $C^{1+\alpha}$ and $d_0 g_x = 0$, there exists a positive constant L, such that $\|d_w g_x\| \leq L\|w\|^\alpha$, and thus

$$\|d_u r_x\| = \|d_u(C_\varepsilon(f(x)) \circ g_x \circ C_\varepsilon(x)^{-1})\|$$
$$\leq L\|C_\varepsilon(f(x))\| \cdot \|C_\varepsilon(x)^{-1}\|^{1+\alpha}\|u\|^\alpha.$$

Hence if $\|w\|$ is sufficiently small, the contribution of the nonlinear part of f_x is negligible. In particular, $\|d_u r_x\| < \varepsilon$ for

$$\|u\| < \delta(x) := \varepsilon^{1/\alpha} L^{-1/\alpha} \|C_\varepsilon(f(x))\|^{-1/\alpha} \|C_\varepsilon(x)^{-1}\|^{-1-1/\alpha}.$$

By the Mean Value Theorem, since $r_x(0) = 0$, we also have

$$\|r_x(u)\| < \varepsilon \quad \text{for } \|u\| \leq \delta(x). \tag{4.31}$$

Since C_ε is tempered, it follows from the definition of $\delta(x)$ that

$$\lim_{m \to +\infty} \frac{1}{m} \log \delta(f^m(x)) = -\lim_{m \to +\infty} \frac{1}{m\alpha} \log \|C_\varepsilon(f^{m+1}(x))\|$$
$$- \lim_{m \to +\infty} \frac{1}{m(1+1/\alpha)} \log \|C_\varepsilon(f^m(x))^{-1}\| = 0.$$

Applying the Tempering Kernel Lemma (see Lemma 4.11) to the function δ, we find a measurable function $q \colon X \to \mathbb{R}$ such that $q(x) \geq \delta(x)$ and

$$e^{-\varepsilon} \leq q(f(x))/q(x) \leq e^\varepsilon.$$

We define the map $\Psi_x \colon B(0, q(x)) \to M$ by $\Psi_x = \exp_x \circ C_\varepsilon(x)$. It is clear that Ψ_x is an embedding for each $x \in \mathcal{R}$, such that $\Psi_x(0) = x$. This proves statements (1) and (2). Statement (3) follows from (4.31) and the identity $d_0 f_x = A_\varepsilon(x)$ (that in turn follows immediately from (4.29), (4.30), and the chain rule).

Now we prove statement (4). It follows from (4.24) that

$$\|y - z\| = \|\Psi_x^{-1}(\Psi_x y) - \Psi_x^{-1}(\Psi_x z)\| \leq 2\|C_\varepsilon(x)^{-1}\|\rho(\Psi_x y, \Psi_x z).$$

On the other hand, we have

$$\rho(\Psi_x y, \Psi_x z) \leq 2\|C_\varepsilon(x)\| \cdot \|y - z\| \leq 2\|y - z\|.$$

This means that if $K_\varepsilon \colon \mathcal{R} \to \mathbb{R}$ is a function as in the Tempering Kernel Lemma (see Lemma 4.11), such that

$$K_\varepsilon(x) \geq \|C_\varepsilon(x)^{-1}\| \quad \text{and} \quad e^{-\varepsilon} \leq K_\varepsilon(f(x))/K_\varepsilon(x) \leq e^\varepsilon,$$

then in statement (4) we can take $A(x) = 2K_\varepsilon(f(x))$. $\square$

We note that for each $x \in \mathcal{R}$ there exists a constant $B(x) \geq 1$ such that for every $y, z \in B(0, q(x))$,
$$B(x)^{-1}\rho(\Psi_x y, \Psi_x z) \leq \rho'_x(\exp_x y, \exp_x z) \leq B(x)\rho(\Psi_x y, \Psi_x z),$$
where $\rho'_x(\cdot, \cdot)$ is the distance in $\exp_x B(0, q(x))$ with respect to the strong Lyapunov norm $\|\cdot\|'_x$. By Luzin's theorem, given $\delta > 0$, there exists a set of measure at least $1 - \delta$ where $x \mapsto B(x)$ as well as $x \mapsto A(x)$ in Theorem 4.13 are bounded.

For each LP-regular point x the set
$$R(x) := \Psi_x(B(0, q(x)))$$
is called a *regular neighborhood* of x or a *strong Lyapunov chart* at x (see Section 5.3.3 and Theorem 5.15 where a weaker version of Lyapunov charts is introduced).

We stress that the existence of regular neighborhoods uses the fact that f is of class $C^{1+\alpha}$ in an essential way.

Chapter 5

Elements of the Nonuniform Hyperbolicity Theory

In this chapter we introduce the notion of nonuniform hyperbolicity which is one of the principal concepts of smooth ergodic theory. While the definition of nonuniform hyperbolicity can be given in the general setting of linear cocycles over measure-preserving transformations, in this book we restrict ourselves to the particular but important case of derivative cocycles generated by diffeomorphisms or flows on compact smooth manifolds. Extension of this definition and related results described in this chapter to the general case is a rather easy task for the curious reader (see [**13**] for details).

Nonuniform hyperbolicity theory is a far-reaching generalization of the classical theory of uniform hyperbolicity. Similarly to the latter, given a diffeomorphism f, any of its nonuniformly hyperbolic trajectories $\{f^m(x)\}_{m \in \mathbb{Z}}$ admits a splitting of the tangent space $T_{f^m(x)}M$ into stable $E^s(f^m(x))$ and unstable $E^u(f^m(x))$ subspaces which are invariant under the action of the differential df. However, there are some principal differences:

(1) the angle $\angle(E^s(f^m(x)), E^u(f^m(x)))$ may vary with m and may become arbitrarily close to zero;

(2) the parameters c, λ, and μ in the hyperbolicity estimates (H4)–(H6) (see Section 5.2) depend on x and may vary uncontrollably (measurably) with x;

(3) the set of all nonuniformly hyperbolic trajectories may occupy a "small" part of the phase space.

In Section 5.2 we will give a precise definition of nonuniform hyperbolicity which addresses all these issues.

We stress that the nonuniform hyperbolicity conditions are quite technical and their direct verification may be a challenging task. To address this problem, we will introduce a class of dynamical systems with nonzero Lyapunov exponents, and we will show that every point with nonzero Lyapunov exponents, which is also LP-regular, is nonuniformly hyperbolic. This gives an efficient way to verify the nonuniform hyperbolicity conditions. Indeed, by the Multiplicative Ergodic Theorem (see Theorem 4.1), given an invariant measure, almost every point is LP-regular, and hence, all we need to do is to check that the system has nonzero Lyapunov exponents on a set of positive measure. There are various techniques that allow one to do this and in Chapter 12 we discuss one of them, known as cone techniques. Furthermore, in Chapter 15 we will consider the problem, known as Anosov rigidity, that roughly speaking claims that if Lyapunov exponents along *every* trajectory of a diffeomorphism of a compact manifold are nonzero, then this diffeomorphism is uniformly hyperbolic.

5.1. Dynamical systems with nonzero Lyapunov exponents

We begin by introducing the class of dynamical systems with nonzero Lyapunov exponents.

5.1.1. Lyapunov exponents for diffeomorphisms.
Let $f \colon M \to M$ be a diffeomorphism of a compact smooth Riemannian p-dimensional manifold M. Consider the derivative cocycle specified by f. Following Chapter 3 we have:

(1) the forward and backward Lyapunov exponents associated with f,
$$\chi^{\pm}(x, v) = \limsup_{m \to \pm\infty} \frac{1}{|m|} \log \|d_x f^m v\|, \quad \text{for } x \in M, \ v \in T_x M;$$

(2) the filtrations $\mathcal{V}_x^{\pm} = \{V_i^{\pm}(x), i = 1, \ldots, s^{\pm}(x)\}$ of $T_x M$ associated to $\chi^{\pm}(x, \cdot)$, where
$$V_i^{\pm}(x) = \{v \in T_x M : \chi^{\pm}(x, v) \le \chi_i^{\pm}(x)\};$$

(3) the Lyapunov spectra of $\chi^{\pm}(x, \cdot)$,
$$\operatorname{Sp}\chi^{\pm}(x) = \{(\chi_i^{\pm}(x), k_i^{\pm}(x)) : 1 \le i \le s^{\pm}(x)\}.$$

We note that the functions $\chi_i^{\pm}(x)$, $s^{\pm}(x)$, and $k_i^{\pm}(x)$ are *invariant* under f and are *Borel measurable* (but not necessarily continuous).

In order to simplify our notation, in what follows, we will often drop the superscript $\pm$ from the notation of the forward and, respectively, backward

5.1. Dynamical systems with nonzero Lyapunov exponents

Lyapunov exponents and the associated quantities if it does not cause any confusion.

Recall that with respect to the derivative cocycle a point $x \in M$ is:

(1) *forward regular* if the Lyapunov exponent χ^+ is forward regular;
(2) *backward regular* if the Lyapunov exponent χ^- is backward regular;
(3) *Lyapunov–Perron regular* or simply *LP-regular* if the Lyapunov exponent χ^+ is forward regular, the Lyapunov exponent χ^- is backward regular, and the filtrations $\mathcal{V}_x^-$ and $\mathcal{V}_x^-$ are coherent (see Sections 2.4 and 2.5).

Note that if x is LP-regular, then so is the point $f(x)$, and thus, one can speak of the whole trajectory $\{f^m(x)\}$ as being forward, backward, or LP-regular.

We state a result characterizing the LP-regularity, which is an application of Theorem 2.29 to the derivative cocycle.

Recall that the cotangent bundle T^*M consists of 1-forms on M. The diffeomorphism f acts on T^*M by its codifferential
$$d_x^*f \colon T_{f(x)}^*M \to T_x^*M$$
defined by
$$d_x^*f\varphi(v) = \varphi(d_xfv), \quad v \in T_xM, \; \varphi \in T_{f(x)}^*M.$$
We denote the inverse map by
$$d_x'f = (d_x^*f)^{-1} \colon T_x^*M \to T_{f(x)}^*M$$
and define the Lyapunov exponent χ^* on T^*M by the formula
$$\chi^*(x,\varphi) = \limsup_{m \to +\infty} \frac{1}{m} \log \|d_x'f^m\varphi\|.$$

Given $\mathbf{v} = (v_1, \ldots, v_k)$, we denote by $\Gamma_k^{\mathbf{v}}(m)$ the volume of the parallelepiped generated by the vectors $d_xf^mv_1, \ldots, d_xf^mv_k$.

Theorem 5.1. *The point $x \in M$ is LP-regular if and only if there exist a decomposition*
$$T_xM = \bigoplus_{i=1}^{s(x)} E_i(x)$$
into subspaces $E_i(x)$ and numbers $\chi_1(x) < \cdots < \chi_{s(x)}(x)$ such that:

(1) *$E_i(x)$ is invariant under d_xf, i.e.,*
$$d_xfE_i(x) = E_i(f(x)),$$
and depends Borel measurably on x;

(2) for $v \in E_i(x) \setminus \{0\}$,
$$\lim_{m \to \pm\infty} \frac{1}{m} \log \|d_x f^m v\| = \chi_i(x)$$
with uniform convergence on $\{v \in E_i(x) : \|v\| = 1\}$;

(3) if $\mathbf{v} = (v_1, \ldots, v_{k_i(x)})$ is a basis of $E_i(x)$, then
$$\lim_{m \to \pm\infty} \frac{1}{m} \log \Gamma^{\mathbf{v}}_{k_i(x)}(m) = \chi_i(x) k_i(x);$$

(4) for any $v, w \in T_x M \setminus \{0\}$ with $\angle(v, w) \neq 0$,
$$\lim_{m \to \pm\infty} \frac{1}{m} \log \angle(d_x f^m v, d_x f^m w) = 0;$$

(5) the Lyapunov exponent $\chi(x, \cdot)$ is exact, that is,
$$\liminf_{m \to \pm\infty} \frac{1}{m} \log \Gamma^{\mathbf{v}}_k(m) = \limsup_{m \to \pm\infty} \frac{1}{m} \log \Gamma^{\mathbf{v}}_k(m)$$
for any $1 \leq k \leq p$ and any vectors $v_1, \ldots, v_k$;

(6) there exists a decomposition of the cotangent bundle
$$T_x^* M = \bigoplus_{i=1}^{s(x)} E_i^*(x)$$
into subspaces $E_i^*(x)$ associated with the Lyapunov exponent χ^*; the subspaces $E_i^*(x)$ are invariant under $d'_x f$, i.e.,
$$d'_x f E_i^*(x) = E_i^*(f(x)),$$
and depend (Borel) measurably on the point x; moreover, if $\{v_i(x) : i = 1, \ldots, p\}$ is a subordinate basis with $v_i(x) \in E_j(x)$ for $n_{j-1}(x) < i \leq n_j(x)$ and if $\{v_i^*(x) : i = 1, \ldots, p\}$ is a dual basis, then $v_i^*(x) \in E_j^*(x)$ for $n_{j-1}(x) < i \leq n_j(x)$.

As an immediate corollary of this theorem we obtain the following result.

Corollary 5.2. *For any LP-regular point $x \in M$ and basis $\mathbf{v} = (v_1, \ldots, v_p)$ of $T_x M$ we have*
$$\lim_{m \to \pm\infty} \frac{1}{m} \log \Gamma^{\mathbf{v}}_p(m) = \sum_{i=1}^{s(x)} \chi_i(x) k_i(x).$$

It is easy to see that the derivative cocycle satisfies condition (3.6) and hence, by the Multiplicative Ergodic Theorem (see Theorem 4.1), we obtain the following result.

Theorem 5.3. *If f is a C^1 diffeomorphism of a compact smooth Riemannian manifold M, then the set of Lyapunov–Perron regular points has full measure with respect to any f-invariant Borel probability measure on M.*

5.1. Dynamical systems with nonzero Lyapunov exponents

We denote by $\mathcal{R} \subset M$ the set of LP-regular points. Although the notion of LP-regularity does not require any invariant measure to be present, the crucial consequence of Theorem 5.3 is that $\mathcal{R}$ is nonempty and indeed has full measure with respect to any f-invariant Borel probability measure on M.

We stress that there may exist trajectories which are both forward and backward regular but *not* LP-regular. However, such trajectories form a negligible set with respect to any f-invariant Borel probability measure. Note that only in some exceptional situations is *every* point in M LP-regular.[1]

Let ν be an f-invariant *ergodic* finite Borel measure. Since the values of Lyapunov exponents are invariant Borel functions, there exist numbers $s = s^\nu$, $\chi_i = \chi_i^\nu$, and $k_i = k_i^\nu$ for $i = 1, \ldots, s$ such that for every $x \in \mathcal{R}$

$$s(x) = s, \quad \chi_i(x) = \chi_i, \quad k_i(x) = k_i. \tag{5.1}$$

The collection of pairs

$$\mathrm{Sp}\,\chi(\nu) = \{(\chi_i, k_i) : 1 \leq i \leq s\} \tag{5.2}$$

is called the *Lyapunov spectrum* of the measure ν.

5.1.2. Diffeomorphisms with nonzero Lyapunov exponents. In this section we consider dynamical systems whose Lyapunov spectrum does not vanish on some subset of M. More precisely, let

$$\mathcal{E} = \big\{x \in \mathcal{R} : \text{there exists } 1 \leq k(x) < s(x)$$
$$\text{such that } \chi_{k(x)}(x) < 0 \text{ and } \chi_{k(x)+1}(x) > 0\big\}. \tag{5.3}$$

This set is f-invariant. We say that f is a *dynamical system with nonzero Lyapunov exponents* if there exists an f-invariant Borel probability measure ν on M such that $\nu(\mathcal{E}) = 1$. In this case we call ν a *hyperbolic measure* for f.

For every $x \in \mathcal{E}$ we define the following subspaces of $T_x M$:

$$E^s(x) = \bigoplus_{i=1}^{k(x)} E_i(x) \quad \text{and} \quad E^u(x) = \bigoplus_{i=k(x)+1}^{s(x)} E_i(x),$$

where $E_i(x)$ are the subspaces in the Oseledets decomposition at the point x given by the Multiplicative Ergodic Theorem. The following theorem from [88] describes some basic properties of these subspaces.

Theorem 5.4. *For every $x \in \mathcal{E}$ the subspaces $E^s(x)$ and $E^u(x)$:*

(L1) *depend Borel measurably on x;*

(L2) *form a splitting of the tangent space, i.e., $T_x M = E^s(x) \oplus E^u(x)$;*

---

[1] This is true, for example, when f is a linear hyperbolic toral automorphism (see Section 1.1).

(L3) *are invariant:*
$$d_x f E^s(x) = E^s(f(x)) \quad \text{and} \quad d_x f E^u(x) = E^u(f(x)).$$

Furthermore, given $x \in \mathcal{E}$, there exist $\varepsilon_0 = \varepsilon_0(x) > 0$ and, for all $0 < \varepsilon \leq \varepsilon_0$, functions $C(x, \varepsilon) > 0$ and $K(x, \varepsilon) > 0$ such that:

(L4) *the subspace $E^s(x)$ is* stable: *if $v \in E^s(x)$ and $n > 0$, then*
$$\|d_x f^n v\| \leq C(x, \varepsilon) e^{(\chi_{k(x)}(x) + \varepsilon) n} \|v\|;$$

(L5) *the subspace $E^u(x)$ is* unstable: *if $v \in E^u(x)$ and $n < 0$, then*
$$\|d_x f^n v\| \leq C(x, \varepsilon) e^{(\chi_{k(x)+1}(x) - \varepsilon) n} \|v\|;$$

(L6) *the angle between $E^s(x)$ and $E^u(x)$ satisfies*
$$\angle(E^s(x), E^u(x)) \geq K(x, \varepsilon);$$

(L7) *the functions $C(x, \varepsilon)$ and $K(x, \varepsilon)$ are Borel measurable in x and for every $m \in \mathbb{Z}$,*
$$C(f^m(x), \varepsilon) \leq C(x, \varepsilon) e^{\varepsilon |m|} \quad \text{and} \quad K(f^m(x), \varepsilon) \geq K(x, \varepsilon) e^{-\varepsilon |m|}.$$

Remark 5.5. Condition (L7) is a manifestation of LP-regularity. Roughly speaking, it means that given $x \in \mathcal{E}$ and an *arbitrarily* small ε, the estimates (L4), (L5), and (L6) may deteriorate along the trajectory $\{f^m(x)\}$ as $m \to \pm \infty$ with the exponential rate $e^{\varepsilon |m|}$. In particular, for sufficiently small ε the rates of contraction along the stable subspaces and of expansion along the unstable subspaces prevail over the rate of deterioration leading to sufficiently strong overall contraction and expansion along these subspaces. We stress that the choice of ε for which this observation holds depends on $x \in \mathcal{E}$ and that for a fixed $x \in \mathcal{E}$ the function $C(\cdot, \varepsilon)$ may increase to ∞ and the function $K(\cdot, \varepsilon)$ may decrease to zero as $\varepsilon \to 0$.

Finally, we emphasize that condition (L7) will play a crucial role in our study of stability of LP-regular trajectories with nonzero Lyapunov exponents as well as of more general nonuniformly hyperbolic trajectories; see the estimates (6.30) and (7.2).

Proof of Theorem 5.4. The first three statements are immediate consequence of Theorem 5.1. Given $q > 0$, consider the sets
$$\mathcal{E}_q = \left\{ x \in \mathcal{E} : \chi_{k(x)}(x) < -\frac{1}{q}, \ \chi_{k(x)+1}(x) > \frac{1}{q} \right\},$$
which are f-invariant, nested, i.e, $\mathcal{E}_q \subset \mathcal{E}_{q+1}$ for each $q > 0$, and exhaust $\mathcal{E}$, i.e., $\bigcup_{q>0} \mathcal{E}_q = \mathcal{E}$. It suffices to prove statements (L4)–(L7) of the theorem for every nonempty set $\mathcal{E}_q$. In what follows, we fix such a $q > 0$ and choose a sufficiently small number $\varepsilon_0 = \varepsilon_0(q)$.

5.1. Dynamical systems with nonzero Lyapunov exponents

Let $A(x,\varepsilon)$ be a Borel function on $X \times [0,\varepsilon_0)$ where $X \subset M$ is an f-invariant Borel subset and $0 < \varepsilon_0 < 1$. We say that $A(x,\varepsilon)$ is ε-tempered if

$$-\varepsilon \leq \liminf_{m \to \pm\infty} \frac{1}{|m|} \log |A(f^m(x),\varepsilon)| \leq \limsup_{m \to \pm\infty} \frac{1}{|m|} \log |A(f^m(x),\varepsilon)| \leq \varepsilon.$$

It follows that there are Borel functions $M_1(x,\varepsilon)$ and $M_2(x,\varepsilon)$ on $X \times [0,\varepsilon_0)$ such that

$$M_1(x,\varepsilon)e^{-\varepsilon|m|} \leq A(f^m(x),\varepsilon) \leq M_2(x,\varepsilon)e^{\varepsilon|m|}.$$

We also say that a function $A(x)$ on X is *tempered* if it is ε-tempered for every $\varepsilon > 0$, that is,

$$\lim_{m \to \pm\infty} \frac{1}{|m|} \log |A(f^m(x))| = 0$$

(compare to (3.2)). We need the following lemma.

Lemma 5.6. *Let $X \subset M$ be an f-invariant Borel subset and let $A(x,\varepsilon)$ be a positive ε-tempered Borel function on $X \times [0,\varepsilon_0)$, $0 < \varepsilon_0 < 1$. Then one can find positive Borel functions $B_1(x,\varepsilon)$ and $B_2(x,\varepsilon)$ such that*

$$B_1(x,\varepsilon) \leq A(x,\varepsilon) \leq B_2(x,\varepsilon), \tag{5.4}$$

and for $m \in \mathbb{Z}$,

$$B_1(x,\varepsilon)e^{-2\varepsilon|m|} \leq B_1(f^m(x),\varepsilon), \quad B_2(x,\varepsilon)e^{2\varepsilon|m|} \geq B_2(f^m(x),\varepsilon). \tag{5.5}$$

Proof of the lemma. It follows from the conditions of the lemma and the definition of ε-tempered function that there exists $m(x,\varepsilon) > 0$ such that if $m \in \mathbb{Z}$ and $|m| > m(x,\varepsilon)$, then

$$-2\varepsilon \leq \frac{1}{|m|} \log A(f^m(x),\varepsilon) \leq 2\varepsilon.$$

Set

$$B_1(x,\varepsilon) = \min_{-m(x,\varepsilon) \leq i \leq m(x,\varepsilon)} \left\{1, A(f^i(x),\varepsilon)e^{2\varepsilon|i|}\right\},$$

$$B_2(x,\varepsilon) = \max_{-m(x,\varepsilon) \leq i \leq m(x,\varepsilon)} \left\{1, A(f^i(x),\varepsilon)e^{-2\varepsilon|i|}\right\}.$$

The functions $B_1(x,\varepsilon)$ and $B_2(x,\varepsilon)$ are Borel functions. Moreover, if $n \in \mathbb{Z}$, then

$$B_1(x,\varepsilon)e^{-2\varepsilon|n|} \leq A(f^n(x),\varepsilon) \leq B_2(x,\varepsilon)e^{2\varepsilon|n|}. \tag{5.6}$$

Furthermore, if $b_1 \leq 1 \leq b_2$ are such that

$$b_1 e^{-2\varepsilon|n|} \leq A(f^n(x),\varepsilon) \tag{5.7}$$

and

$$b_2 e^{2\varepsilon|n|} \geq A(f^n(x),\varepsilon) \tag{5.8}$$

for all $n \in \mathbb{Z}$, then $b_1 \leq B_1(x,\varepsilon)$ and $b_2 \geq B_1(x,\varepsilon)$. In other words,

$$B_1(x,\varepsilon) = \sup\{b \leq 1 : \text{inequality (5.7) holds for all } n \in \mathbb{Z}\}, \\ B_2(x,\varepsilon) = \inf\{b \geq 1 : \text{inequality (5.8) holds for all } n \in \mathbb{Z}\}. \tag{5.9}$$

Inequalities (5.4) follow from (5.6) (with $n = 0$). We also have

$$A(f^{n+m}(x),\varepsilon) \leq B_2(x,\varepsilon)e^{2\varepsilon|n+m|} \leq B_2(x,\varepsilon)e^{2\varepsilon|n|+2\varepsilon|m|},$$

$$A(f^{n+m}(x),\varepsilon) \geq B_1(x,\varepsilon)e^{-2\varepsilon|n+m|} \geq B_1(x,\varepsilon)e^{-2\varepsilon|n|-2\varepsilon|m|}.$$

Comparing these inequalities with (5.6) written at the point $f^m(x)$ and taking (5.9) into account, we obtain (5.5). The proof of the lemma is complete. $\square$

We use Lemma 5.6 to construct the function $K(x,\varepsilon)$. Fix $0 < \varepsilon < \varepsilon_0(q)$ and for $x \in \mathcal{E}_q$ consider the function

$$\gamma(x) = \angle(E^s(x), E^u(x)).$$

By Theorem 5.1, $\gamma(x)$ is an ε-tempered function for every $\varepsilon > 0$. Therefore, applying Lemma 5.6 to the function $A(x,\varepsilon) = \gamma(x)$, we conclude that the function $K(x,\varepsilon) = B_1(x, \frac{1}{2}\varepsilon)$ satisfies conditions (L6) and (L7) of the theorem provided the number $\varepsilon_0(q)$ is sufficiently small.

We will now show how to construct the function $C(x,\varepsilon)$.

Lemma 5.7. *There exist $\varepsilon_0 = \varepsilon_0(q)$ and a positive Borel function $D(x,\varepsilon)$ (where $x \in \mathcal{E}_q$ and $0 < \varepsilon < \varepsilon_0$) such that for $m \in \mathbb{Z}$ and $1 \leq i \leq s$,*

$$D(f^m(x),\varepsilon) \leq D(x,\varepsilon)^2 e^{2\varepsilon|m|} \tag{5.10}$$

and for any $n \geq 0$,

$$\|df_{ix}^n\| \leq D(x,\varepsilon)e^{(\chi_i+\varepsilon)n}, \quad \|df_{ix}^{-n}\| \geq D(x,\varepsilon)^{-1}e^{-(\chi_i+\varepsilon)n}$$

where $\chi_i = \chi_i(x)$ and $df_{ix}^n = d_x f^n | E_i(x)$.

Proof of the lemma. Let $x \in \mathcal{E}_q$. By Theorem 5.1 (we use the notation of that theorem), there exists $\varepsilon_0 = \varepsilon_0(q)$ such that for every $0 < \varepsilon < \varepsilon_0$ one can find a number $n(x,\varepsilon) \in \mathbb{N}$ such that for $n \geq n(x,\varepsilon)$,

$$\chi_i - \varepsilon \leq \frac{1}{n}\log\|df_{ix}^n\| \leq \chi_i + \varepsilon, \quad -\chi_i - \varepsilon \leq \frac{1}{n}\log\|df_{ix}^{-n}\| \leq -\chi_i + \varepsilon$$

and

$$-\chi_i - \varepsilon \leq \frac{1}{n}\log\|d'f_{ix}^n\| \leq -\chi_i + \varepsilon, \quad \chi_i - \varepsilon \leq \frac{1}{n}\log\|d'f_{ix}^{-n}\| \leq \chi_i + \varepsilon,$$

5.1. Dynamical systems with nonzero Lyapunov exponents

where $d'f_{ix}^n = d'_x f^n | E_i^*(x)$ (recall that $E_i^*(x) \subset T_x^* M$ is the dual space to $E_i(x)$ and $d'_x f$ is the inverse of the codifferential). Set

$$D_1^+(x,\varepsilon) = \min_{1 \leq i \leq s} \min_{0 \leq j \leq n(x,\varepsilon)} \left\{ 1, \|df_{ix}^j\| e^{(-\chi_i + \varepsilon)j}, \|d'f_{ix}^j\| e^{(\chi_i + \varepsilon)j} \right\},$$

$$D_1^-(x,\varepsilon) = \min_{1 \leq i \leq s} \min_{-n(x,\varepsilon) \leq j \leq 0} \left\{ 1, \|df_{ix}^j\| e^{(-\chi_i - \varepsilon)j}, \|d'f_{ix}^j\| e^{(\chi_i - \varepsilon)j} \right\},$$

$$D_2^+(x,\varepsilon) = \max_{1 \leq i \leq s} \max_{0 \leq j \leq n(x,\varepsilon)} \left\{ 1, \|df_{ix}^j\| e^{(-\chi_i - \varepsilon)j}, \|d'f_{ix}^j\| e^{(\chi_i - \varepsilon)j} \right\},$$

$$D_2^-(x,\varepsilon) = \max_{1 \leq i \leq s} \max_{-n(x,\varepsilon) \leq j \leq 0} \left\{ 1, \|df_{ix}^j\| e^{(-\chi_i + \varepsilon)j}, \|d'f_{ix}^j\| e^{(\chi_i + \varepsilon)j} \right\},$$

and

$$D_1(x,\varepsilon) = \min\{D_1^+(x,\varepsilon), D_1^-(x,\varepsilon)\},$$
$$D_2(x,\varepsilon) = \max\{D_2^+(x,\varepsilon), D_2^-(x,\varepsilon)\},$$
$$D(x,\varepsilon) = \max\{D_1(x,\varepsilon)^{-1}, D_2(x,\varepsilon)\}.$$

The function $D(x,\varepsilon)$ is measurable, and if $n \geq 0$ and $1 \leq i \leq s$, then

$$D(x,\varepsilon)^{-1} e^{(\chi_i - \varepsilon)n} \leq \|df_{ix}^n\| \leq D(x,\varepsilon) e^{(\chi_i + \varepsilon)n},$$
$$D(x,\varepsilon)^{-1} e^{(-\chi_i - \varepsilon)n} \leq \|df_{ix}^{-n}\| \leq D(x,\varepsilon) e^{(-\chi_i + \varepsilon)n},$$
$$D(x,\varepsilon)^{-1} e^{(-\chi_i - \varepsilon)n} \leq \|d'f_{ix}^n\| \leq D(x,\varepsilon) e^{(-\chi_i + \varepsilon)n}, \quad (5.11)$$
$$D(x,\varepsilon)^{-1} e^{(\chi_i - \varepsilon)n} \leq \|d'f_{ix}^{-n}\| \leq D(x,\varepsilon) e^{(\chi_i + \varepsilon)n}.$$

Moreover, if $d \geq 1$ is a number for which inequalities (5.11) hold for all $n \geq 0$ and $1 \leq i \leq s$ with $D(x,\varepsilon)$ replaced by d, then $d \geq D(x,\varepsilon)$. Therefore,

$$D(x,\varepsilon) = \inf\{d \geq 1 : \text{inequalities (5.11) hold for all } n \geq 0 \\ \text{and } 1 \leq i \leq s \text{ with } D(x,\varepsilon) \text{ replaced by } d\}. \quad (5.12)$$

We wish to compare the values of the function $D(x,\varepsilon)$ at the points x and $f^m(x)$ for $m \in \mathbb{Z}$. Notice that for every $x \in M$, $v \in T_x M$, and $\varphi \in T_x^* M$ with $\varphi(v) = 1$ we have

$$(d'_x f \varphi)(d_x f v) = \varphi((d_x f)^{-1} d_x f v) = \varphi(v) = 1. \quad (5.13)$$

Using the Riemannian metric on the manifold M, we introduce the identification map $\tau_x \colon T_x^* M \to T_x M$ such that $\langle \tau_x(\varphi), v \rangle = \varphi(v)$ where $v \in T_x M$ and $\varphi \in T_x^* M$.

Let $\{v_k^n : k = 1, \ldots, p\}$ be a basis of $E_i(f^n(x))$ and let $\{w_k^n : k = 1, \ldots, p\}$ be the dual basis of $E_i^*(f^n(x))$. We have $\tau_{f^n(x)}(w_k^n) = v_k^n$. Denote by $A_{n,m}^i$ and $B_{n,m}^i$ the matrices corresponding to the linear maps $df_{if^m(x)}^n$ and $d'f_{if^m(x)}^n$ with respect to the above bases. It follows from (5.13) that

$$A_{m,0}^i (B_{m,0}^i)^* = \mathrm{Id}$$

where * stands for matrix transposition. Hence, for every $n > 0$ the matrix corresponding to the map $df^n_{if^m(x)}$ is

$$A^i_{n,m} = A^i_{n+m,0}(A^i_{m,0})^{-1} = A^i_{n+m,0}(B^i_{m,0})^*.$$

Therefore, in view of (5.11), we obtain the following:

(1) if $n > 0$, then

$$\|df^n_{if^m(x)}\| \leq D(x,\varepsilon)^2 e^{(\chi_i+\varepsilon)(n+m)+(-\chi_i+\varepsilon)m}$$
$$= D(x,\varepsilon)^2 e^{2\varepsilon m} e^{(\chi_i+\varepsilon)n},$$
$$\|df^n_{if^m(x)}\| \geq D(x,\varepsilon)^{-2} e^{(\chi_i-\varepsilon)(n+m)+(-\chi_i-\varepsilon)m}$$
$$= D(x,\varepsilon)^{-2} e^{-2\varepsilon m} e^{(\chi_i-\varepsilon)n},$$

(2) if $n > 0$ and $m - n \geq 0$, then

$$\|df^{-n}_{if^m(x)}\| \leq D(x,\varepsilon)^2 e^{(\chi_i+\varepsilon)(m-n)+(-\chi_i+\varepsilon)m}$$
$$= D(x,\varepsilon)^2 e^{2\varepsilon m} e^{(-\chi_i+\varepsilon)n},$$
$$\|df^{-n}_{if^m(x)}\| \geq D(x,\varepsilon)^{-2} e^{(\chi_i-\varepsilon)(m-n)+(-\chi_i-\varepsilon)m}$$
$$= D(x,\varepsilon)^{-2} e^{-2\varepsilon m} e^{(-\chi_i-\varepsilon)n},$$

(3) if $n > 0$ and $n - m \geq 0$, then

$$\|df^{-n}_{if^m(x)}\| \leq D(x,\varepsilon)^2 e^{(\chi_i+\varepsilon)(n-m)+(-\chi_i+\varepsilon)m}$$
$$= D(x,\varepsilon)^2 e^{2\varepsilon m} e^{(-\chi_i+\varepsilon)n},$$
$$\|df^{-n}_{if^m(x)}\| \geq D(x,\varepsilon)^{-2} e^{(\chi_i-\varepsilon)(n-m)+(-\chi_i-\varepsilon)m}$$
$$= D(x,\varepsilon)^{-2} e^{-2\varepsilon m} e^{(-\chi_i-\varepsilon)n}.$$

Similar inequalities hold for the maps $d'f^n_{if^m(x)}$ for each $n, m \in \mathbb{Z}$. Comparing this with inequalities (5.11) applied to the point $f^m(x)$ and using (5.12), we conclude that if $m \geq 0$, then

$$D(f^m(x),\varepsilon) \leq D(x,\varepsilon)^2 e^{2\varepsilon m}. \tag{5.14}$$

Similar arguments show that if $m \leq 0$, then

$$D(f^{-m}(x),\varepsilon) \leq D(x,\varepsilon)^2 e^{-2\varepsilon m}. \tag{5.15}$$

It follows from (5.14) and (5.15) that if $m \in \mathbb{Z}$, then

$$D(f^m(x),\varepsilon) \leq D(x,\varepsilon)^2 e^{2\varepsilon|m|}.$$

This completes the proof of the lemma. $\square$

We now proceed with the proof of the theorem. Replacing in (5.10) m by $-m$ and x by $f^m(x)$, we obtain

$$D(f^m(x),\varepsilon) \geq \sqrt{D(x,\varepsilon)}e^{-\varepsilon|m|}. \tag{5.16}$$

Consider two disjoint subsets $\sigma_1, \sigma_2 \subset [1,s] \cap \mathbb{N}$ and set

$$L_1(x) = \bigoplus_{i \in \sigma_1} E_i(x), \quad L_2(x) = \bigoplus_{i \in \sigma_2} E_i(x),$$

and $\gamma_{\sigma_1\sigma_2}(x) = \angle(L_1(x), L_2(x))$. By Theorem 5.1, the function $\gamma_{\sigma_1\sigma_2}(x)$ is tempered, and hence, in view of Lemma 5.6, one can find a function $K_{\sigma_1\sigma_2}(x)$ satisfying condition (L7) such that

$$\gamma_{\sigma_1\sigma_2}(x) \geq K_{\sigma_1\sigma_2}(x).$$

Set

$$T(x,\varepsilon) = \min K_{\sigma_1\sigma_2}(x),$$

with the minimum taken over all pairs of disjoint subsets $\sigma_1, \sigma_2 \subset [1,s] \cap \mathbb{N}$. The function $T(x,\varepsilon)$ satisfies condition (L7).

Let $v \in E^s(x)$. Write $v = \sum_{i=1}^{k(x)} v_i$ where $v_i \in E_i(x)$. We have

$$\|v\| \leq \sum_{i=1}^{k(x)} \|v_i\| \leq LT^{-1}(x,\varepsilon)\|v\|,$$

where $L > 1$ is a constant. Let us set

$$C'(x,\varepsilon) = LD(x,\varepsilon)T(x,\varepsilon)^{-1}.$$

It follows from (5.10) and (5.16) that the function $C'(x,\varepsilon)$ satisfies the condition of Lemma 5.6 with

$$M_1(x,\varepsilon) = \frac{2}{\pi}L\sqrt{D(x,\varepsilon)} \quad \text{and} \quad M_2(x,\varepsilon) = LD(x,\varepsilon)^2 T(x,\varepsilon)^{-1}.$$

Therefore, there exists a function $C_1(x,\varepsilon) \geq C'(x,\varepsilon)$ for which the statements of Lemma 5.6 hold.

Applying the above arguments to the inverse map f^{-1} and the subspace $E^u(x)$, one can construct a function $C_2(x,\varepsilon)$ for which the statements of Lemma 5.6 hold. The desired function $C(x,\varepsilon)$ is defined by

$$C(x,\varepsilon) = \max\{C_1(x,\varepsilon/2), C_2(x,\varepsilon/2)\}.$$

This completes the proof of Theorem 5.4. □

5.2. Nonuniform complete hyperbolicity

In this section we introduce one of the principal concepts of smooth ergodic theory—the notion of nonuniform hyperbolicity—and we discuss its relation to dynamical systems with nonzero Lyapunov exponents introduced in the previous section.

5.2.1. Definition of nonuniform hyperbolicity. Let $f\colon M \to M$ be a diffeomorphism of a compact smooth Riemannian manifold M of dimension p and let $Y \subset M$ be an f-invariant nonempty measurable subset. We say that the set Y is *nonuniformly (completely) hyperbolic* if there are a number $\varepsilon_0 > 0$, measurable functions $\lambda, \mu, C, K\colon Y \to (0, \infty)$, and $\varepsilon\colon Y \to [0, \varepsilon_0]$, and subspaces $E^s(x)$ and $E^u(x)$, which depend measurably on $x \in Y$ such that for every $x \in Y$ the following *complete hyperbolicity conditions* hold:

(H1) the functions λ, μ, ε are f-invariant, i.e.,
$$\lambda(f(x)) = \lambda(x), \quad \mu(f(x)) = \mu(x), \quad \varepsilon(f(x)) = \varepsilon(x), \tag{5.17}$$
and satisfy
$$\lambda(x)e^{\varepsilon(x)} < 1 < \mu(x)e^{-\varepsilon(x)}; \tag{5.18}$$

(H2) for $m \in \mathbb{Z}$,
$$C(f^m(x)) \leq C(x)e^{\varepsilon(x)|m|}, \quad K(f^m(x)) \geq K(x)e^{-\varepsilon(x)|m|}; \tag{5.19}$$

(H3) $T_x M = E^s(x) \oplus E^u(x)$ and
$$d_x f E^s(x) = E^s(f(x)), \quad d_x f E^u(x) = E^u(x); \tag{5.20}$$

(H4) for $v \in E^s(x)$ and $n \geq 0$,
$$\|d_x f^n v\| \leq C(x) \lambda(x)^n e^{\varepsilon(x)n} \|v\|; \tag{5.21}$$

(H5) for $v \in E^u(x)$ and $n \leq 0$,
$$\|d_x f^n v\| \leq C(x) \mu(x)^n e^{-\varepsilon(x)n} \|v\|; \tag{5.22}$$

(H6) the angle
$$\angle(E^s(x), E^u(x)) \geq K(x). \tag{5.23}$$

If a subset $Y \subset M$ is nonuniformly (completely) hyperbolic for a diffeomorphism f, then we call the map $f|Y$ *nonuniformly (completely) hyperbolic*. The number ε_0 and the functions $\{\lambda(x), \mu(x), \varepsilon(x), C(x), K(x)\}$ are called *parameters of hyperbolicity*.

Conditions (H1), (H4), and (H5) mean that vectors in $E^s(x)$ (respectively, $E^u(x)$) contract with an exponential rate under the dynamics as the time $n \to \infty$ (respectively, $n \to -\infty$). This justifies calling $E^s(x)$ and $E^u(x)$, respectively, *stable* and *unstable* subspaces.

Condition (H3) means that the subspaces $E^s(x)$ and $E^u(x)$ form an invariant splitting of the tangent bundle over the set Y.

We stress that the functions C and K are Borel measurable and may not be continuous. This means that these functions may jump arbitrarily near a given point $x \in Y$ in an uncontrollable way. Condition (H2), however, provides some control over these functions along the trajectory $\{f^n(x)\}$ for $x \in Y$ by requiring that the function $C(x)$ increases and the function $K(x)$

decreases with at most small exponential rate. Along with conditions (H4)–(H6) this ensures the hyperbolic behavior of f along the trajectory of the point x.

If ν is an invariant Borel probability measure for which $\nu(Y) > 0$, then given a sufficiently small $\varepsilon > 0$, there exists a subset $A \subset Y$ with $\nu(A) \geq \nu(Y) - \varepsilon > 0$ such that the function $C(x)$ is bounded from above on A. Moreover, due to the Poincaré Recurrence Theorem almost every point $x \in A$ returns to A infinitely often. Therefore, the function $C(x)$ indeed oscillates along the trajectory $f^n(x)$, for almost every $x \in A$, but may still become arbitrarily large. A similar observation holds for the function $K(x)$.

Note that the dimensions of $E^s(x)$ and $E^u(x)$ are measurable f-invariant functions, and hence, the set Y can be decomposed into finitely many disjoint invariant measurable subsets on which these dimensions are constant.

Exercise 5.8. Show that if Y is nonuniformly (completely) hyperbolic, then for every $x \in Y$:

(1) $d_x f^m E^s(x) = E^s(f^m(x))$ and $d_x f^m E^u(x) = E^u(f^m(x))$;

(2) for $v \in E^s(x)$ and $m < 0$,
$$\|d_x f^m v\| \geq C(f^m(x))^{-1} \lambda(x)^m e^{\varepsilon(x)m} \|v\|;$$

(3) for $v \in E^u(x)$ and $m > 0$,
$$\|d_x f^m v\| \geq C(f^m(x))^{-1} \mu(x)^m e^{-\varepsilon(x)m} \|v\|.$$

We summarize the discussion in the previous section by saying that the set $\mathcal{E}$ (see (5.3)) of LP-regular points with nonzero Lyapunov exponents is nonuniformly (completely) hyperbolic with
$$\lambda(x) = e^{\chi_{k(x)}(x)}, \quad \mu(x) = e^{\chi_{k(x)+1}(x)},$$
$$C(x) = C(x, \varepsilon), \quad K(x) = K(x, \varepsilon)$$
for *any* fixed $0 < \varepsilon \leq \varepsilon_0(x)$ with sufficiently small $\varepsilon_0(x)$ (see conditions (L1)–(L7) in Section 5.1). In fact, finding trajectories with nonzero Lyapunov exponents seems to be a universal approach in establishing nonuniform hyperbolicity.

5.2.2. Nonuniformly hyperbolic flows. We introduce the notion of nonuniform (complete) hyperbolicity for dynamical systems with continuous time. Consider a smooth flow φ_t on a compact smooth Riemannian manifold M which is generated by a vector field $X(x)$. We say that the set Y is *nonuniformly (completely) hyperbolic* if there are a number $\varepsilon_0 > 0$, measurable functions $\lambda, \mu, C, K \colon Y \to (0, \infty)$, $\varepsilon \colon Y \to [0, \varepsilon_0]$, and subspaces $E^s(x)$ and $E^u(x)$, which depend measurably on $x \in Y$ such that conditions (H1),

(H2), (H4), (H5), and (H6)[2] hold and condition (H3) is replaced with the following condition:

(H3′) $T_xM = E^s(x) \oplus E^u(x) \oplus E^0(x)$, where
$$E^0(x) = \{\alpha X(x) : \alpha \in \mathbb{R}\}$$
and for every $t \in \mathbb{R}$,
$$d_x\varphi_t E^s(x) = E^s(\varphi_t(x)) \quad \text{and} \quad d_x\varphi_t E^u(x) = E^u(\varphi_t(x)).$$

We say that a dynamical system (with discrete or continuous time) is *nonuniformly (completely) hyperbolic* if it possesses an invariant nonuniformly hyperbolic subset.

Remark 5.9. For any point x in a nonuniformly hyperbolic set $Y \subset M$ (with parameter ε_0) we have that $\chi^+(x,v) < 0$ for any $v \in E^s(x)$ and $\chi^+(x,v) > 0$ for any $v \in E^u(x)$. It follows that if ν is an invariant measure for f and $\nu(Y) = 1$, then ν is a measure with nonzero Lyapunov exponents and by the Multiplicative Ergodic Theorem (see Theorem 4.1), ν-almost every point $x \in Y$ is LP-regular. In particular, for any $x \in \mathcal{R} \cap Y$ the irregularity coefficient of the Lyapunov exponent χ^+ at x is zero while for any $x \in Y \setminus \mathcal{R}$ the irregularity coefficient at x may not be zero but does not exceed ε_0. We will use this observation in Chapters 6 and 7 where we study stability of nonuniformly hyperbolic trajectories for diffeomorphisms and flows.

5.3. Regular sets

By Luzin's theorem every measurable function on a measurable space X is "nearly" continuous with respect to a finite measure μ, that is, it is continuous outside a set of arbitrarily small measure. In other words, X can be exhausted by an increasing sequence of measurable subsets on which the function is continuous. In line with this idea, the regular sets are built to exhaust an invariant nonuniformly (completely) hyperbolic set Y by an increasing sequence of (not necessarily invariant) *uniformly* (completely) hyperbolic subsets, demonstrating that nonuniform (complete) hyperbolicity is "nearly" uniform.

5.3.1. Definition of regular sets. Let f be a diffeomorphism of a compact smooth Riemannian manifold M and let λ, μ, ε be positive numbers satisfying
$$0 < \lambda e^\varepsilon < 1 < \mu e^{-\varepsilon}. \tag{5.24}$$

---
[2] In conditions (H4) and (H5) the discrete time m should be replaced with the continuous time t.

5.3. Regular sets

In this section, given integers j, $1 \leq j < n$, and $\ell \geq 1$, we introduce a special collection of subsets $\{\Lambda_\iota^\ell\}$ depending on the *index set* $\iota = \{\lambda, \mu, \varepsilon, j\}$. Each of these subsets consists of points $x \in M$ with the following properties: for $x \in \Lambda_\iota^\ell$ there exists a decomposition $T_x M = E_x^1 \oplus E_x^2$ such that for every $k \in \mathbb{Z}$ and $m > 0$ we have

(1) $\dim E_x^1 = j$ (and hence, $\dim E_x^2 = p - j$, $p = \dim M$);
(2) if $v \in d_x f^k E_x^1$, then
$$\|d_{f^k(x)} f^m v\| \leq \ell \lambda^m e^{\varepsilon(m+|k|)} \|v\|; \tag{5.25}$$
(3) if $v \in d_x f^k E_x^2$, then
$$\|d_{f^k(x)} f^{-m} v\| \leq \ell \mu^{-m} e^{\varepsilon(m+|k|)} \|v\|; \tag{5.26}$$
(4) the angle
$$\angle(d_x f^k E_x^1, d_x f^k E_x^2) \geq \ell^{-1} e^{-\varepsilon|k|}. \tag{5.27}$$

The set Λ_ι^ℓ is called a *regular set* (or a *Pesin set*). We stress that regular sets need not be invariant nor compact. Moreover, the estimates (5.25)–(5.27) are uniform in $x \in \Lambda_\iota^\ell$.

We also introduce the *level set*
$$\Lambda_\iota := \bigcup_{\ell \geq 1} \Lambda_\iota^\ell.$$

We describe the main properties of regular and level sets.

Theorem 5.10. *The following statements hold:*

(1) *if $\ell < \ell'$, then $\Lambda_\iota^\ell \subset \Lambda_\iota^{\ell'}$, i.e., regular sets are nested; in particular, the subspaces E_x^1 and E_x^2 do not depend on the choice of ℓ;*
(2) *for $x \in \Lambda_\iota^\ell$ the splitting $T_x M = E_x^1 \oplus E_x^2$ is uniquely defined, i.e., if $v \in T_x M$ is such that (5.25) or (5.26) holds for $k = 0$ and all $m \geq 0$, then $v \in E_x^1$ or, respectively, $v \in E_x^2$;*
(3) *if $n \in \mathbb{Z}$, then $f^n(\Lambda_\iota^\ell) \subset \Lambda_\iota^{\ell'}$, where $\ell' = \ell \exp(|n|\varepsilon)$; in particular, the subspaces E_x^1 and E_x^2 are invariant under df and the set Λ_ι is f-invariant.*
(4) *the subspaces E_x^1 and E_x^2 can be extended to the closure $\overline{\Lambda_\iota^\ell}$ of Λ_ι^ℓ and the subspaces E_x^1 and E_x^2 vary continuously with x on $\overline{\Lambda_\iota^\ell}$.*

Proof. To prove the first statement choose $x \in \Lambda_\iota^\ell$, $\ell' > \ell$, and observe that for every $k \in \mathbb{Z}$ and $m > 0$ the inequalities (5.25)–(5.27) hold with ℓ replaced by ℓ'. This implies that $x \in \Lambda_\iota^{\ell'}$ with the same splitting of the tangent space at x.

To prove the second statement we follow the line of argument in the proof of Theorem 1.10. Fix $x \in \Lambda_\iota^\ell$ and let $v \in T_x M$ be such that (5.25) holds for $k = 0$ and every $m > 0$. We wish to show that $v \in E_x^1$. Otherwise, write $v = v^1 + v^2$ where $v^1 \in E_x^1$, $v^2 \in E_x^2$, and $v^2 \neq 0$. Fix $m > 0$ and let $w = d_x f^m v^2$. By (5.26),

$$\|v^2\| = \|d_x f^{-m} w\| \leq \ell \mu^{-m} e^{\varepsilon m} \|w\|.$$

In view of (5.24) this implies that for all sufficiently large m

$$\|w\| \geq \ell^{-1} \mu^m e^{-\varepsilon m} \|v^2\| = \ell^{-1} (\mu e^{-\varepsilon})^m \|v^2\| > 1.$$

On the other hand, since $d_x f^m v = d_x f^m v^1 + d_x f^m v^2$, by (5.24) and (5.25), for all sufficiently large m

$$\|w\| = \|d_x f^m v^2\| \leq \|d_x f^m v\| + \|d_x f^m v^1\|$$
$$\leq \ell \lambda^m e^{\varepsilon m}(\|v\| + \|v^1\|) = \ell(\lambda e^\varepsilon)^m (\|v\| + \|v^1\|) < 1,$$

leading to a contradiction. Let now $v \in T_x M$ be such that (5.26) holds for $k = 0$ and every $m > 0$. Using an argument similar to the above, one can show that $v \in E^2(x)$, thus completing the proof of the second statement.

To prove the third statement for every $k, n \in \mathbb{Z}$, $m > 0$, and $x \in \Lambda_\iota^\ell$, for $v \in E_x^1$ we have

$$\|d_{f^{k+n}(x)} f^m v\| \leq \ell \lambda^m e^{\varepsilon(m+|k+n|)} \|v\| \leq \ell e^{\varepsilon|n|} \lambda^m e^{\varepsilon(m+|k|)} \|v\|,$$

and for $v \in E_x^2$ we have

$$\|d_{f^{k+n}(x)} f^{-m} v\| \leq \ell \mu^{-m} e^{\varepsilon(m+|k+n|)} \|v\| \leq \ell e^{\varepsilon|n|} \mu^{-m} e^{\varepsilon(m+|k|)} \|v\|.$$

Finally, the angle

$$\angle(d_x f^{k+n} E_x^1, d_x f^{k+n} E_x^2) \geq \ell^{-1} e^{-\varepsilon|k+n|} \geq \ell^{-1} e^{-\varepsilon|n|} e^{-\varepsilon|k|},$$

implying the third statement.

To prove the last statement we again follow the line of argument in the proof of Theorem 1.10 and consider a sequence $(x_q)_{q \geq 1}$ of points in Λ_ι^ℓ converging to a point $x \in \overline{\Lambda_\iota^\ell}$. Since the Grassmannian manifold over $\overline{\Lambda_\iota^\ell}$ is compact, we can choose a subsequence q_t such that the sequence of subspaces $E^1(x_{q_t})$ converges to a subspace $E'(x) \subset T_x M$ and the sequence of subspaces $E^2(x_{q_t})$ converges to a subspace $E''(x) \subset T_x M$. Let us fix $k \in \mathbb{Z}$ and $m > 0$. Since the inequalities (5.25)–(5.27) hold for vectors in $E^1(x_{q_t})$ and $E^2(x_{q_t})$ for every q_t, they also hold for vectors in $E'(x)$ and $E''(x)$, and since $k \in \mathbb{Z}$ and $m > 0$ are arbitrary, we obtain that these inequalities hold for the subspaces $E'(x)$ and $E''(x)$. Observe now that

$$\dim E^s(x_{m_k}) = \dim E'(x) =: a, \quad \dim E^u(x_{m_k}) = \dim E''(x) =: b$$

and that $a + b = \dim M$. By (5.27), $E'(x) \cap E''(x) = \{0\}$, meaning that the subspaces $E'(x)$ and $E''(x)$ form a splitting of $T_x M$ which satisfies (5.25)–(5.27). Statement (2) now yields that $E'(x) = E^1(x)$ and $E''(x) = E^2(x)$.

5.3. Regular sets

In fact, the above argument implies continuous dependence of subspaces $E^1(x)$ and $E^2(x)$ on $x \in \overline{\Lambda_\iota^\ell}$. This completes the proof of the theorem. □

Exercise 5.11. Show that

(1) if $\varepsilon < \log(1 + 1/\ell)$, then
$$\Lambda_\iota^{\ell-1} \subset f(\Lambda_\iota^\ell) \subset \Lambda_\iota^{\ell+1} \quad \text{and} \quad \Lambda_\iota^{\ell-1} \subset f^{-1}(\Lambda_\iota^\ell) \subset \Lambda_\iota^{\ell+1};$$

(2) for every $x \in \Lambda_\iota^\ell$, $k \in \mathbb{Z}$, and $m > 0$, if $v \in d_x f^k E_x^1$, then
$$\|d_{f^k(x)} f^{-m} v\| \geq \ell^{-1} \lambda^{-m} e^{-\varepsilon(|k-m|+m)} \|v\|$$
and if $v \in d_x f^k E_x^2$, then
$$\|d_{f^k(x)} f^m v\| \geq \ell^{-1} \mu^m e^{-\varepsilon(|k+m|+m)} \|v\|.$$

In view of statement (4) of Theorem 5.10 from now on we will always assume that the regular sets Λ_ι^ℓ are compact. In Section 5.5 we will show that the subspaces $E^1(x)$ and $E^2(x)$ depend Hölder continuously on $x \in \Lambda_\iota^\ell$.

5.3.2. Regular subsets of nonuniformly hyperbolic sets. Consider a nonuniformly (completely) hyperbolic set Y for a diffeomorphism f whose parameters are the number ε_0 and functions $\{\lambda(x), \mu(x), \varepsilon(x), C(x), K(x)\}$. Given positive numbers λ, μ, ε satisfying (5.24) and an integer j, $1 \leq j < n < p$ (where $p = \dim M$), consider the measurable set

$$Y_\iota = \{x \in Y : \lambda(x) \leq \lambda < 1 < \mu \leq \mu(x),\ \varepsilon(x) \leq \varepsilon,\ \dim E^s(x) = j\}$$

where we recall that $\iota = \{\lambda, \mu, \varepsilon, j\}$ is the index set. Clearly, Y_ι is invariant under f and is nonempty if the numbers λ, μ, ε, and j are chosen appropriately. For each integer $\ell \geq 1$, consider the measurable subset

$$Y_\iota^\ell = \{x \in Y_\iota : C(x) \leq \ell,\ K(x) \geq \ell^{-1}\}.$$

We have
$$Y_\iota^\ell \subset Y_\iota^{\ell+1} \quad \text{and} \quad Y_\iota = \bigcup_{\ell \geq 1} Y_\iota^\ell.$$

Note that the set Y_ι^ℓ need not be invariant nor compact. We therefore obtain that every nonuniformly (completely) hyperbolic set Y can be exhausted by a nested sequence of subset Y_ι^ℓ on which hyperbolicity estimates are *uniform*.

Exercise 5.12. Let $\{\Lambda_\iota^\ell\}$ be the collection of regular sets for f. Show that:

(1) $Y_\iota^\ell \subset \Lambda_\iota^\ell$ for every $\ell \geq 1$;
(2) $E_x^1 = E^s(x)$ and $E_x^2 = E^u(x)$ for every $x \in Y$.

It follows that to the set Y one can associate: (1) the family of nonempty f-invariant level sets

$$\{\Lambda_\iota : \lambda, \mu, \varepsilon \text{ satisfy } (5.24) \text{ and } 1 \leq j = \dim E^s(x) < p \text{ for } x \in \Lambda_\iota\} \quad (5.28)$$

and (2) for each ι, the collection of nonempty *compact* regular sets

$$\{\Lambda^\ell = \Lambda_\iota^\ell : \ell \geq 1\}. \tag{5.29}$$

Note that f is nonuniformly (completely) hyperbolic on each level set Λ_ι with parameters

$$\lambda(x) = \lambda, \quad \mu(x) = \mu, \quad \varepsilon(x) = \varepsilon, \quad C(x) = \ell, \quad K(x) = \ell^{-1}$$

as well as on the set $\Lambda = \bigcup \Lambda_\iota$ (here the union is taken over index sets $\iota = \{\lambda, \mu, \varepsilon\}$ with the numbers λ, μ, ε satisfying (5.24) and $1 \leq j < p$) that can be viewed as an "extension" of the "original" nonuniformly (completely) hyperbolic set Y. We stress that the rates of exponential contraction $\lambda(x)$ and of exponential expansion $\mu(x)$ are uniformly bounded away from 1 on each level set Λ_ι but may be arbitrarily close to 1 on Λ.

In what follows, when we talk about nonuniformly hyperbolic sets, we always assume a set Λ with an associated family of nonempty f-invariant level sets (5.28) and a collection of nonempty compact regular sets (5.29).

5.3.3. The Lyapunov inner product. We will construct a special inner product in the tangent bundle $T\Lambda$ which extends to maps on nonuniformly hyperbolic sets the Lyapunov inner product introduced in Section 1.3 for maps on uniformly hyperbolic sets. We will do this on each level set Λ_ι, with the index set $\iota = \{\lambda, \mu, \varepsilon, j\}$ where the numbers λ, μ, ε, and j satisfy (5.28).

Choose numbers λ' and μ' such that

$$\lambda e^\varepsilon < \lambda' < 1 < \mu' < \mu e^{-\varepsilon} \tag{5.30}$$

and define a new inner product $\langle \cdot, \cdot \rangle'_x$, called a *weak Lyapunov inner product* or simply a *Lyapunov inner product*, by (1.9).

Exercise 5.13. Using the Cauchy–Schwarz inequality, conditions (5.19), (5.21)–(5.23), and (5.30), show that each of the above series converges; indeed show that if $v, w \in E^s(x)$, then

$$|\langle v, w \rangle'_x| \leq \sum_{k=0}^\infty C(f^k(x))^2 \lambda^{2k} (\lambda')^{-2k} \|v\|_x \|w\|_x$$
$$\leq C(x)^2 (1 - (\lambda e^\varepsilon / \lambda')^2)^{-1} \|v\|_x \|w\|_x < \infty,$$

and if $v, w \in E^u(x)$, then

$$|\langle v, w \rangle'_x| \leq \sum_{k=0}^\infty C(f^{-k}(x))^2 \mu^{-2k} (\mu')^{2k} \|v\|_x \|w\|_x$$
$$\leq C(x)^2 (1 - (\mu'/(\mu e^{-\varepsilon}))^2)^{-1} \|v\|_x \|w\|_x < \infty.$$

5.3. Regular sets

We extend $\langle \cdot, \cdot \rangle'_x$ to all vectors in $T_x M$ by declaring the subspaces $E^s(x)$ and $E^u(x)$ to be mutually orthogonal with respect to $\langle \cdot, \cdot \rangle'_x$, i.e., we set
$$\langle v, w \rangle'_x = \langle v^s, w^s \rangle'_x + \langle v^u, w^u \rangle'_x,$$
where $v = v^s + v^u$ and $w = w^s + w^u$ with $v^s, w^s \in E^s(x)$ and $v^u, w^u \in E^u(x)$.

The norm in $T_x M$ induced by the Lyapunov inner product is called a *weak Lyapunov norm* or simply a *Lyapunov norm* and is denoted by $\|\cdot\|'_x$.[3] We emphasize that the Lyapunov inner product and, hence, the corresponding norm depend on the choice of numbers λ' and μ'.

It is worth mentioning that in general the Lyapunov inner product $\langle \cdot, \cdot \rangle'_x$ is *not* a Riemannian metric. While in the case of uniform hyperbolicity it depends continuously (in fact, Hölder continuously but, in general, not smoothly) in x, in the case of nonuniform hyperbolicity it depends Borel measurably in x.

Exercise 5.14. For the Lyapunov inner product, show that for each $x \in \Lambda_\iota$:

(1) the angle between the subspaces $E^s(x)$ and $E^u(x)$ with respect to the inner product $\langle \cdot, \cdot \rangle'_x$ is $\pi/2$;

(2) $\|d_x f v\|'_{f(x)} \leq \lambda' \|v\|_x$ for any $v \in E^s(x)$ and $\|(d_x f)^{-1} v\|'_{f^{-1}(x)} \leq (\mu')^{-1} \|v\|_x$ for any $v \in E^u(x)$;

(3) the relation between the Lyapunov inner product and the Riemannian inner product is given by
$$\frac{1}{\sqrt{2}} \|v\|_x \leq \|v\|'_x \leq D(x) \|v\|_x,$$
where $v \in T_x M$ and
$$D(x) = C(x) K(x)^{-1} \left[(1 - (\lambda e^\varepsilon / \lambda')^2)^{-1} + (1 - (\mu'/(\mu e^{-\varepsilon}))^2)^{-1} \right]^{1/2}$$
is a measurable function satisfying
$$D(f^m(x)) \leq D(x) e^{2\varepsilon |m|}, \quad m \in \mathbb{Z}. \tag{5.31}$$

Compare with Theorem 1.14 and Exercise 4.9.

Properties (1) and (2) in Exercise 5.14 show that the action of the differential df on $T\Lambda_\iota$ is *uniformly* hyperbolic with respect to the Lyapunov inner product.

Using the Lyapunov inner product, one can construct a particular family of local coordinate charts around points $x \in \Lambda_\iota$ which are similar to those constructed in Section 4.4 (compare to Theorem 4.13). These coordinate charts are called *(weak) Lyapunov charts*.

The proof of the following theorem is left to the reader.

---

[3] Compare this weak Lyapunov inner product and the associated weak Lyapunov norm with the one introduced in (1.9) in the case of a diffeomorphism on a uniformly hyperbolic set.

Theorem 5.15. *For every sufficiently small $\varepsilon > 0$ the following properties hold:*

(1) *there exist a function $q\colon \Lambda_\iota \to (0,1]$ and a collection of embeddings $\Psi_x\colon B(0, q(x)) \subset \mathbb{R}^p \to M$ for each $x \in \Lambda_\iota$ such that:*
   (a) *$\Psi_x(0) = x$ and $\Psi_x = \exp_x \circ C(x)$, where $C(x)$ is the Lyapunov change of coordinates;*
   (b) *$e^{-\varepsilon} < q(f(x))/q(x) < e^\varepsilon$ and $q(x) \leq \|C(x)^{-1}\|^{-1}$;*

(2) *if $f_x := \Psi_{f(x)}^{-1} \circ f \circ \Psi_x\colon B(0, q(x)) \to \mathbb{R}^p$, then*
$$d_0 f_x = (d_x f | E^s(x), d_x f | E^u(x));$$

(3) *the C^1 distance $d_{C^1}(f_x, d_0 f_x) < \varepsilon$ in $B(0, q(x))$;*

(4) *there exist a constant $K > 0$ and a measurable function $a\colon \Lambda_\iota \to \mathbb{R}$ such that for every $y, z \in B(0, q(x))$,*
$$K^{-1}\rho(\Psi_x y, \Psi_x z) \leq \|y - z\| \leq a(x)\rho(\Psi_x y, \Psi_x z)$$
with $e^{-\varepsilon} < a(f(x))/a(x) < e^\varepsilon$;

(5) *the matrix function $C(x)\colon \mathbb{R}^p \to T_x M$ depends measurably on x and the matrix $A(x) = C(f(x))^{-1} d_x f C(x)$ has the block form*
$$A(x) = \begin{pmatrix} A^1(x) & 0 \\ 0 & A^2(x) \end{pmatrix},$$
where $\|A^1(x)\| \leq \lambda'$ and $\|A^2(x)^{-1}\| \leq (\mu')^{-1}$.

This theorem means that the action of the restriction of f to the Lyapunov chart is uniformly hyperbolic.

Remark 5.16. In Section 4.3.1 we introduced the strong Lyapunov inner product (see (4.18)) and the associated strong Lyapunov norms and strong Lyapunov charts (or regular neighborhoods; see Theorem 4.13) for cocycles over measurable transformations. In particular, given a $C^{1+\alpha}$ diffeomorphism of a compact smooth Riemannian manifold M, we obtain two special inner products associated with f—the strong Lyapunov inner product associated with the derivative cocycle generated by f and the (weak) Lyapunov inner product constructed in this section. Both inner products capture a great deal of information about the infinitesimal behavior of trajectories but the weak one is in a sense more "relaxed" and is easier to work with. Here we discuss and compare some properties of these two inner products.

(1) While the weak Lyapunov inner product is defined on the tangent spaces over a set of nonuniformly hyperbolic points (which necessarily have nonzero Lyapunov exponents), the strong Lyapunov inner product is defined on the tangent spaces over the set of Lyapunov–Perron regular points which may or may not have nonzero Lyapunov exponents.

(2) The weak Lyapunov inner product takes into account the stable $E^s(x)$ and unstable $E^u(x)$ subspaces, so that in this inner product they are orthogonal and the contraction and expansion rates along these subspaces are bounded by constants λ' and μ' independent of $x \in \Lambda_\iota$ (see Exercise 5.14; this is why in the definition of the weak Lyapunov inner product, the sum is taken over all $m \in \mathbb{N}$; see (1.9)).

This inner product provides an important technical tool in our construction of local stable $V^s(x)$ and unstable $V^u(x)$ manifolds (see Section 7.1.3).

It is also used in an essential way in the work of Sarig [**104**] and Ben Ovadia [**17**] on constructing countable Markov partitions and the corresponding symbolic dynamics for $C^{1+\alpha}$ diffeomorphisms on nonuniformly hyperbolic sets. We outline this construction in Section 11.3.

(3) The strong Lyapunov inner product, on the other hand, takes into account the entire collection of Oseledets subspaces, $\{E_i(x)\}_{i=1,\ldots,s(x)}$, so that in this inner product any two of them are orthogonal and the contraction and expansion rates along every $E_i(x)$ are determined by the Lyapunov exponent $\chi_i(x)$ up to a small error term $e^{\pm\varepsilon}$ (see (4.22); this is why in the definition of the strong Lyapunov inner product along $E_i(x)$, the sum is taken over all $m \in \mathbb{Z}$; see (4.17)).

This manifests itself in two ways: if x is an LP-regular point, then:

(a) there are two stable $\{V_i^-(x)\}$ and unstable $\{V_i^+(x)\}$ filtrations associated with the Oseledets decomposition (see Section 3.2) which give rise to two flags of stable $V_i^s(x)$ and unstable $V_i^u(x)$ local manifolds (see [**13**, Theorem 8.4.1]); thus one obtains a sufficiently detailed description of the structure of the strong Lyapunov chart at x;

(b) starting with the ball $B(0,r) \subset T_x M$ centered at zero of small radius r, for any $n \in \mathbb{Z}$, the image $d_x f^n(B(0,r))$ is an ellipsoid whose axes in the direction of $E_i(x)$ have length in the strong Lyapunov norm equal to $e^{\chi_i n} e^{\pm \varepsilon n}$, $i = 1, \ldots, s(x)$.

We note that these two properties play a crucial role in the proof of the Oseledets–Pesin Reduction Theorem (see Theorem 4.10) and in the proof of the Ledrappier–Young entropy formula. The latter establishes a fundamental relation between the three arguably most important invariants in smooth dynamics—entropy, dimension, and Lyapunov exponents; see Section 9.3.4 for more details.

5.4. Nonuniform partial hyperbolicity

In Section 5.1 we studied diffeomorphisms whose values of the Lyapunov exponent are *all* nonzero on a nonempty set $\mathcal{E}$ (with some of the values being negative and the remaining ones being positive; see (5.3)). As we saw in Section 5.2, the set $\mathcal{E}$ is nonuniformly hyperbolic and the hyperbolicity is complete. In this section we discuss the more general case of partial hyperbolicity. It deals with the situation when *some* of the values of the Lyapunov exponent are negative and *some* among the remaining ones may be zero.

While for dynamical systems that are nonuniformly completely hyperbolic one can obtain a sufficiently complete description of their ergodic properties (with respect to smooth invariant measures; see Chapter 9), dynamical systems that are nonuniformly partially hyperbolic may not possess "nice" ergodic properties. However, some principal results describing local behavior of systems that are nonuniformly completely hyperbolic can be extended without much extra work to systems that are only nonuniformly partially hyperbolic. This includes constructing local stable manifolds (see Section 7.3.5) and establishing their absolute continuity (see Remark 8.19).

We call an f-invariant nonempty measurable subset $Z \subset M$ *nonuniformly partially hyperbolic* if there are a number $\varepsilon_0 > 0$, measurable functions $\lambda_1, \lambda_2, \mu_1, \mu_2, C, K \colon Z \to (0, \infty)$, and $\varepsilon \colon Z \to [0, \varepsilon_0]$, and subspaces $E^s(x)$, $E^u(x)$, and $E^0(x)$, which depend measurably on $x \in Z$, such that the following *partial hyperbolicity conditions* hold:

(PH1) the functions λ_1, λ_2, μ_1, μ_2, ε are f-invariant (see (5.17)) and satisfy for $x \in Z$ (see (5.18))
$$\lambda_1(x)e^{\varepsilon(x)} < \lambda_2(x)e^{-\varepsilon(x)} < 1 < \mu_2(x)e^{\varepsilon(x)} < \mu_1(x)e^{-\varepsilon(x)};$$

(PH2) the functions $C(x)$ and $K(x)$ satisfy condition (5.19);

(PH3) $T_x M = E^s(x) \oplus E^0(x) \oplus E^u(x)$ and for $a = s, u, 0$,
$$d_x f E^a(x) = E^a(f(x));$$

(PH4) for $v \in E^s(x)$ and $n \geq 0$, condition (5.21) holds (with $\lambda(x)$ replaced by $\lambda_1(x)$) and for $v \in E^u(x)$ and $n \leq 0$, condition (5.22) holds (with $\mu(x)$ replaced by $\mu_1(x)$);

(PH5) for $v \in E^0(x)$
$$C(x)^{-1}\lambda_2(x)^n e^{-\varepsilon(x)n}\|v\| \leq \|d_x f^n v\| \leq C(x)\mu_2(x)^n e^{\varepsilon(x)n}\|v\|, \quad n \geq 0,$$
$$C(x)^{-1}\mu_2(x)^n e^{\varepsilon(x)n}\|v\| \leq \|d_x f^n v\| \leq C(x)\lambda_2(x)^n e^{-\varepsilon(x)n}\|v\|, \quad n \leq 0;$$

(PH6) the angles $\angle(E^a(x), E^b(x)) \geq K(x)$ where $a, b \in \{s, u, 0\}$ and $a \neq b$.

If a subset $Z \subset M$ is nonuniformly partially hyperbolic, then we call the map $f|Z$ *nonuniformly partially hyperbolic*.

Let $\iota = \{\lambda_1, \lambda_2, \mu_1, \mu_2, \varepsilon, j_1, j_2\}$ be a collection of numbers satisfying
$$\begin{aligned} 0 < \lambda_1 e^\varepsilon &< \lambda_2 e^{-\varepsilon} < 1 < \mu_2 e^\varepsilon < \mu_1 e^{-\varepsilon}, \\ 1 \leq j_1 &< p, \quad 1 \leq j_2 < p, \quad 1 < j_1 + j_2 < p. \end{aligned} \quad (5.32)$$

As in Section 5.3, to each nonuniformly partially hyperbolic set Z one can associate a collection of level sets

$$\Lambda_\iota = \bigcup_{\ell \geq 1} \Lambda_\iota^\ell, \tag{5.33}$$

where Λ_ι^ℓ are regular sets given by

$$\Lambda_\iota^\ell = \{x \in M : \lambda_1(x) < \lambda_1 < \lambda_2 < \lambda_2(x), \ \mu_2(x) < \mu_2 < \mu_1 < \mu_1(x),$$
$$\varepsilon(x) = \varepsilon, \ j_1 = \dim E^s(x), \ j_2 = \dim E^u(x), \ C(x) \leq \ell, \ K(x) \geq \ell\}. \tag{5.34}$$

The level sets are invariant and the regular sets are nested. The set $\Lambda = \bigcup_\iota \Lambda_\iota$ (here the union is taken over all collections ι satisfying (5.32)) is nonuniformly partially hyperbolic and $Z \subset \Lambda$. Observe that each regular set is *compact* and *uniformly* partially hyperbolic but not necessarily invariant. One can show that the subspaces $E^a(x)$, $a = s, u, 0$, depend (Hölder) continuously on x.

Let $f \colon M \to M$ be a diffeomorphism of a compact smooth Riemannian manifold M of dimension p. Consider the f-invariant set

$$\mathcal{F} = \{x \in \mathcal{R} : \text{there exist } 1 \leq k_1(x) < k_2(x) < s(x)$$
$$\text{such that } \chi_{k_1(x)}(x) < 0 \text{ and } \chi_{k_2(x)}(x) > 0\}.$$

Repeating the arguments in the proof of Theorem 5.4, one can show that f is nonuniformly partially hyperbolic on $\mathcal{F}$.

The time-one map of a nonuniformly hyperbolic flow is an example of a nonuniformly partially hyperbolic diffeomorphism.

5.5. Hölder continuity of invariant distributions

As we saw in Section 1.1, the stable and unstable subspaces of an Anosov diffeomorphism f depend continuously on the point in the manifold. Since these subspaces at a point x are determined by the whole positive and, respectively, negative semitrajectory through x, their dependence on the point may not be differentiable even if f is real analytic. However, one can show that they depend Hölder continuously on the point.[4]

We remind the reader of the definition of Hölder continuous distribution. A k-dimensional *distribution* E on a smooth manifold M is a family of k-dimensional subspaces $E(x) \subset T_x M$. A Riemannian metric on M naturally induces distances in TM and in the space of k-dimensional distributions on TM. The Hölder continuity of a distribution E can be defined using these

---

[4]This result was proved by Anosov in [4] and is a corollary of Theorem 5.18 below.

distances. More precisely, for a subspace $A \subset \mathbb{R}^p$ (where $p = \dim M$) and a vector $v \in \mathbb{R}^p$, set
$$\rho(v, A) = \min_{w \in A} \|v - w\|.$$
In other words, $\rho(v, A)$ is the distance from v to its orthogonal projection on A. For subspaces A and B in $\mathbb{R}^p$, define
$$d(A, B) = \max \left\{ \max_{v \in A, \|v\|=1} \rho(v, B), \max_{w \in B, \|w\|=1} \rho(w, A) \right\}.$$
Let $D \subset \mathbb{R}^p$ be a subset and let E be a k-dimensional distribution. The distribution E is said to be *Hölder continuous* with *Hölder exponent* $\alpha \in (0, 1]$ and *Hölder constant* $L > 0$ if there exists $\varepsilon_0 > 0$ such that
$$d(E(x), E(y)) \leq L \|x - y\|^\alpha$$
for every $x, y \in D$ with $\|x - y\| \leq \varepsilon_0$.

Now let E be a continuous distribution on M. Choose a small number $\varepsilon > 0$ and an atlas $\{U_i\}$ of M. We say that E is *Hölder continuous* if the restriction $E|U_i$ is Hölder continuous for every i.

Exercise 5.17. (1) Show that if a distribution E on M is Hölder continuous with respect to an atlas $\{U_i\}$ of M, then it is also Hölder continuous with respect to any other atlas of M with the same Hölder exponent (but the Hölder constant may be different).

(2) Show that if a distribution E on M is Hölder continuous, then it remains Hölder continuous with the same Hölder exponent if the Riemannian metric is replaced by an equivalent smooth metric.

(3) Show that a distribution E on M is Hölder continuous if and only if there are positive constants C, α, and ε such that for every two points x and y with $\rho(x, y) \leq \varepsilon$ we have
$$d(E(x), \tilde{E}(x)) \leq C \rho(x, y)^\alpha,$$
where $\tilde{E}(x)$ is the subspace of $T_x M$ that is the parallel transport of the subspace $E(y) \subset T_y M$ along the unique geodesic connecting x and y.[5]

Finally, given a subset $\Lambda \subset M$, we say that a distribution $E(x)$ on Λ is Hölder continuous if for an atlas $\{U_i\}$ of M, the restriction $E|U_i \cap \Lambda$ is Hölder continuous for every i.

Let f be a $C^{1+\alpha}$ diffeomorphism of a compact smooth manifold M and let Λ be a nonuniformly (completely) hyperbolic set for f. Consider the corresponding collection $\{\Lambda_\iota\}$ of level sets (see (5.28)) and for each λ, μ, and ε the corresponding collection of regular sets $\{\Lambda^\ell = \Lambda^\ell_\iota : \ell \geq 1\}$ (see (5.29)).

---
[5] The geodesic connecting x and y is unique if the number ε is sufficiently small.

5.5. Hölder continuity of invariant distributions

As we know, the stable $E^s(x)$ and unstable $E^u(x)$ subspaces depend only (Borel) measurably on $x \in \Lambda$. However, by Theorem 5.10, these subspaces vary continuously on the point in a regular set, and in this section we show that they depend Hölder continuously on the point. It should be stressed that the Hölder continuity property requires higher regularity of the system, i.e., that f is of class $C^{1+\alpha}$. In what follows, we consider only the stable subspaces; the Hölder continuity of the unstable subspaces follows by reversing the time.

Theorem 5.18. *For every index set $\{\iota = \lambda, \mu, \varepsilon, j\}$ and every $\ell \geq 1$ the stable and unstable distributions $E^s(x)$ and $E^u(x)$ depend Hölder continuously on $x \in \Lambda_\iota^\ell$.*

We shall prove a more general statement of which Theorem 5.18 is an easy corollary. It applies to the cases of complete as well as of partial hyperbolicity. By the Whitney Embedding Theorem, every manifold M can be embedded in the Euclidean space $\mathbb{R}^N$ for a sufficiently large N. If M is compact, the Riemannian metric on M is equivalent to the distance $\|x - y\|$ induced by the embedding. We assume in Theorem 5.20, without loss of generality, that the manifold is embedded in $\mathbb{R}^N$.

Given a number $\kappa > 0$, we say that two subspaces $E_1, E_2 \subset \mathbb{R}^N$ are κ-*transverse* if $\|v_1 - v_2\| \geq \kappa$ for all unit vectors $v_1 \in E_1$ and $v_2 \in E_2$.

Exercise 5.19. Show that the subspaces $E^s(x)$ and $E^u(x)$ are κ-transverse for $x \in \Lambda^\ell$ and some $\kappa > 0$ which is independent of x.

Theorem 5.20. *Let M be a compact m-dimensional C^2 submanifold of $\mathbb{R}^N$ for some $m < N$, and let $f \colon M \to M$ be a $C^{1+\beta}$ map for some $\beta \in (0,1)$. Assume that there exist a set $D \subset M$ and real numbers $0 < \lambda < \mu$, $c > 0$, and $\kappa > 0$ such that for each $x \in D$ there are κ-transverse subspaces $E_1(x)$, $E_2(x) \subset T_x M$ such that:*

(1) $T_x M = E_1(x) \oplus E_2(x)$;

(2) *for every $n > 0$ and every $v_1 \in E_1(x)$, $v_2 \in E_2(x)$ we have*

$$\|d_x f^n v_1\| \leq c\lambda^n \|v_1\| \quad \text{and} \quad \|d_x f^n v_2\| \geq c^{-1} \mu^n \|v_2\|.$$

Then for every $a > \max_{z \in M} \|d_z f\|^{1+\beta}$, the distribution E_1 is Hölder continuous with exponent

$$\alpha = \frac{\log \mu - \log \lambda}{\log a - \log \lambda} \beta.$$

Proof. We follow the argument in [28] and we begin with two technical lemmas.

Lemma 5.21. *Let A_n and B_n, for $n = 0, 1, \ldots$, be two sequences of real $N \times N$ matrices such that for some $\Delta \in (0,1)$,*
$$\|A_n - B_n\| \le \Delta a^n$$
for every positive integer n. Assume that there exist subspaces $E_A, E_B \subset \mathbb{R}^N$ and numbers $0 < \lambda < \mu$ and $C > 1$ such that $\lambda < a$ and for each $n \ge 0$,

$\|A_n v\| \le C\lambda^n \|v\|$ *if* $v \in E_A$, $\qquad \|A_n w\| \ge C^{-1}\mu^n \|w\|$ *if* $w \in E_A^\perp$,

$\|B_n v\| \le C\lambda^n \|v\|$ *if* $v \in E_B$, $\qquad \|B_n w\| \ge C^{-1}\mu^n \|w\|$ *if* $w \in E_B^\perp$.

Then
$$\operatorname{dist}(E_A, E_B) \le 3C^2 \frac{\mu}{\lambda} \Delta^{\frac{\log \mu - \log \lambda}{\log a - \log \lambda}}.$$

Proof of the lemma. Set
$$Q_A^n = \{v \in \mathbb{R}^N : \|A_n v\| \le 2C\lambda^n \|v\|\}$$
and
$$Q_B^n = \{v \in \mathbb{R}^N : \|B_n v\| \le 2C\lambda^n \|v\|\}.$$
For each $v \in \mathbb{R}^N$, write $v = v_1 + v_2$, where $v_1 \in E_A$ and $v_2 \in E_A^\perp$. If $v \in Q_A^n$, then
$$\|A_n v\| = \|A_n(v_1 + v_2)\| \ge \|A_n v_2\| - \|A_n v_1\| \ge C^{-1}\mu^n \|v_2\| - C\lambda^n \|v_1\|,$$
and hence,
$$\|v_2\| \le C\mu^{-n}(\|A_n v\| + C\lambda^n \|v_1\|) \le 3C^2 \left(\frac{\lambda}{\mu}\right)^n \|v\|.$$
Therefore,
$$\operatorname{dist}(v, E_A) \le 3C^2 \left(\frac{\lambda}{\mu}\right)^n \|v\|. \tag{5.35}$$
Set $\gamma = \lambda/a < 1$. There exists a unique nonnegative integer n such that $\gamma^{n+1} < \Delta \le \gamma^n$. If $w \in E_B$, then
$$\|A_n w\| \le \|B_n w\| + \|A_n - B_n\| \cdot \|w\|$$
$$\le C\lambda^n \|w\| + \Delta a^n \|w\|$$
$$\le (C\lambda^n + (\gamma a)^n)\|w\| \le 2C\lambda^n \|w\|.$$
It follows that $w \in Q_A^n$, and hence, $E_B \subset Q_A^n$. By symmetry, $E_A \subset Q_B^n$. By (5.35) and the choice of n, we obtain
$$\operatorname{dist}(E_A, E_B) \le 3C^2 \left(\frac{\lambda}{\mu}\right)^n \le 3C^2 \frac{\mu}{\lambda} \Delta^{\frac{\log \mu - \log \lambda}{\log a - \log \lambda}}.$$

This completes the proof of the lemma. □

We now consider a diffeomorphism $f \colon M \to M$ of class $C^{1+\beta}$.

5.5. Hölder continuity of invariant distributions

Exercise 5.22. Show that there are positive constants $L > 0$ and $r > 0$ such that for any two points $x, y \in M$ for which $\|x - y\| < r$ we have that
$$\|d_x f - d_y f\| \leq L \|x - y\|^\beta. \tag{5.36}$$

Since M is compact, covering M by finitely many balls of radius r, we obtain that (5.36) holds for any $x, y \in M$ (with an appropriately chosen constant L). The following result extends the estimate (5.36) to powers of the map f.

Lemma 5.23. Let $f: M \to M$ be a $C^{1+\beta}$ map of a compact m-dimensional C^2 submanifold $M \subset \mathbb{R}^N$. Then for every $a > \max_{z \in M} \|d_z f\|^{1+\beta}$ there exists $D > 1$ such that for every $n \in \mathbb{N}$ and every $x, y \in M$ we have
$$\|d_x f^n - d_y f^n\| \leq D a^n \|x - y\|^\beta.$$

Proof of the lemma. Let D' be such that
$$\|d_x f - d_y f\| \leq D' \|x - y\|^\beta.$$
Set $b = \max_{z \in M} \|d_z f\| \geq 1$ and observe that for every $x, y \in M$,
$$\|f^n(x) - f^n(y)\| \leq b^n \|x - y\|.$$
Fix $a > b$. Then the lemma holds true for $n = 1$ and any $D \geq D'$. For the inductive step we note that
$$\|d_x f^{n+1} - d_y f^{n+1}\| \leq \|d_{f^n(x)} f\| \cdot \|d_x f^n - d_y f^n\|$$
$$+ \|d_{f^n(x)} f - d_{f^n(y)} f\| \cdot \|d_y f^n\|$$
$$\leq b D a^n \|x - y\|^\beta + D' (b^n \|x - y\|)^\beta b^n$$
$$\leq D a^{n+1} \|x - y\|^\beta \left(\frac{b}{a} + \frac{D'}{D} \frac{(b^{1+\beta})^n}{a^{n+1}} \right).$$
If $a > b^{1+\beta}$, then for some $D \geq D'$ the factor in parentheses is ≤ 1. □

We proceed with the proof of the theorem. For $x \in M$, let $(T_x M)^\perp$ denote the orthogonal complement to the tangent plane $T_x M$ in $\mathbb{R}^N$. Since the distribution $(TM)^\perp$ is smooth, it is sufficient to prove that the distribution $F = E_1 \oplus (TM)^\perp$ is Hölder continuous. Since $E_1(x)$ and $E_2(x)$ are κ-transverse and of complementary dimensions in $T_x M$, there exists $d > 1$ such that $\|d_x f^n w\| \geq d^{-1} \mu^n \|w\|$ for every $x \in D$ and $w \perp E_1(x)$. For $x, y \in D$ and a positive integer n, let A_n and B_n be $N \times N$ matrices such that
$$A_n v = d_x f^n v \text{ if } v \in T_x M \quad \text{and} \quad A_n w = 0 \text{ if } w \in (T_x M)^\perp,$$
$$B_n v = d_y f^n v \text{ if } v \in T_y M \quad \text{and} \quad B_n w = 0 \text{ if } w \in (T_y M)^\perp.$$
By Lemma 5.23, $\|A_n - B_n\| \leq D a^n \|x - y\|^\beta$ and Theorem 5.20 follows from Lemma 5.21 with $\Delta = D \|x - y\|^\beta$, $E_A = F(x)$, $E_B = F(y)$, and $C = \max\{c, d\}$. □

Chapter 6

Lyapunov Stability Theory of Nonautonomous Equations

The stability theory of differential equations is centered around the problem of whether a given solution $x(t) = x(t, x_0) \in \mathbb{C}^p$ of the equation

$$\dot{x} = f(t, x) \tag{6.1}$$

is stable under small perturbations of either the initial condition $x_0 = x(0)$ or the function f (in the topology of the space of C^r functions, $r \geq 0$; it is assumed that the solution $x(t)$ is well-defined for all $t \geq 0$). The latter is important in applications since the function f is usually known only up to a given precision.

Stability under small perturbations of initial conditions means that every solution $y(t)$, whose initial condition $y(0)$ lies in the δ-ball around x_0, stays within the ε-neighborhood of the solution $x(t)$ for all sufficiently small ε and some $\delta = \delta(\varepsilon)$ chosen appropriately (in particular, it is assumed that the solution $y(t)$ is well-defined for all $t \geq 0$). One says that the stability is exponential if the solutions $y(t)$ approach $x(t)$ with an exponential rate and that it is uniformly exponential if the convergence is uniform over the initial condition. The uniform exponential stability survives under small perturbations of the right-hand side function f and, therefore, is "observable in practical situations".

In order to study the stability of a solution $x(t)$ of the nonlinear system (6.1) under small perturbations of the initial conditions, one linearizes this system along $x(t)$, i.e., considers the linear system of differential equations known as the system of *variational equations*:

$$\dot{v} = A(t)v, \text{ where } A(t) = \frac{df}{dt}(t, x(t)). \tag{6.2}$$

Moving back from the linear system (6.2) to the nonlinear system (6.1) can be viewed as solving the problem of stability under small perturbations of the right-hand side function in (6.2), thus reducing the study of stability under small perturbations of the initial condition to the study of stability under small perturbations of the right-hand side function.

To characterize stability of the zero solution $v(t) = 0$ of the linear system (6.2), one introduces the Lyapunov exponent associated with the system (6.2), which is a function of the initial condition v_0 (and, hence, of the corresponding solution $v(t) = v(t, v_0)$), given by

$$\chi(v_0) = \limsup_{t \to +\infty} \frac{1}{t} \log \|v(t, v_0)\|,$$

with the convention that $\log 0 = -\infty$.

One can show that if all values of the Lyapunov exponent are negative, then the zero solution is exponentially (albeit possibly nonuniformly) stable. If only some of these values are negative, then the zero solution is conditionally exponentially stable, i.e., it is stable along some directions in the space. These directions form a subspace called *stable*.

One of the main results in the classical stability theory asserts that if the zero solution $v(t) = 0$ of the linear system (6.2) is uniformly exponentially stable, then so is the solution $x(t)$ of the nonlinear system (6.1). In the case when the zero solution is conditionally exponentially stable, the solution $x(t)$ is stable along a local smooth submanifold—called *local stable manifold*— which is tangent to the stable subspace. If the solution $v(t) = 0$ is only (nonuniformly) exponentially stable, the situation becomes much more subtle and requires the additional assumption that the solution is *LP-regular*.

In this chapter we present some core results in the Lyapunov–Perron absolute and conditional stability theory.

6.1. Stability of solutions of ordinary differential equations

We first introduce the notions of stability and asymptotic stability for solutions of ordinary differential equations considering the general case of nonautonomous equations.

6.1. Stability of solutions of ordinary differential equations

Let $f\colon \mathbb{R} \times \mathbb{R}^n \to \mathbb{R}^n$ be a continuous function such that the differential equation
$$x' = f(t, x) \tag{6.3}$$
has unique solutions. For each pair $(t_0, x_0) \in \mathbb{R} \times \mathbb{R}^n$ let $x(t, t_0, x_0)$ be the solution of the initial value problem
$$\begin{cases} x' = f(t, x), \\ x(t_0) = x_0. \end{cases}$$
A solution $x(t, t_0, \bar{x}_0)$ of equation (6.3) defined for every $t > t_0$ is said to be *(Lyapunov) stable* if for each $\varepsilon > 0$ there exists $\delta > 0$ such that if $\|x_0 - \bar{x}_0\| < \delta$, then:

(1) $x(t, t_0, x_0)$ is defined for $t > t_0$;

(2) $\|x(t, t_0, x_0) - x(t, t_0, \bar{x}_0)\| < \varepsilon$ for $t > t_0$.

Otherwise, the solution $x(t, t_0, \bar{x}_0)$ is said to be *unstable*. We also introduce the stronger notion of asymptotic stability. A solution $x(t, t_0, \bar{x}_0)$ of equation (6.3) defined for every $t > t_0$ is said to be *asymptotically stable* if:

(1) $x(t, t_0, \bar{x}_0)$ is stable;

(2) there exists $\alpha > 0$ such that if $\|x_0 - \bar{x}_0\| < \alpha$, then
$$\|x(t, t_0, x_0) - x(t, t_0, \bar{x}_0)\| \to 0 \quad \text{when} \quad t \to +\infty.$$

An asymptotically stable solution $x(t, t_0, \bar{x}_0)$ is said to be *exponentially stable* if there exist $\beta, c, \lambda > 0$ such that if $\|x_0 - \bar{x}_0\| < \beta$, then
$$\|x(t, t_0, x_0) - x(t, t_0, \bar{x}_0)\| \le ce^{-\lambda t}, \quad t > t_0.$$

We note that in general a solution satisfying the second condition in the definition of asymptotic stability need not be stable.

Exercise 6.1. For the equation in polar coordinates
$$r' = r(1 - r), \quad \theta' = \sin^2(\theta/2),$$
show that the fixed point $(1, 0)$ satisfies the second condition in the notion of asymptotic stability but not the first one.

Now we consider an autonomous differential equation
$$x' = f(x), \tag{6.4}$$
where $f\colon \mathbb{R}^n \to \mathbb{R}^n$ is a continuous bounded function such that equation (6.4) has unique and global solutions. It induces a flow φ_t on $\mathbb{R}^n$ such that the solution of equation (6.4) satisfying the initial condition $x(0) = x_0$ is given by $x(t) = \varphi_t(x_0)$. The vector field f can be recovered from the equation
$$f(x) = \frac{\partial}{\partial t}\varphi_t(x)|_{t=0}.$$

Given a trajectory $\{\varphi_t(x) : t \in \mathbb{R}\}$ of a point $x \in M$, we consider the *system of variational equations*

$$v' = A(x,t)v, \tag{6.5}$$

where

$$A(x,t) = d_{\varphi_t(x)}f.$$

In general this is a nonautonomous linear differential equation. It turns out that the stability of a given trajectory of the flow φ_t can be described by studying small perturbations $g(t,v)$ of its system of variational equations (6.5). We thus consider the nonautonomous linear equation

$$v' = A(t)v \tag{6.6}$$

in $\mathbb{R}^n$, where $A(t)$ is an $n \times n$ matrix with real entries which depend continuously on $t \in \mathbb{R}$ and are bounded. For each $v_0 \in \mathbb{R}^n$ there exists a unique global solution $v(t)$ of equation (6.6) satisfying the initial condition $v(0) = v_0$. One can easily verify that for equation (6.6) a given solution is stable (respectively, asymptotically stable, unstable) if and only if all solutions are stable (respectively, asymptotically stable, unstable). Thus we only need to consider the trivial solution $v(t) = 0$.

Let $V(t)$ be a monodromy matrix, that is, an $n \times n$ matrix whose columns form a basis of the n-dimensional space of solutions of equation (6.6).

Exercise 6.2. Show that the trivial solution is:

(1) stable if and only if $\sup\{\|V(t)\| : t > 0\} < +\infty$;

(2) asymptotically stable if and only if

$$\|V(t)\| \to 0 \quad \text{when} \quad t \to +\infty;$$

(3) exponentially stable if and only if there exist $c, \lambda > 0$ such that

$$\|V(t)\| \le c e^{-\lambda t}, \quad t > 0,$$

or equivalently if

$$\limsup_{t \to +\infty} \frac{1}{t} \log \|V(t)\| < 0. \tag{6.7}$$

Let χ^+_{s+} be the largest value of the Lyapunov exponent χ^+ in (2.9). Condition (6.7) is equivalent to

$$\chi^+_{s+} < 0. \tag{6.8}$$

Exercise 6.3. Assume that the matrix function $A(t)$ is constant, that is, $A(t) = A$ for every $t \in \mathbb{R}$ and some $n \times n$ matrix A.

(1) Show that the trivial solution is stable if and only if A has no eigenvalues with positive real part and each eigenvalue with zero real part has a diagonal Jordan block (in the Jordan canonical form).

(2) Show that the trivial solution is asymptotically stable (in which case it is also exponentially stable) if and only if A has only eigenvalues with negative real part.

One can obtain a similar characterization of stability in the case when the matrix function $A(t)$ is periodic. Namely, assume that

$$A(t+T) = A(t), \quad t \in \mathbb{R},$$

for some constant $T > 0$ (this includes the constant case). In this case any monodromy matrix is of the form $P(t)E^{Bt}$ for some $n \times n$ matrices $P(t)$ and B with $P(t+T) = P(t)$ for $t \in \mathbb{R}$. The eigenvalues of the matrix B (which are well-defined $\mod(2\pi/T)$) are called *characteristic exponents*. Since the matrix $P(t)$ is periodic, the stability of solutions depends only on the characteristic exponents. Thus the trivial solution is stable if and only if there are no characteristic exponents with positive real part and for each characteristic exponent with zero real part the corresponding Jordan block of B is diagonal. Moreover, the trivial solution is asymptotically stable (in which case it is also exponentially stable) if and only if there are only characteristic exponents with negative real part.

We consider the problem of the stability under nonlinear perturbations. Under condition (6.8) any solution of equation (6.6) is exponentially stable.

Now we consider a nonlinear differential equation

$$u' = A(t)u + f(t, u), \tag{6.9}$$

which is a *perturbation* of equation (6.6). We assume that $f(t, 0) = 0$, and hence, $u(t) = 0$ is a solution of equation (6.9). We also assume that there exists a neighborhood U of 0 in $\mathbb{R}^n$ such that f is continuous on $[0, +\infty) \times U$ and for every $u, v \in U$ and $t \geq 0$,

$$\|f(t, u) - f(t, v)\| \leq K\|u - v\|(\|u\| + \|v\|)^{q-1} \tag{6.10}$$

for some constants $K > 0$ and $q > 1$. This means that the perturbation is sufficiently small in U. The number q is called the *order of the perturbation*.

One can ask whether condition (6.8) implies that the solution $u(t) = 0$ of the perturbed equation (6.9) is exponentially stable. Perron [86] showed that in general this may not be true.

Example 6.4. Consider the nonlinear system of differential equations in $\mathbb{R}^2$ given by
$$\begin{aligned} u_1' &= [-\omega - a(\sin\log t + \cos\log t)]u_1, \\ u_2' &= [-\omega + a(\sin\log t + \cos\log t)]u_2 + |u_1|^{\lambda+1}, \end{aligned} \quad (6.11)$$
for some positive constants ω, a, and λ. It is a perturbation of the system of linear equations
$$\begin{aligned} v_1' &= [-\omega - a(\sin\log t + \cos\log t)]v_1, \\ v_2' &= [-\omega + a(\sin\log t + \cos\log t)]v_2. \end{aligned} \quad (6.12)$$
We assume that
$$a < \omega < (2e^{-\pi}+1)a \quad \text{and} \quad 0 < \lambda < \frac{2a}{\omega-a} - e^\pi. \quad (6.13)$$
Notice that the perturbation $f(t,(u_1,u_2)) = (0,|u_1|^{\lambda+1})$ satisfies condition (6.10) with $q = \lambda+1 > 1$.

The general solution of (6.11) is given by
$$\begin{aligned} u_1(t) &= c_1 e^{-\omega t - at\sin\log t}, \\ u_2(t) &= c_2 e^{-\omega t + at\sin\log t} \\ &\quad + |c_1|^{\lambda+1} e^{-\omega t + at\sin\log t} \int_{t_0}^{t} e^{-(2+\lambda)a\tau\sin\log\tau - \omega\lambda\tau}\,d\tau, \end{aligned} \quad (6.14)$$
while the general solution of (6.12) is given by
$$v_1(t) = d_1 e^{-\omega t - at\sin\log t}, \quad v_2(t) = d_2 e^{-\omega t + at\sin\log t},$$
where c_1, c_2, d_1, d_2, and t_0 are arbitrary numbers.

Exercise 6.5. Show that the values of the Lyapunov exponent associated with (6.12) are $\chi_1^+ = \chi_2^+ = -\omega + a < 0$.

Let $u(t) = (u_1(t), u_2(t))$ be a solution of the system of nonlinear equations (6.11). In view of (6.14) it is also a solution of the system of linear equations
$$\begin{aligned} u_1' &= [-\omega - a(\sin\log t + \cos\log t)]u_1, \\ u_2' &= [-\omega + a(\sin\log t + \cos\log t)]u_2 + \delta(t)u_1, \end{aligned} \quad (6.15)$$
where
$$\delta(t) = \operatorname{sgn} c_1 |c_1|^\lambda e^{-\omega\lambda t - a\lambda t \sin\log t}.$$
We have
$$|\delta(t)| \le |c_1|^\lambda e^{(-\omega+a)\lambda t},$$
and hence, by (6.13), condition (2.8) holds for (6.15). Fix $0 < \varepsilon < \pi/4$ and for each $k \in \mathbb{N}$ set
$$t_k = e^{2k\pi - \frac{1}{2}\pi}, \quad t_k' = e^{2k\pi - \frac{1}{2}\pi - \varepsilon}.$$

Clearly, $0 < t'_k < t_k$ and $t_k \to \infty$ as $k \to \infty$. We have that
$$\int_{t_0}^{t_k} e^{-(2+\lambda)a\tau \sin\log\tau - \omega\lambda\tau}\, d\tau > \int_{t'_k}^{t_k} e^{-(2+\lambda)a\tau \sin\log\tau - \omega\lambda\tau}\, d\tau$$
and for every $\tau \in [t'_k, t_k]$,
$$2k\pi - \frac{\pi}{2} - \varepsilon \leq \log\tau \leq 2k\pi - \frac{\pi}{2},$$
$$(2+\lambda)a\tau\cos\varepsilon \leq -(2+\lambda)a\tau\sin\log\tau.$$
This implies that
$$\int_{t'_k}^{t_k} e^{-(2+\lambda)a\tau \sin\log\tau - \omega\lambda\tau}\, d\tau \geq \int_{t'_k}^{t_k} e^{(2+\lambda)a\tau\cos\varepsilon - \omega\lambda\tau}\, d\tau.$$
Set $r = (2+\lambda)a\cos\varepsilon - \omega\lambda$ and if necessary choose ε so small that $r > 0$. It follows that if $k \in \mathbb{N}$ is sufficiently large, then
$$\int_{t_0}^{t_k} e^{-(2+\lambda)a\tau \sin\log\tau - \omega\lambda\tau}\, d\tau > \int_{t'_k}^{t_k} e^{r\tau}\, d\tau > ce^{rt_k},$$
where $c = [1 - \exp(e^{-\varepsilon} - 1)]/r$. Set $t_k^* = t_k e^\pi = e^{2k\pi + \frac{1}{2}\pi}$. We obtain
$$e^{at_k^* \sin\log t_k^*} \int_{t_0}^{t_k^*} e^{-(2+\lambda)a\tau \sin\log\tau - \omega\lambda\tau}\, d\tau > e^{at_k^*} \int_{t_0}^{t_k} e^{-(2+\lambda)a\tau \sin\log\tau - \omega\lambda\tau}\, d\tau$$
$$> ce^{at_k^* + rt_k} = ce^{(a + re^{-\pi})t_k^*}.$$

It follows from (6.13) that if $c_1 \neq 0$ and ε is sufficiently small, then the Lyapunov exponent of any solution $u(t)$ of (6.11) (that is also a solution of (6.15)) satisfies
$$\chi^+(u) \geq -\omega + a + re^{-\pi} = -\omega + a + [(2+\lambda)a\cos\varepsilon - \omega\lambda]e^{-\pi} > 0.$$
Therefore, the solution $u(t)$ is not asymptotically stable. This completes the construction of the example.

Lyapunov introduced regularity conditions which guarantee the exponential stability of the solution $u(t) = 0$ of the perturbed equation (6.9). Although there are many different ways to state the regularity conditions, for a given differential equation the regularity is often difficult to verify.

6.2. Lyapunov absolute stability theorem

We now state the *Lyapunov Stability Theorem*. It claims that under an additional assumption known as *forward regularity*, condition (6.8) indeed implies the exponential stability of the trivial solution $u(t) = 0$ of equation (6.9). The theorem was essentially established by Lyapunov.

Theorem 6.6. *Assume that the matrix function $A(t)$ satisfies condition (2.8). Assume also that the Lyapunov exponent χ^+ associated with equation (2.7) is forward regular and satisfies condition (6.8). Then the solution $u(t) = 0$ of the perturbed equation (6.9) is exponentially stable.*

We present some more general results on exponential stability of equation (6.9) of which the above theorem is an immediate corollary. Consider the perturbed differential equation (6.9) and assume that the function $f(t, u)$ satisfies (6.10). Let $V(t)$ be a monodromy matrix (the fundamental solution matrix) for which $V(0) = \mathrm{Id}$. Denote by $\mathcal{V}(t, s)$ the Cauchy matrix of equation (2.7) defined by $\mathcal{V}(t, s) = V(t)V(s)^{-1}$.

Theorem 6.7. *Assume that there are two continuous functions R and r on $[0, +\infty)$ such that for every $t \geq 0$,*

$$\int_0^t R(\tau)\,d\tau \leq D_1 < +\infty, \tag{6.16}$$

$$\int_0^\infty e^{(q-1)\int_0^s (R+r)(\tau)\,d\tau}\,ds = D_2 < +\infty, \tag{6.17}$$

and for every $0 \leq s \leq t$,

$$\|\mathcal{V}(t, s)\| \leq D_3 e^{\int_s^t R(\tau)\,d\tau + (q-1)\int_0^s r(\tau)\,d\tau}. \tag{6.18}$$

Then there exist $D > 0$ and $\delta > 0$ such that for every $u_0 \in \mathbb{R}^n$ with $\|u_0\| < \delta$, there is a unique solution u of equation (6.9) which satisfies:

(1) *u is well-defined on the interval $[0, \infty)$, and $u(0) = u_0$;*
(2) *for every t,*

$$\|u(t)\| \leq \|u_0\| D e^{\int_0^t R(\tau)\,d\tau}. \tag{6.19}$$

Proof. Equation (6.9) is equivalent to the integral equation

$$u(t) = \mathcal{V}(t, 0)u_0 + \int_0^t \mathcal{V}(t, s) f(s, u(s))\,ds. \tag{6.20}$$

Moreover, by (6.10), for every $u_0 \in \mathbb{R}^n$ with $u_0 \in U$ there exists a unique solution $u(t)$ satisfying the initial condition $u(0) = u_0$.

We consider the linear space $\mathcal{B}_C$ of $\mathbb{R}^n$-valued continuous functions $x = x(t)$ on the interval $[0, \infty)$ such that

$$\|x(t)\| \leq C e^{\int_0^t R(\tau)\,d\tau}$$

for every $t \geq 0$ where $C > 0$ is a constant. We endow this space with the norm

$$\|x\|_R = \sup\{\|x(t)\| e^{-\int_0^t R(\tau)\,d\tau} : t \geq 0\}.$$

6.2. Lyapunov absolute stability theorem

It follows from (6.16) that $\|x(t)\| \leq \|x\|_R e^{D_1}$. Therefore, if $\|x\|_R$ is sufficiently small, then $x(t) \in U$ for every $t \geq 0$. This means that the operator

$$(Jx)(t) = \int_0^t \mathcal{V}(t,s) f(s, x(s)) \, ds$$

is well-defined on $\mathcal{B}_C$. Let

$$L = K(\|x_1\|_R + \|x_2\|_R)^{q-1} \|x_1 - x_2\|_R.$$

By (6.10), (6.17), and (6.18), we have

$$\|(Jx_1)(t) - (Jx_2)(t)\| \leq \int_0^t D_3 e^{\int_s^t R(\tau) \, d\tau + (q-1) \int_0^s r(\tau) \, d\tau} L e^{q \int_0^s R(\tau) \, d\tau} \, ds$$

$$= D_3 L e^{\int_0^t R(\tau) \, d\tau} \int_0^t e^{(q-1) \int_0^s (R+r)(\tau) \, d\tau} \, ds$$

$$\leq D_3 D_2 L e^{\int_0^t R(\tau) \, d\tau}.$$

It follows that

$$\|Jx_1 - Jx_2\|_R \leq D_3 D_2 K (\|x_1\|_R + \|x_2\|_R)^{q-1} \|x_1 - x_2\|_R.$$

Now choose $\varepsilon > 0$ such that

$$\theta := D_3 D_2 K (2\varepsilon)^{q-1} < 1. \tag{6.21}$$

Whenever $x_1, x_2 \in \mathcal{B}_\varepsilon$, we have

$$\|Jx_1 - Jx_2\|_R \leq \theta \|x_1 - x_2\|_R. \tag{6.22}$$

Choose a point $u_0 \in \mathbb{R}^n$ and set $\xi(t) = \mathcal{V}(t, 0) u_0$. It follows from (6.18) that

$$\|\xi(t)\| \leq \|u_0\| D_3 e^{\int_0^t R(\tau) \, d\tau},$$

and hence, $\|\xi\|_R \leq \|u_0\| D_3$. Consider the operator $\tilde{J}$ defined by

$$(\tilde{J} x)(t) = \xi(t) + (Jx)(t)$$

on the space $\mathcal{B}_\varepsilon$. Note that $\mathcal{B}_\varepsilon$ is a complete metric space with the distance $\rho(x, y) = \|x - y\|_R$. We have

$$\|\tilde{J} x\|_R \leq D_3 \|u_0\| + \theta \|x\|_R < D_3 \delta + \theta \varepsilon < \varepsilon \tag{6.23}$$

provided that δ is sufficiently small. Therefore, $\tilde{J}(\mathcal{B}_\varepsilon) \subset \mathcal{B}_\varepsilon$. By (6.21) and (6.22), the operator $\tilde{J}$ is a contraction, and hence, there exists a unique function $u \in \mathcal{B}_\varepsilon$ which is a solution of (6.20). Inequality (6.19) now follows readily from (6.23). □

An immediate consequence of this theorem is the following statement due to Malkin (see [**79**]).

Theorem 6.8. *Assume that the Cauchy matrix of* (2.7) *admits the estimate*

$$\|\mathcal{V}(t,s)\| \le De^{\alpha(t-s)+\beta s}, \qquad (6.24)$$

for any $0 \le s \le t$ *and some constants* α, β, D, *such that*

$$(q-1)\alpha + \beta < 0. \qquad (6.25)$$

Then the trivial solution of equation (6.9) *is exponentially stable.*

Proof. Set

$$R(t) = \alpha \quad \text{and} \quad r(t) = \frac{\beta}{q-1}.$$

It follows from (6.24) and (6.25) that conditions (6.17) and (6.18) hold. In order to verify condition (6.16), note that by setting $t = s$ in (6.18) we obtain

$$1 = \|\mathcal{V}(t,t)\| \le D_3 e^{(q-1)\int_0^t r(\tau)\,d\tau}.$$

It follows that for every t,

$$\frac{\beta t}{q-1} = \int_0^t r(\tau)\,d\tau \ge -C, \qquad (6.26)$$

where $C \ge 0$ is a constant. Therefore, by (6.25),

$$\alpha t = \int_0^t R(\tau)\,d\tau \le C + \int_0^t (R(\tau)+r(\tau))\,d\tau \le C \qquad (6.27)$$

for all t. This implies condition (6.16). Therefore, by (6.19), any solution u of (6.9) satisfies

$$\|u(t)\| \le \|u_0\|De^{\alpha t} \le \|u_0\|De^C. \qquad (6.28)$$

This shows that the trivial solution of (6.9) is asymptotically stable. By (6.26) and (6.27), we have $\beta \ge 0$ and $\alpha \le 0$. In view of (6.25), $\alpha < 0$, and in view of (6.28), the solution $u(t) = 0$ is exponentially stable. The desired result follows. □

Another important consequence of Theorem 6.7 is the following statement due to Lyapunov (see [78]). It shows that exponential stability may hold even for some nonregular systems, provided that the irregularity coefficient is sufficiently small.

Theorem 6.9. *Assume that the maximal value* $\chi_{\max}$ *of the Lyapunov exponent of* (2.7) *is strictly negative (see* (6.8)*) and that*

$$(q-1)\chi_{\max} + \gamma < 0,$$

where γ *is the irregularity coefficient. Then the trivial solution of equation* (6.9) *is exponentially stable.*

Proof. The proof is an elaboration for continuous time of the argument in the proof of Lemma 5.7. Consider a subordinate basis $(v_1(t), \ldots, v_n(t))$ of the space of solutions of (2.7) such that the numbers

$$\chi'_1 \leq \cdots \leq \chi'_n = \chi_{\max}$$

are the values of the Lyapunov exponent χ^+ of $v_1(t), \ldots, v_n(t)$. We may assume that the matrix $V(t)$ whose columns are $v_1(t), \ldots, v_n(t)$ is such that the columns $w_1(t), \ldots, w_n(t)$ of the matrix $W(t) = [V(t)^*]^{-1}$ form a subordinate basis of the space of solutions of the dual equation $\dot{w} = -A(t)^* w$.

Let $\mu_1, \ldots, \mu_n$ be the values of the Lyapunov exponent $\tilde{\chi}^+$ of the vectors $w_1(t), \ldots, w_n(t)$. For every $\varepsilon > 0$ there exists $D_\varepsilon > 0$ such that

$$\|v_j(t)\| \leq D_\varepsilon e^{(\chi'_j + \varepsilon)t} \quad \text{and} \quad \|w_j(t)\| \leq D_\varepsilon e^{(\mu_j + \varepsilon)t}$$

for every t. Note that there is a pair of dual bases $\mathbf{v}(t) = (v_1(t), \ldots, v_n(t))$ and $\mathbf{w}(t) = (w_1(t), \ldots, w_n(t))$ for which

$$\gamma = \gamma(\chi, \chi^*) = \max\{\chi(v_j) + \chi^*(w_j) : j = 1, \ldots, n\}.$$

Therefore,

$$\gamma = \max\{\chi'_j + \mu_j : j = 1, \ldots, n\}.$$

It follows that $\chi'_j + \mu_j \leq \gamma$ for every $j = 1, \ldots, n$. Consider the Cauchy matrix

$$\mathcal{V}(t, s) = V(t) V(s)^{-1} = V(t) W(s)^*.$$

Its entries are

$$v_{ik}(t, s) = \sum_{j=1}^{n} v_{ij}(t) \overline{w_{kj}(s)},$$

where $v_{ij}(t)$ is the ith coordinate of the vector $v_j(t)$ and $w_{kj}(s)$ is the kth coordinate of the vector $w_j(s)$. It follows that

$$|v_{ik}(t, s)| \leq \sum_{j=1}^{n} |v_{ij}(t)| \cdot |\overline{w_{kj}(s)}| \leq \sum_{j=1}^{n} \|v_j(t)\| \cdot \|w_j(s)\|$$

$$\leq n D_\varepsilon^2 e^{(\chi'_j + \varepsilon)t + (\mu_j + \varepsilon)s} = n D_\varepsilon^2 e^{(\chi'_j + \varepsilon)(t-s) + (\chi'_j + \mu_j + 2\varepsilon)s}.$$

Therefore, there exists $D > 0$ such that

$$\|\mathcal{V}(t, s)\| \leq D e^{(\chi_{\max} + \varepsilon)(t-s) + (\gamma + 2\varepsilon)s}. \tag{6.29}$$

Since ε can be chosen arbitrarily small, the desired result follows from Theorem 6.8 by setting $\alpha = \chi_{\max} + \varepsilon$ and $\beta = \gamma + 2\varepsilon$. $\square$

We conclude by observing that the Lyapunov Stability Theorem (see Theorem 6.6) is an immediate corollary of Theorem 6.9. Moreover, if χ^+ is forward regular, setting $\gamma = 0$ in (6.29), we obtain that for every solution $u(t)$ of equation (6.9) and every $s \in \mathbb{R}$,

$$\|u(t)\| \leq De^{(\chi_{\max}+\varepsilon)(t-s)+2\varepsilon s}\|u(s)\| = De^{2\varepsilon s}e^{(\chi_{\max}+\varepsilon)(t-s)}\|u(s)\|. \quad (6.30)$$

In other words the contraction constant $De^{2\varepsilon s}$ may deteriorate along the solution of equation (6.9). This implies that the "size" of the neighborhood at time s, where the exponential stability of the trivial solution is guaranteed, may decay with "small" exponential rate. Note that the same conclusion holds even if the Lyapunov exponent χ^+ is not forward regular but its irregularity coefficient $\gamma \leq \varepsilon$. One can view the estimate (6.30) as a continuous-time analog of the first inequality in statement (L7) of Theorem 5.4. In the next chapter we will observe the crucial role it plays in constructing local stable and unstable manifolds in the nonuniform hyperbolicity theory.

6.3. Lyapunov conditional stability theorem

In this section we discuss the concept of conditional stability that is a foundation for hyperbolicity theory and we present a version of Theorem 6.6 that deals with conditional stability.

A solution $x(t, t_0, \bar{x}_0)$ of equation (6.3) defined for every $t > t_0$ is said to be *(conditionally) stable* along a set $V \subset \mathbb{R}^n$ if for each $\varepsilon > 0$ there exists $\delta > 0$ such that if $x_0, \bar{x}_0 \in V$ and $\|x_0 - \bar{x}_0\| < \delta$, then:

(1) $x(t, t_0, x_0)$ is defined for $t > t_0$;
(2) $\|x(t, t_0, x_0) - x(t, t_0, \bar{x}_0)\| < \varepsilon$ for $t > t_0$.

A solution $x(t, t_0, \bar{x}_0)$ of equation (6.3) defined for every $t > t_0$ is said to be *(conditionally) asymptotically stable* along a set $V \subset \mathbb{R}^n$ if:

(1) $x(t, t_0, \bar{x}_0)$ is stable along V;
(2) there exists $\alpha > 0$ such that if $x_0, \bar{x}_0 \in V$ and $\|x_0 - \bar{x}_0\| < \alpha$, then

$$\|x(t, t_0, x_0) - x(t, t_0, \bar{x}_0)\| \to 0 \quad \text{when} \quad t \to +\infty.$$

An asymptotically stable solution $x(t, t_0, \bar{x}_0)$ along V is said to be *(conditionally) exponentially stable* along V if there exist $\beta, c, \lambda > 0$ such that if $x_0, \bar{x}_0 \in V$ and $\|x_0 - \bar{x}_0\| < \beta$, then

$$\|x(t, t_0, x_0) - x(t, t_0, \bar{x}_0)\| \leq ce^{-\lambda t}, \quad t > t_0.$$

The classical Hadamard–Perron Theorem claims that a hyperbolic fixed point is exponentially stable along its stable manifold.

6.3. Lyapunov conditional stability theorem

We now describe a generalization of the Lyapunov Stability Theorem when *not all* Lyapunov exponents are negative. Namely, let $\chi_1^+ < \cdots < \chi_s^+$ be the distinct values of the Lyapunov exponent χ^+ for equation (2.7). Assume that

$$\chi_k^+ < 0, \tag{6.31}$$

where $1 \le k < s$ and the number χ_{k+1}^+ can be negative, positive, or equal to 0 (compare to (6.8)).

We recall that a *local smooth submanifold* of $\mathbb{R}^n$ is the graph of a smooth function $\psi \colon B \to \mathbb{R}^n$, where B is the open unit ball in $\mathbb{R}^k$ for $k \le n$.

Theorem 6.10. *Assume that the Lyapunov exponent χ^+ for equation (2.7) with the matrix function $A(t)$ satisfying (2.8) is forward regular and satisfies condition (6.31). Then there exists a local smooth submanifold $V^s \subset \mathbb{R}^n$ which passes through 0, is tangent at 0 to the linear subspace V_k (defined by (2.1)), and is such that the trivial solution of (6.9) is exponentially stable along V^s.*

This statement is an extension of the classical Hadamard–Perron Theorem to the case of conditional stability and is a particular case of a general stable manifold theorem that we discuss in Section 7.1.

Following [16], we give an outline of the proof of the theorem restricting ourselves to showing the existence of *Lipschitz* stable manifolds along which the trivial solution is exponentially stable. The stable manifolds are obtained as graphs of Lipschitz functions. Let

$$E(t) = V(t)\bigoplus_{i=1}^{k} E_i \quad \text{and} \quad F(t) = V(t)\bigoplus_{i=k+1}^{s} E_i,$$

where $V(t)$ is a monodromy matrix of equation (6.6). Take $\varepsilon > 0$ such that

$$\chi_k + \alpha + \varepsilon < 0 \quad \text{and} \quad \chi_k + 2\varepsilon < \chi_{k+1},$$

where $\alpha = 2\varepsilon(1 + 2/q)$. Given $\delta > 0$, consider the set

$$X_\alpha = \{(s,\xi) : s \ge 0 \text{ and } \xi \in B(\delta e^{-\alpha s})\},$$

where $B(\delta)$ is the open ball in $E(s)$ of radius $\delta > 0$ centered at zero. Let $\mathfrak{X}_\alpha$ be the space of continuous functions $\varphi \colon X_\alpha \to \mathbb{R}^n$ such that for each $s \ge 0$,

$$\varphi(s, B(\delta e^{-\alpha s})) \subset F(s), \quad \varphi(s,0) = 0,$$

and

$$\|\varphi(s,x) - \varphi(s,y)\| \le \|x - y\| \quad \text{for } x, y \in B(\delta e^{-\alpha s}).$$

The graph of a function $\varphi \in \mathfrak{X}_\alpha$ is given by

$$\mathcal{V} = \{(s, \xi, \varphi(s,\xi)) : (s,\xi) \in X_\alpha\} \subset \mathbb{R}_0^+ \times \mathbb{R}^n. \tag{6.32}$$

We will show that there exists $\varphi \in \mathfrak{X}_\alpha$ such that for every initial condition $(s,\xi) \in X_\beta \subset X_\alpha$, with $\beta = \alpha + 2\varepsilon$, the corresponding solution is entirely contained in $\mathcal{V}$. In order for $\mathcal{V}$ to be forward invariant, any solution with initial condition in $\mathcal{V}$ must remain in $\mathcal{V}$ for all time and thus must be of the form
$$(x(t), y(t)) = (x(t), \varphi(t, x(t)))$$
with components in $E(t)$ and $F(t)$, respectively. Equation (6.9) is then equivalent to
$$x(t) = U(t,s)x(s) + \int_s^t U(t,\tau)f(\tau, x(\tau), \varphi(\tau, x(\tau)))\, d\tau, \qquad (6.33)$$
$$\varphi(t, x(t)) = V(t,s)\varphi(s, x(s)) + \int_s^t V(t,\tau)f(\tau, x(\tau), \varphi(\tau, x(\tau)))\, d\tau, \quad (6.34)$$
where
$$U(t,s) = V(t)V(s)^{-1}|E(s) \quad \text{and} \quad V(t,s) = V(t)V(s)^{-1}|F(s).$$
Let
$$X_\alpha^* = \{(s,\xi) : s \geq 0 \text{ and } \xi \in E(s)\}.$$
Consider the space $\mathfrak{X}_\alpha^*$ of continuous functions $\varphi \colon X_\alpha^* \to \mathbb{R}^n$ for which $\varphi|X_\alpha \in \mathfrak{X}_\alpha$ and
$$\varphi(s,\xi) = \varphi(s, \delta e^{-\alpha s}\xi/\|\xi\|), \quad s \geq 0,\ \xi \notin B(\delta e^{-\alpha s}).$$
Arguing similarly as in the proof of Theorem 6.7, one can show that for each $\varphi \in \mathfrak{X}_\alpha^*$ there exists a unique function $x(t) = x_\varphi(t)$ satisfying (6.33).

Lemma 6.11. *There exists $R > 0$ such that for all sufficiently small $\delta > 0$, $\varphi \in \mathfrak{X}_\alpha^*$, and $(s,\xi) \in X_\alpha$, one can find a unique continuous function $x = x_\varphi \colon [s, +\infty) \to \mathbb{R}^n$ satisfying:*

(1) *$x_\varphi(s) = \xi$, $x_\varphi(t) \in E(t)$, and (6.33) holds for $t \geq s$;*
(2) *$\|x_\varphi(t)\| \leq Re^{(\chi_k + \varepsilon)(t-s) + 2\varepsilon s}\|\xi\|$, $t \geq s$.*

It remains to find a function φ satisfying identity (6.34) with $x = x_\varphi$. We first observe that this identity can be reduced to the identity
$$\varphi(s,\xi) = -\int_s^\infty V(\tau,s)^{-1} f(\tau, x_\varphi(\tau), \varphi(\tau, x_\varphi(\tau)))\, d\tau \qquad (6.35)$$
for every $(s,\xi) \in X_\alpha$. This means that we need to find a fixed point of the operator Φ defined for each $\varphi \in \mathfrak{X}_\alpha^*$ by
$$(\Phi\varphi)(s,\xi) = -\int_s^\infty V(\tau,s)^{-1} f(\tau, x_\varphi(\tau), \varphi(\tau, x_\varphi(\tau)))\, d\tau$$

6.3. Lyapunov conditional stability theorem

with $(s,\xi) \in X_\alpha$. Here x_φ is the unique continuous function given by Lemma 6.11 if $x_\varphi(s) = \xi$ and by

$$(\Phi\varphi)(s,\xi) = (\Phi\varphi)(s, \delta e^{-\alpha s}\xi/\|\xi\|)$$

otherwise. One can show that Φ is a contraction in the space $\mathcal{X}_\alpha^*$, and thus, given a sufficiently small $\delta > 0$, there exists a unique function $\varphi \in \mathcal{X}_\alpha^*$ such that (6.35) holds for every $(s,\xi) \in X_\alpha$. The desired Lipschitz stable manifold $\mathcal{V}$ in (6.32) is the restriction of this function φ to X_α.

Chapter 7

Local Manifold Theory

One of the goals of hyperbolicity theory is to describe the behavior of trajectories near a hyperbolic trajectory. In the case of uniform hyperbolicity such a description is provided by the classical Hadamard–Perron theorem, which effectively reduces the study of a nonlinear system in a small neighborhood of a hyperbolic trajectory to the study of the corresponding system of variational equations (in the continuous-time case) or the study of the action of the differential (in the discrete-time case) along the hyperbolic trajectory. A crucial manifestation of the Hadamard–Perron theorem is a construction of a local stable (and unstable) manifold which consists of all points whose trajectories start near the hyperbolic one and approach it with an exponential rate.

Moving to the case of the nonuniform hyperbolicity, one can generalize the Hadamard–Perron theorem and construct local stable (and unstable) manifolds for trajectories which are Lyapunov–Perron regular. However, local manifolds depend "wildly" on the base points: their sizes are measurable functions of the base point (that in general are not continuous) and may decrease along the trajectory with a subexponential rate. In this chapter we carry out a detailed construction of local stable (and unstable) manifolds for nonuniformly hyperbolic systems by substantially generalizing the Perron method in the stability theory of nonlinear differential equations. In particular, we obtain local stable manifolds in the setting of nonlinear maps of Banach spaces. We also describe some basic properties of local manifolds.

We stress that constructing local stable and unstable manifolds requires a higher regularity of the system, i.e., that f is of class $C^{1+\alpha}$. This reflects a principal difference between uniform and nonuniform hyperbolicity—while the local manifold theory of Anosov diffeomorphisms can be developed under the assumption that the systems are of class C^1, the corresponding results for nonuniformly hyperbolic C^1 diffeomorphisms fail to be true. See Chapter 16 for some "pathological" phenomena in the C^1 topology.

7.1. Local stable manifolds

Let f be a $C^{1+\alpha}$ diffeomorphism of a compact smooth Riemannian manifold M. Recall that a *local smooth submanifold* of M is the graph of a smooth injective function $\psi\colon B \to M$, where B is the open unit ball in $\mathbb{R}^k$ for $k \le \dim M$.

Consider a nonuniformly completely hyperbolic f-invariant set Λ and let $\{\Lambda_\iota\}$ be the corresponding collection of level sets (see (5.28)) where $\iota = \{\lambda, \mu, \varepsilon, j\}$ is the index set, the numbers λ, μ, ε satisfy (5.24), and $1 \le j < p$ (recall that $j = E^s(x)$ for every $x \in \Lambda_\iota$ and $p = \dim M$). In what follows, we fix an index set ι and assume that the level set Λ_ι is not empty. We stress that we allow the case when this set consists of a single trajectory.

7.1.1. Stable Manifold Theorem. In this section we prove the following theorem, which is one of the key results in hyperbolicity theory.

Theorem 7.1 (Stable Manifold Theorem). *For every $x \in \Lambda_\iota$ there exists a local stable manifold $V^s(x)$ such that $x \in V^s(x)$, $T_x V^s(x) = E^s(x)$, and for every $y \in V^s(x)$ and $n \ge 0$,*

$$\rho(f^n(x), f^n(y)) \le T(x)(\lambda')^n \rho(x,y), \tag{7.1}$$

where ρ is the distance in M induced by the Riemannian metric, λ' is a number satisfying $0 < \lambda e^\varepsilon < \lambda' < 1$, and $T\colon \Lambda_\iota \to (0, \infty)$ is a Borel function satisfying

$$T(f^m(x)) \le T(x) e^{10\varepsilon|m|}, \quad m \in \mathbb{Z}. \tag{7.2}$$

Furthermore, $f^m(V^s(x)) \subset V^s(f^m(x))$ for every $m \in \mathbb{Z}$.

Inequality (7.2) illustrates the crucial fact that while the estimate (7.1) may deteriorate along the trajectory $\{f^n(x)\}$, this only happens with a sufficiently small exponential rate. This fact is essentially due to condition (5.19).

The proof of the Stable Manifold Theorem (see Theorem 7.1) presented below is due to Pesin [**88**] and is an elaboration of Perron's approach [**86**].

7.1.2. The functional equation. We obtain the local stable manifold in the form

$$V^s(x) = \exp_x\{(v, \psi_x^s(v)) : v \in B^s(r)\}, \tag{7.3}$$

where $\psi_x^s\colon B^s(r) \to E^u(x)$ is a smooth map satisfying

$$\psi_x^s(0) = 0 \quad \text{and} \quad d_0 \psi_x^s = 0.$$

7.1. Local stable manifolds

Here $B^s(r) \subset E^s(x)$ is the ball of radius r centered at the origin; $r = r(x)$ is called the *size* of the local stable manifold. We now describe how to construct the function ψ_x^s. Fix $x \in \Lambda_\iota$ and consider the map

$$\tilde{f}_x = \exp_{f(x)}^{-1} \circ f \circ \exp_x \colon B^s(r) \times B^u(r) \to T_{f(x)}M, \tag{7.4}$$

which is well-defined if r is sufficiently small.[1] Here $B^u(r)$ is the ball of radius r in $E^u(x)$ centered at the origin. Since the stable and unstable subspaces are invariant under the action of the differential df (see condition (5.20)), the map $\tilde{f}$ can be written in the following form:

$$\tilde{f}_x(v) = (A_x^1 v_1 + g_x^1(v), A_x^2 v_2 + g_x^2(v)), \tag{7.5}$$

where $v = (v_1, v_2)$, $v_1 \in B^s(x)$, $v_2 \in B^u(x)$,

$$A_x^1 \colon E^s(x) \to E^s(f(x)), \quad A_x^2 \colon E^u(x) \to E^u(f(x))$$

are linear maps, and

$$g_x^1 \colon B^s(r) \times B^u(r) \to E^s(f(x)), \quad g_x^2 \colon B^s(r) \times B^u(r) \to E^u(f(x))$$

are nonlinear maps. In view of conditions (5.21), (5.22), and (5.23), the map A_x^1 acts along the trajectory $\{f^n(x)\}$ as a contraction while the map A_x^2 acts as an expansion. Moreover, for every $x \in \Lambda_\iota$ the maps g_x^i, $i = 1, 2$, satisfy: $g_x^i(0) = 0$, $d_0 g_x^i = 0$, and for any $v, w \in B^s(r) \times B^u(r)$, $v = (v_1, v_2)$, $w = (w_1, w_2)$,[2]

$$\|d_v g_x^i - d_w g_x^i\|_x \leq C(\|v - w\|_x)^\alpha, \tag{7.6}$$

where $C > 0$ is a constant independent of x and

$$\|v - w\|_x = \|v_1 - w_1\|_x + \|v_2 - w_2\|_x.$$

Note that in (7.6) we use the fact that the map f is of class $C^{1+\alpha}$, and hence, for every $x \in \Lambda_\iota$, so is the map $\tilde{f}_x$.

In other words, the map $\tilde{f}_x$ can be viewed as a small perturbation of the linear map $(v_1, v_2) \mapsto (A_x^1 v_1, A_x^2 v_2)$ by the map $(g_x^1(v_1, v_2), g_x^2(v_1, v_2))$ satisfying condition (7.6), which is analogous to condition (6.10).

The local stable manifold must be invariant under the action of the map f. In other words, one should effectively construct a collection of local stable manifolds $V^s(f^m(x))$ along the trajectory of x in such a way that

$$f(V^s(f^m(x))) \subset V^s(f^{m+1}(x)). \tag{7.7}$$

---

[1] The number $r = r(x)$ should be less than the radius of injectivity of the exponential map $\exp_x$ at the point x.

[2] Given a point $x \in \Lambda_\iota$ and a linear map $L_x \colon T_x M \to T_{f(x)} M$, the norm $\|L_x\|_x$ is defined by $\|L_x\|_x = \sup_{v \in T_x M \setminus \{0\}} \{\|L_x v\|_{f(x)} / \|v\|_x\}$.

Exercise 7.2. Using the relations (7.3) and (7.5), show that (7.7) yields the following relation on the collection of functions $\psi^s_{f^m(x)}$: for any $y \in B^s(r)$

$$\psi^s_{f^{m+1}(x)}(A^1_{f^m(x)}y + g^1_{f^m(x)}(y, \psi^s_{f^m(x)}(y))) \\ = A^2_{f^m(x)}\psi^s_{f^m(x)}(y) + g^2_{f^m(x)}(y, \psi^s_{f^m(x)}(y)). \tag{7.8}$$

This means that constructing local stable manifolds $V^s(f^m(x))$ boils down to solving the functional equation (7.8) for the collection of locally defined functions $\psi^s_{f^m(x)}$ along the trajectory of x.

To construct local stable manifolds, one can now use either the Hadamard method or the Perron method, which are well known in the uniform hyperbolicity theory. Note that the "perturbation map" $\tilde{f}_x$ (see (7.4) and (7.5)) satisfies condition (7.6) in a neighborhood U_x of the point x whose size depends on x. Moreover, the size of U_x decays along the trajectory of x with small exponential rate (see (5.31)). This requires a substantial modification of the classical Hadamard–Perron approach.

7.1.3. The family of local nonlinear maps. Fix $x \in \Lambda_\iota$ and consider the positive semitrajectory $\{f^m(x)\}_{m\geq 0}$ and the family of maps

$$\{\tilde{F}_m = \tilde{f}_{f^m(x)}\}_{m\geq 0}.$$

We identify the tangent spaces $T_{f^m(x)}M$ with $\mathbb{R}^p = \mathbb{R}^k \times \mathbb{R}^{p-k}$ (recall that $p = \dim M$ and $1 \leq k < p$) via an isomorphism τ_m such that

$$\tau_m(E^s(f^m(x))) = \mathbb{R}^k \quad \text{and} \quad \tau_m(E^u(f^m(x))) = \mathbb{R}^{p-k}.$$

In view of (7.5) the map $F_m = \tau_{m+1} \circ \tilde{F}_m \circ \tau_m^{-1}$ is of the form

$$F_m(v) = (A^1_m v_1 + g^1_m(v), A^2_m v_2 + g^2_m(v)), \tag{7.9}$$

where $v = (v_1, v_2)$, $v_1 \in \mathbb{R}^k$, $v_2 \in \mathbb{R}^{p-k}$,

$$A^1_m \colon \mathbb{R}^k \to \mathbb{R}^k \quad \text{and} \quad A^2_m \colon \mathbb{R}^{p-k} \to \mathbb{R}^{p-k}$$

are linear maps, and

$$g^1_m \colon B^s(r_0) \times B^u(r_0) \to \mathbb{R}^k \quad \text{and} \quad g^2_m \colon B^s(r_0) \times B^u(r_0) \to \mathbb{R}^{p-k}$$

are nonlinear maps (here $B^s(r_0)$ and $B^u(r_0)$ are balls centered at 0 of radius r_0). Furthermore, we will choose the maps τ_m such that for every $v \in T_{f^m(x)}M$

$$\|v\|'_{f^m(x)} = \|\tau_m v\|,$$

where $\|\cdot\|'$ is the weak Lyapunov norm (see Section 5.3.3 for the definition of this norm.) Therefore, in view of Exercise 5.14, we have

$$\|A^1_m\| \leq \lambda' \quad \text{and} \quad \|(A^2_m)^{-1}\| \leq (\mu')^{-1}, \tag{7.10}$$

7.1. Local stable manifolds

where the numbers λ' and μ' satisfy (5.30). Moreover, by (7.6), for $i = 1, 2$

$$g_m^i(0) = 0, \quad d_0 g_m^i = 0, \tag{7.11}$$

and

$$\|d_v g_m^i - d_w g_m^i\| \le C\gamma^{-m}\|v - w\|^\alpha, \tag{7.12}$$

where $0 < \alpha \le 1$, $C > 0$, and the number γ is chosen such that

$$(\lambda')^\alpha < \gamma < 1. \tag{7.13}$$

Here $v = (v_1, v_2)$, $w = (w_1, w_2)$ with $v_1, w_1 \in \mathbb{R}^k$ and $v_2, w_2 \in \mathbb{R}^{p-k}$.

For $m \ge 1$ set

$$\mathcal{F}_m = F_{m-1} \circ \cdots \circ F_0 \tag{7.14}$$

and

$$\prod_{j=0}^{m-1} A_j^i = A_{m-1}^i \circ \cdots \circ A_0^i.$$

Exercise 7.3. Writing

$$\mathcal{F}_m(v) = \bigl(\mathcal{F}_m^1(v), \mathcal{F}_m^2(v)\bigr), \ v = (v_1, v_2),$$

show that for $m \ge 1$

$$\mathcal{F}_m^1(v) = \left(\prod_{i=0}^{m-1} A_i^1\right) v_1 + \sum_{n=0}^{m-1} \left(\prod_{i=n+1}^{m-1} A_i^1\right) g_n^1(\mathcal{F}_n(v)) \tag{7.15}$$

and for $m < 0$

$$\mathcal{F}_m^2(v) = -\sum_{n=0}^{\infty} \left(\prod_{i=0}^{n} A_{i+m}^2\right)^{-1} g_{n+m}^2(\mathcal{F}_{m+n}(v)). \tag{7.16}$$

Hint: Observe that $\mathcal{F}_{m+1} = F_m \circ \mathcal{F}_m$, and hence,

$$\mathcal{F}_{m+1}^1(v) = A_m^1 \mathcal{F}_m^1(v) + g_m^1(\mathcal{F}_m(v)),$$

$$\mathcal{F}_{m+1}^2(v) = A_m^2 \mathcal{F}_m^2(v) + g_m^2(\mathcal{F}_m(v)).$$

To show the relation (7.15), iterate the first equality "forward". To obtain (7.16), rewrite the second equality in the form

$$\mathcal{F}_m^2(v) = (A^2)_m^{-1} \mathcal{F}_{m+1}^2(v) - (A^2)_m^{-1} g_m^2(\mathcal{F}_m(v))$$

and iterate it "backward".

7.2. An abstract version of the Stable Manifold Theorem

Let E be a Banach space with norm $\|\cdot\|$ and let $E^1, E^2 \subset E$ be linear spaces such that $E = E^1 \times E^2$. We identify $E^1 \times \{0\}$ with E^1 and $\{0\} \times E^2$ with E^2, and we assume that $\|v\| = \|v_1\| + \|v_2\|$ for every $v = (v_1, v_2) \in E$. Let also $B^i(r_0) \subset E^i$ be a ball centered at zero of radius $r_0 > 0$ and let $B(r_0) = B^1(r_0) \times B^2(r_0)$. Choose numbers $0 < \alpha \leq 1$, λ, μ, and γ such that

$$0 < \lambda < 1, \quad \lambda < \mu, \quad \lambda^\alpha < \gamma < 1. \tag{7.17}$$

We assume that there is a sequence of maps

$$\{F_m \colon B(r_0) \to E\}_{m \in \mathbb{N}}$$

of the form (7.9) where $v_i \in E^i$, $A_m^i \colon E^i \to E^i$ are linear maps, and $g_m^i \colon B(r_0) \to E^i$ are nonlinear maps, $i = 1, 2$, satisfying (see (7.10))

$$\|A_m^1\| \leq \lambda \quad \text{and} \quad \|(A_m^2)^{-1}\| \leq \mu^{-1}, \tag{7.18}$$

as well as (7.11) and (7.12) where $v = (v_1, v_2), w = (w_1, w_2) \in E$.

We are now ready to state a general version of the Stable Manifold Theorem (see Theorem 7.1).

Theorem 7.4 (Abstract Stable Manifold Theorem). *Let κ be any number satisfying*

$$\lambda < \kappa < \gamma^{1/\alpha}. \tag{7.19}$$

Then there exist constants $D > 0$ and $r_0 > r > 0$ and a map $\psi \colon B^1(r) \to E^2$ such that:

(1) *smoothness: ψ is of class C^1, $\psi(0) = 0$, $d_0\psi = 0$, and for any $v, w \in B^1(r)$,*

$$\|d_v\psi - d_w\psi\| \leq D\|v - w\|^\alpha;$$

(2) *stability: if $m \geq 0$ and $v \in B^1(r)$, then $\mathcal{F}_m(v, \psi(v)) \in B(r)$ and*

$$\|\mathcal{F}_m(v, \psi(v))\| \leq D\kappa^m \|(v, \psi(v))\|,$$

where the maps $\mathcal{F}_m$ are given by (7.14);

(3) *uniqueness: given $v_i \in B^i(r)$, $i = 1, 2$, if there is a number $K > 0$ such that*

$$\mathcal{F}_m(v_1, v_2) \in B(r) \quad \text{and} \quad \|\mathcal{F}_m(v_1, v_2)\| \leq K\kappa^m$$

for every $m \geq 0$, then $v_2 = \psi(v_1)$;

(4) *the numbers D and r depend only on the numbers λ, μ, γ, α, κ, and C (and do not depend on m).*

7.2. An abstract version of the Stable Manifold Theorem

Proof. We present an approach based upon the Perron method. Consider the linear space Γ_κ of sequences of vectors

$$z = \{z(m) \in E\}_{m \in \mathbb{N}},$$

satisfying the following condition:

$$\|z\|_\kappa = \sup_{m \geq 0}\{\kappa^{-m}\|z(m)\|\} < \infty.$$

Γ_κ is a Banach space with the norm $\|z\|_\kappa$. Given $r > 0$, let

$$W = \{z \in \Gamma_\kappa : z(m) \in B(r) \text{ for every } m \in \mathbb{N}\}.$$

Since $0 < \kappa < 1$, the set W is open. Consider the map $\Phi_\kappa \colon B^1(r_0) \times W \to \Gamma_\kappa$ given by[3]

$$\Phi_\kappa(y, z)(0) = \left(y, -\sum_{k=0}^{\infty} \left(\prod_{i=0}^{k} A_i^2\right)^{-1} g_k^2(z(k))\right),$$

and for $m > 0$ by

$$\Phi_\kappa(y, z)(m) = -z(m) + \left(\left(\prod_{i=0}^{m-1} A_i^1\right) y, 0\right)$$

$$+ \left(\sum_{n=0}^{m-1} \left(\prod_{i=n+1}^{m-1} A_i^1\right) g_n^1(z(n)), -\sum_{n=0}^{\infty} \left(\prod_{i=0}^{n} A_{i+m}^2\right)^{-1} g_{n+m}^2(z(n+m))\right).$$

Lemma 7.5. *There are constants $C, M > 0$ independent of the numbers λ, μ, γ, α, and κ such that for every $y \in B^1(r_0)$ and $z \in W$*

$$\|\Phi_\kappa(y, z)\|_\kappa \leq \|z\|_\kappa + \|y\| + CM(\|z\|_\kappa)^{1+\alpha}. \tag{7.20}$$

Proof of the lemma. By (7.11) and (7.12), for $z \in B^1(r_0) \times B^2(r_0)$, $i = 1, 2$, and $n \geq 0$,

$$\begin{aligned}\|g_n^i(z)\| &= \|g_n^i(z) - g_n^i(0)\| \leq \|d_\xi g_n^i\| \cdot \|z\| \\ &= \|d_\xi g_n^i - d_0 g_n^i\| \cdot \|z\| \\ &\leq C\gamma^{-n}\|\xi\|^\alpha \|z\| \leq C\gamma^{-n}\|z\|^{1+\alpha},\end{aligned} \tag{7.21}$$

where ξ lies on the interval joining the points 0 and z.

---

[3] The formulae in the definition of the operator Φ_κ are motivated by Exercise 7.3 and will guarantee the invariance relations (7.8) (see (7.38) below).

Using (7.18) and (7.21), we obtain

$$\|\Phi_\kappa(y,z)\|_\kappa = \sup_{m\geq 0}\{\kappa^{-m}\|\Phi_\kappa(y,z)(m)\|\}$$
$$\leq \sup_{m\geq 0}\{\kappa^{-m}\|z(m)\|\}$$
$$+ \sup_{m\geq 0}\left\{\kappa^{-m}\left[\prod_{i=0}^{m-1}\|A_i^1\|\cdot\|y\| + \sum_{n=0}^{m-1}\left(\prod_{i=n+1}^{m-1}\|A_i^1\|\right)\right.\right.$$
$$\times C\gamma^{-n}\|z(n)\|^{1+\alpha}$$
$$\left.\left.+ \sum_{n=0}^{\infty}\left(\prod_{i=0}^{n}\|(A_{i+m}^2)^{-1}\|\right)C\gamma^{-(m+n)}\|z(n+m)\|^{1+\alpha}\right]\right\} \quad (7.22)$$
$$\leq \|z\|_\kappa + \sup_{m\geq 0}(\kappa^{-m}\lambda^m)\|y\|$$
$$+ \sup_{m\geq 0}\left(\kappa^{-m}C\|z\|_\kappa^{1+\alpha}\left[\sum_{n=0}^{m-1}\lambda^{m-n-1}\gamma^{-n}\kappa^{(1+\alpha)n}\right.\right.$$
$$\left.\left.+ \sum_{n=0}^{\infty}\mu^{-(n+1)}\gamma^{-(m+n)}\kappa^{(1+\alpha)(m+n)}\right]\right).$$

In view of (7.19), we have

$$\sup_{m\geq 0}(\kappa^{-m}\lambda^m) = 1.$$

Since the function $x \mapsto xa^x$ reaches its maximum $-1/e\log a$ at $x = -1/\log a$, we obtain

$$\sup_{m\geq 0}\left[\kappa^{-m}\lambda^{m-1}\sum_{n=0}^{m-1}\left(\lambda^{-1}\gamma^{-1}\kappa^{(1+\alpha)}\right)^n\right]$$
$$\leq \begin{cases}\lambda^{-1}(\kappa^{-1}\lambda)^m m & \text{if } \lambda^{-1}\gamma^{-1}\kappa^{(1+\alpha)} \leq 1, \\ \lambda^{-1}(\gamma^{-1}\kappa^\alpha)^m m & \text{if } \lambda^{-1}\gamma^{-1}\kappa^{(1+\alpha)} \geq 1\end{cases}$$
$$\leq \lambda^{-1}e^{-1}\left(\log\max\left\{\frac{\kappa^\alpha}{\gamma},\frac{\lambda}{\kappa}\right\}\right)^{-1} =: M_1.$$

Furthermore, since $\mu^{-1}\gamma^{-1}\kappa^{1+\alpha} = (\mu^{-1}\kappa)(\kappa^\alpha\gamma^{-1}) < 1$, we have

$$\sup_{m\geq 0}\left(\kappa^{-m}\gamma^{-m}\kappa^{(1+\alpha)m}\mu^{-1}\sum_{n=0}^{m-1}\left(\mu^{-1}\gamma^{-1}\kappa^{(1+\alpha)}\right)^n\right) \quad (7.23)$$
$$= \frac{1}{\mu - \gamma^{-1}\kappa^{1+\alpha}} =: M_2.$$

Setting $M = M_1 + M_2$ and using (7.22), we obtain (7.20), thus completing the proof of the lemma. □

7.2. An abstract version of the Stable Manifold Theorem

In particular, we conclude that the map Φ_κ is well-defined. We also have that $\Phi_\kappa(0,0) = (0,0)$.

Lemma 7.6. *For any $y \in B^1(r_0)$ and $z \in W$ we have:*

(1) *for any $m \geq 1$ the partial derivative*

$$\partial^1_{(y,z)} \Phi_\kappa(m) = \left(\prod_{i=0}^{m-1} A_i^1, 0 \right); \tag{7.24}$$

(2) *the partial derivative*

$$\partial^2_{(y,z)} \Phi_\kappa = \mathcal{A}_\kappa(z) - \mathrm{Id}, \tag{7.25}$$

where Id *is the identity map and $\mathcal{A}_\kappa \colon W \to \Gamma_\kappa$ satisfies*

$$\|\mathcal{A}_\kappa(z_1) - \mathcal{A}_\kappa(z_2)\|_\kappa \leq CM(\|z_1 - z_2\|_\kappa)^\alpha. \tag{7.26}$$

Proof of the lemma. Choose any $y \in B^1(r_0)$ and $z \in W$. If $s \in B^1(r_0)$ is such that $y + s \in B^1(r_0)$, then for any $m \geq 0$ we have that

$$\Phi_\kappa(y+s,z)(m) - \Phi_\kappa(y,z)(m) = \left(\left(\prod_{i=0}^{m-1} A_i^1\right) s, 0 \right),$$

implying (7.24).

Now let $t \in \Gamma_\kappa$ be such that $z + t \in W$. Write

$$\Phi_\kappa(y, z+t) - \Phi_\kappa(y,z) = (\mathcal{A}_\kappa(z) - \mathrm{Id})t + o(z,t),$$

where

$$(\mathcal{A}_\kappa(z))t(m) = \left(\sum_{n=0}^{m-1} \left(\prod_{i=n+1}^{m-1} A_i^1 \right) d_{z(n)} g_n^1 \, t(n), \right.$$
$$\left. - \sum_{n=0}^{\infty} \left(\prod_{i=0}^{n} A_{i+m}^2 \right)^{-1} d_{z(n+m)} g_{n+m}^2 \, t(m+n) \right),$$

$o(z,t)(m)$
$$= \left(\sum_{n=0}^{m-1} \left(\prod_{i=n+1}^{m-1} A_i^1 \right) o_1(z,t)(n), -\sum_{n=0}^{\infty} \left(\prod_{i=0}^{n} A_{i+m}^2 \right)^{-1} o_2(z,t)(m+n) \right).$$

Here the $o_i(z,t)(m)$ for $i = 1,2$ are defined by

$$o_i(z,t)(m) = g_m^i((z+t)(m)) - g_m^i(z(m)) - d_{z(m)} g_m^i \, t(m). \tag{7.27}$$

For $z_1, z_2 \in W$ and $t \in \Gamma_\kappa$ we have that
$$\|(\mathcal{A}_\kappa(z_1) - \mathcal{A}_\kappa(z_2))t\|_\kappa$$
$$\leq \sup_{m \geq 0}\left\{\kappa^{-m}\left[\sum_{n=0}^{m-1}\left(\prod_{i=n+1}^{m-1}\|A_i^1\|\right)C\gamma^{-n}\|z_1(n) - z_2(n)\|^\alpha \|t(n)\|\right.\right.$$
$$+ \sum_{n=0}^\infty \left(\prod_{i=0}^n \|(A_{i+m}^2)^{-1}\|\right) C\gamma^{-(n+m)}\|z_1(n+m) - z_2(n+m)\|^\alpha$$
$$\left.\left. \times \|t(n+m)\|\right]\right\}$$
$$\leq \sup_{m \geq 0}\left\{\kappa^{-m} C\left[\sum_{n=0}^{m-1} \lambda^{m-n-1}\gamma^{-n}\kappa^{(1+\alpha)n}\right.\right.$$
$$\left.\left. + \sum_{n=0}^\infty \mu^{-(n+1)}\gamma^{-(n+m)}\kappa^{(1+\alpha)(n+m)}\right]\right\}(\|z_1 - z_2\|_\kappa)^\alpha \|t\|_\kappa.$$

The estimate (7.26) follows from (7.22) and (7.23).

Applying the Mean Value Theorem to (7.27) and using (7.12), (7.22), and (7.23), it is easy to show that
$$\|o_i(z,t)(m)\| \leq C\gamma^{-m}\|t(m)\|^{1+\alpha},$$
implying (7.25). □

It is easy to derive from Lemma 7.6 that the map Φ_κ is of class C^1 and
$$d_{(y,z)}\Phi_\kappa(m)(s,t) = \partial^1_{(y,z)}\Phi_\kappa(m)s + \partial^2_{(y,z)}\Phi_\kappa(m)t.$$
Indeed, the map Φ_κ is of class $C^{1+\alpha}$ since
$$\kappa^{-m}\|(d_{(y_1,z_1)}\Phi_\kappa - d_{(y_2,z_2)}\Phi_\kappa)(m)(s,t)\| \leq \|\mathcal{A}_\kappa(z_1) - \mathcal{A}_\kappa(z_2)\|_\kappa \|t\|,$$
which, by (7.26), gives
$$\|d_{(y_1,z_1)}\Phi_\kappa - d_{(y_2,z_2)}\Phi_\kappa\|_\kappa \leq \|\mathcal{A}_\kappa(z_1) - \mathcal{A}_\kappa(z_2)\|_\kappa \leq CM(\|z_1 - z_2\|_\kappa)^\alpha.$$
Moreover, we have from (7.25) and (7.26) that $\partial^2_{(y,0)}\Phi_\kappa = -\operatorname{Id}$. Therefore, the map Φ_κ satisfies the conditions of the Implicit Function Theorem, and hence, there exist a number $r \leq r_0$ and a map $\varphi\colon B^1(r) \to W$ of class C^1 with
$$\varphi(0) = 0 \quad \text{and} \quad \Phi_\kappa(y, \varphi(y)) = 0. \tag{7.28}$$
We need a special version of the Implicit Function Theorem which will allow us to (1) show that the function φ is Hölder continuous and (2) obtain an explicit estimate of the number r which will demonstrate that this number depends only on λ, μ, γ, α, κ, and C. More precisely, the following statement holds.

7.2. An abstract version of the Stable Manifold Theorem

Lemma 7.7 (Implicit Function Theorem). *Let E_1, E_2, and G be Banach spaces and let $h\colon B_1 \times B_2 \to G$ be a C^1 map, where $B_i = B_i(r_i) \subset E_i$ is a ball centered at 0 of radius r_i, for $i = 1, 2$. Assume that $h(0,0) = 0$ and that the partial derivative (with respect to the second coordinate) $\partial^2_{(0,0)} h\colon E_2 \to G$ is a linear homeomorphism. Assume also that dh is Hölder continuous in $B_1 \times B_2$ with Hölder constant a and Hölder exponent α. Let B be the ball in E_1 centered at 0 of radius*

$$\rho_1 = \min\left\{r_1, r_2, \frac{r_2}{2bc}, \frac{1}{(1+2bc)(2ac)^{1/\alpha}}\right\}, \tag{7.29}$$

where

$$b = \sup_{x \in B_1} \|\partial^1_{(x,0)} h\|, \quad c = \|(\partial^2_{(0,0)} h)^{-1}\|.$$

Then there exists a unique map $u\colon B \to B_2$ satisfying the following properties:

(1) *u is of class $C^{1+\alpha}$, $h(x, u(x)) = 0$ for every $x \in B$, and $u(0) = 0$;*
(2) *if $x \in B$, then*

$$\left\|\frac{du}{dx}(x)\right\| \le 1 + 2bc;$$

(3) *if $x_1, x_2 \in B$, then*

$$\left\|\frac{du}{dx}(x_1) - \frac{du}{dx}(x_2)\right\| \le 8ac(1+bc)^2 \|x_1 - x_2\|^\alpha.$$

Proof of the lemma. Let $\rho_2 = \max\{\rho_1, 2bc\rho_1\}$. By (7.29), we have that $\rho_1 \le r_1$, $\rho_2 \le r_2$. Also let $S_i \subset E_i$ be the open balls centered at zero of radii ρ_i, for $i = 1, 2$. We equip the space $E_1 \times E_2$ with the norm

$$\|(x,y)\| = \|x\| + \|y\|,$$

for $(x, y) \in E_1 \times E_2$. By (7.29), if $x \in S_1$ and $y_1, y_2 \in S_2$, then

$$\|h(x, y_1) - h(x, y_2) - \partial^2_{(0,0)} h(y_1 - y_2)\| \le a(\rho_1 + \rho_2)^\alpha \|y_1 - y_2\| \le \frac{1}{2c}\|y_1 - y_2\|.$$

Define a function $H\colon S_1 \times S_2 \to E_2$ by

$$H(x, y) = y - (\partial^2_{(0,0)} h)^{-1} h(x, y).$$

We obtain

$$\begin{aligned}
\|H(x, y_1) - H(x, y_2)\| &= \|(\partial^2_{(0,0)} h)^{-1} [\partial^2_{(0,0)} h(y_1 - y_2) - (h(x, y_1) - h(x, y_2))]\| \\
&\le \frac{1}{2}\|y_1 - y_2\|.
\end{aligned} \tag{7.30}$$

Furthermore,
$$\begin{aligned}\|H(x,0)\| &= \|-(\partial^2_{(0,0)}h)^{-1}h(x,0)\| \\ &= \|-(\partial^2_{(0,0)}h)^{-1}[h(x,0)-h(0,0)]\| \\ &\le cb\rho_1 \le \frac{\rho_2}{2}.\end{aligned} \quad (7.31)$$

We need the following statement.

Exercise 7.8. Let E be a Banach space, let $S \subset E$ be the open ball of radius r centered at 0, and let $t\colon S \to E$ be a transformation. Assume that there exists $0 < \beta < 1$ such that
$$\|t(y_1) - t(y_2)\| \le \beta \|y_1 - y_2\|$$
for any $y_1, y_2 \in S$ and $\|t(0)\| \le r(1-\beta)$. Then there exists a unique $y \in S$ such that $t(y) = y$.

Applying this exercise to the map $t = H$ (for which $E = E_1 \times E_2$ and $S = S_1 \times S_2$) with $r = \rho_2$ and $\beta = 1/2$ (see (7.30) and (7.31)), there exists a unique function $u\colon S_1 \to S_2$ such that $u(0) = 0$ and
$$h(x, u(x)) = 0$$
for every $x \in S_1$. Write $\tau_x(y) = H(x, y)$. Then
$$\|(\tau_x)^n(0) - (\tau_x)^{n-1}(0)\| \le \frac{\rho_2}{2^n}.$$
This implies that the series
$$\sum_{n=0}^{\infty}[\tau_x^n(0) - \tau_x^{n-1}(0)] = u(x)$$
converges uniformly and defines a continuous function.

We now show that u is differentiable. Choose $x, x+s \in S_1$ and set $t = u(x+s) - u(x)$. For any $\delta > 0$ there exists $r > 0$ such that
$$\begin{aligned}\|\partial^1_{(x,u(x))}hs &+ \partial^2_{(x,u(x))}ht\| \\ &= \|h(x+s, u(x)+t) - h(x,u(x)) - \partial^1_{(x,u(x))}hs - \partial^2_{(x,u(x))}ht\| \\ &\le \delta\|(s,t)\|\end{aligned}$$
whenever $\|s\| < r$. Set
$$\sigma = 2\|(\partial^2_{(x,u(x))}h)^{-1}\partial^1_{(x,u(x))}h\| + 1.$$
Choosing δ so small that $\delta\|(\partial^2_{(x,u(x))}h)^{-1}\| \le 1/2$, we find that
$$\|t\| - \frac{\sigma-1}{2}\|s\| \le \|t + (\partial^2_{(x,u(x))}h)^{-1}\partial^1_{(x,u(x))}hs\| \le \frac{\|s\|+\|t\|}{2}.$$

7.2. An abstract version of the Stable Manifold Theorem

This implies that $\|t\| \le \sigma\|s\|$, and thus,

$$\|t + (\partial^2_{(x,u(x))}h)^{-1}\partial^1_{(x,u(x))}hs\| \le \delta(\sigma+1)\|(\partial^2_{(x,u(x))}h)^{-1}\| \cdot \|s\|.$$

We conclude that u is differentiable and its derivative is given by

$$\frac{du}{dx}(x) = -(\partial^2_{(x,u(x))}h)^{-1}\partial^1_{(x,u(x))}h. \tag{7.32}$$

Therefore, u is of class C^1. Since dh is Hölder continuous, it follows from (7.32) that u has a Hölder continuous derivative.

One can also verify that

$$\begin{aligned}\|\partial^1_{(x,y)}h\| &\le \|\partial^1_{(x,y)}h - \partial^1_{(0,0)}h\| + \|\partial^1_{(0,0)}h\| \\ &\le a(\rho_1+\rho_2)^\alpha + b \le \frac{1}{2c} + b.\end{aligned} \tag{7.33}$$

Furthermore,

$$\begin{aligned}\|\partial^2_{(x,y)}h\| &\ge \|\partial^2_{(0,0)}h\| - \|\partial^2_{(x,y)}h - \partial^2_{(0,0)}h\| \\ &\ge \frac{1}{c} - a(\rho_1+\rho_2)^\alpha \ge \frac{1}{2c}.\end{aligned}$$

This implies that $\partial^2_{(x,y)}h$ is invertible for every $(x,y) \in S_1 \times S_2$ and that

$$\|(\partial^2_{(x,y)}h)^{-1}\| \le 2c. \tag{7.34}$$

It follows from (7.33) and (7.34) that for every $x \in S_1$,

$$\begin{aligned}\left\|\frac{du}{dx}(x)\right\| &= \|-(\partial^2_{(x,u(x))}h)^{-1}\partial^1_{(x,u(x))}h\| \\ &\le 2c\left(\frac{1}{2c}+b\right) = 1 + 2bc.\end{aligned} \tag{7.35}$$

In view of (7.32), we obtain that

$$\partial^1_{(x,u(x))}h + \partial^2_{(x,u(x))}h\frac{du}{dx}(x) = 0,$$

and thus,

$$\begin{aligned}\partial^1_{(x_1,u(x_1))}h - \partial^1_{(x_2,u(x_2))}h &+ [\partial^2_{(x_1,u(x_1))}h - \partial^2_{(x_2,u(x_2))}h]\frac{du}{dx}(x_1) \\ &+ \partial^2_{(x_2,u(x_2))}h\left(\frac{du}{dx}(x_1) - \frac{du}{dx}(x_2)\right) = 0\end{aligned}$$

for every $x_1, x_2 \in S_1$. Using (7.33) and (7.35), we conclude that
$$\left\| \frac{du}{dx}(x_1) - \frac{du}{dx}(x_2) \right\| \leq \|(\partial^2_{(x_2, u(x_2))} h)^{-1}\|$$
$$\times a\|(x_1, u(x_1)) - (x_2, u(x_2))\|^\alpha (2 + 2bc)$$
$$\leq 4ac(1 + bc)(2 + 2bc)^\alpha \|x_1 - x_2\|^\alpha$$
$$\leq 8ac(1 + bc)^2 \|x_1 - x_2\|^\alpha.$$

This completes the proof of Lemma 7.7. □

The map $h = \Phi_\kappa$ satisfies the conditions of Lemma 7.7 with
$$c = 1, \quad b = 1, \quad a = CM. \tag{7.36}$$

To confirm this, we need only to show that the derivatives $\partial^1 \Phi_\kappa$ and $\partial^2 \Phi_\kappa$ are Hölder continuous. This holds true for $\partial^1 \Phi_\kappa$ in view of (7.24). Furthermore, by (7.25) and (7.26),
$$\|\partial^2_{(y_1, z_1)} \Phi_\kappa - \partial^2_{(y_2, z_2)} \Phi_\kappa\| = \|\mathcal{A}_\kappa(z_1) - \mathcal{A}_\kappa(z_2)\|$$
$$\leq CM \|z_1 - z_2\|_\kappa{}^\alpha.$$

We now describe some properties of the map φ. Differentiating the second equality in (7.28) with respect to y, we obtain
$$d_y \varphi = -[\partial^2_{(y, \varphi(y))} \Phi_\kappa]^{-1} \partial^1_{(y, \varphi(y))} \Phi_\kappa.$$

Setting $y = 0$ in this equality yields
$$d_0 \varphi(m) = \left(\prod_{i=0}^{m-1} A_i^1, 0 \right). \tag{7.37}$$

One can write the vector $\varphi(y)(m)$ in the form
$$\varphi(y)(m) = (\varphi_1(y)(m), \varphi_2(y)(m)),$$
where $\varphi_1(y)(m) \in E^1$ and $\varphi_2(y)(m) \in E^2$. The desired map ψ is now defined by
$$\psi(v) = \varphi_2(v)(0)$$
for each $v \in B^1(r)$. Note that $\varphi_1(v)(0) = v$. It follows from (7.28) that $\varphi_1(y)(0) = 0$ and
$$\varphi_1(y)(m) = \left(\prod_{i=0}^{m-1} A_i^1 \right) y + \sum_{n=0}^{m-1} \left(\prod_{i=n+1}^{m-1} A_i^1 \right) g_n^1(\varphi(y)(n)), \quad m > 0,$$
$$\varphi_2(y)(m) = -\sum_{n=0}^{\infty} \left(\prod_{i=0}^{n} A_{i+m}^2 \right)^{-1} g_{n+m}^2(\varphi(y)(n+m)), \quad m \geq 0.$$

7.2. An abstract version of the Stable Manifold Theorem

By Exercise 7.3, these equalities imply that

$$\varphi_1(y)(m+1) = A_m^1 \varphi_1(y)(m) + g_m^1(\varphi_1(y)(m), \varphi_2(y)(m)),$$
$$\varphi_2(y)(m+1) = A_m^2 \varphi_2(y)(m) + g_m^2(\varphi_1(y)(m), \varphi_2(y)(m)).$$

Thus, we obtain that the function $\varphi(y)$ is invariant under the family of maps F_m, i.e.,

$$F_m(\varphi(y)(m)) = \varphi(y)(m+1) \tag{7.38}$$

(compare to (7.8)). It follows from (7.38) that

$$\mathcal{F}_m(v, \psi(v)) = \mathcal{F}_m(\varphi_1(v)(0), \varphi_2(v)(0))$$
$$= \mathcal{F}_m(\varphi(v)(0)) = \varphi(v)(m).$$

Applying Lemma 7.7 and (7.36), we find that

$$\|\mathcal{F}_m(v, \psi(v))\| \leq \kappa^m \|\varphi(v)\|_\kappa = \kappa^m \|\varphi(v) - \varphi(0)\|_\kappa$$
$$\leq \kappa^m \sup_{\xi \in B^s(r)} \|d_\xi \varphi\|_\kappa \|v\|$$
$$\leq 3\kappa^m \|v\| \leq 3\kappa^m \|(v, \psi(v))\|$$

(we use here the fact that $\|v_1\|, \|v_2\| \leq \|v\|$ for every $v = (v_1, v_2) \in \mathbb{R}^2$). This proves statement (2) of Theorem 7.4. Furthermore, by (7.36) and Lemma 7.7, for any $v_1, v_2 \in B^1(r)$, we have

$$\|d_{v_1}\psi - d_{v_2}\psi\| = \|d_{v_1}\varphi_2(0) - d_{v_1}\varphi_2(0)\|$$
$$\leq \|d_{v_1}\varphi(0) - d_{v_1}\varphi(0)\|$$
$$\leq 32CM\|v_1 - v_2\|^\alpha.$$

Statement (1) of the theorem now follows from (7.28) and (7.37) with the numbers D and r depending only on the numbers λ, μ, γ, α, κ, and C. This implies statement (4) of the theorem. Take $(v, w) \in B^1(r) \times B^2(r)$ satisfying the assumptions of statement (4) of the theorem. Set

$$z(m) = \mathcal{F}_m(v, w).$$

It follows that $z \in \Gamma_\kappa$ (with $\|z\|_\kappa \leq K$) and that $\Phi_k(v, z) = 0$. The uniqueness of the map φ implies that $z = \varphi(v)$, and hence,

$$w = \varphi_2(v)(0) = \psi(v).$$

This establishes statement (3) of the theorem and completes its proof. $\square$

Now the Stable Manifold Theorem (see Theorem 7.1) follows from Theorem 7.4.

7.3. Basic properties of stable and unstable manifolds

In hyperbolicity theory there is a symmetry between the objects marked by the index "s" and those marked by the index "u". Namely, when the time direction is reversed, the statements concerning objects with index "s" become the statements about the corresponding objects with index "u". In particular, this allows one to define a *local unstable manifold* $V^u(x)$ at a point x in a nonuniformly completely hyperbolic set Y as a local stable manifold for f^{-1}. In this section we describe some basic properties of local stable and unstable manifolds.

Recall that as in the beginning of Section 7.1 we fix an index set $\iota = \{\lambda, \mu, \varepsilon, j\}$ and consider the level set $\{\Lambda_\iota\}$. Let $\{\Lambda_\iota^\ell : \ell \geq 1\}$ be the collection of regular sets (see (5.29)). Note that the regular sets are compact. In what follows, we fix the number ℓ.

7.3.1. Sizes of local manifolds.
We discuss how the sizes of local stable $V^s(x)$ and unstable $V^u(x)$ manifolds vary with $x \in \Lambda_\iota$. Recall that the size of a local stable manifold $V^s(x)$ at a point $x \in \Lambda_\iota$ is the number $r = r(x)$ that is determined by Theorem 7.4 and such that (7.3) holds. Similarly, one defines the size of a local unstable manifold $V^u(x)$ at x. Note that the size of local stable and unstable manifolds is not uniquely defined as, for example, the number $r = \frac{1}{2}r(x)$ can be used as well. However, unless stated otherwise, we will always assume that the size of local stable and unstable manifold at a point $x \in \Lambda_\iota$ is the number $r(x)$ given by Theorem 7.4.

It follows from statement (4) of Theorem 7.4 and statement (3) in Exercise 5.14 that the sizes of the local stable and unstable manifolds at a point $x \in \Lambda_\iota$ and at any point $y = f^m(x)$ for any $m \in \mathbb{Z}$ are related by

$$r(f^m(x)) \geq Ke^{-\varepsilon|m|}r(x), \tag{7.39}$$

where $K > 0$ is a constant.

Let μ be an ergodic invariant Borel measure for f such that $\mu(\Lambda_\iota) > 0$. Then for all sufficiently large ℓ the regular set Λ_ι^ℓ has positive measure. Therefore, the trajectory of almost every point visits Λ_ι^ℓ infinitely many times in the future and in the past. It follows that for generic points x the function $r(f^m(x))$ is oscillating in m and is of the same order as $r(x)$ for infinitely many values of m. Nevertheless, for some integers m the value $r(f^m(x))$ may become as small as is allowed by (7.39). Let us emphasize that the rate with which the sizes of the local stable manifolds $V^s(f^m(x))$ decrease as $m \to +\infty$ is smaller than the rate with which the trajectories $\{f^m(x)\}$ and $\{f^m(y)\}$, $y \in V^s(x)$, approach each other.

The Stable Manifold Theorem (see Theorem 7.1) applies to the case when the nonuniformly hyperbolic set Λ_ι consists of a *single* nonuniformly hyperbolic trajectory $\{f^n(x)\}$. Thus it allows us to describe the behavior of nearby trajectories without the need to have any other nonuniformly hyperbolic trajectories in the vicinity of the given one. Note also that one can replace the assumption that the manifold M is compact by the assumption that for every $x \in \Lambda_\iota$ the map $\tilde{f}_x$ given by (7.4) satisfies conditions (7.5) and (7.6).

Fix a regular set Λ_ι^ℓ.

Exercise 7.9. Using Theorems 5.10, 5.18, and 7.4 and statement (3) in Exercise 5.14 show that:

(1) the sizes of local manifolds are bounded from below on Λ_ι^ℓ, i.e., there exists a number $r_\ell > 0$ that depends only on ℓ such that

$$r(x) \geq r_\ell \quad \text{for } x \in \Lambda_\iota^\ell; \tag{7.40}$$

(2) local manifolds depend uniformly continuously on $x \in \Lambda_\iota^\ell$ in the C^1 topology, i.e., if $x_n \in \Lambda_\iota^\ell$ is a sequence of points converging to a point $x \in \Lambda_\iota^\ell$, then

$$d_{C^1}(V^s(x_n), V^s(x)) \to 0, \quad d_{C^1}(V^u(x_n), V^u(x)) \to 0$$

as $n \to \infty$; here for two given local smooth submanifolds V_1 and V_2 in M of the same size r (i.e., V_i is of the form (7.3) for some r and some functions ψ_i, $i = 1, 2$), we have

$$d_{C^1}(V_1, V_2) = \sup_{x \in B^s(r)} \|\psi_1(x) - \psi_2(x)\| + \sup_{x \in B^s(r)} \|d_x\psi_1 - d_x\psi_2\|;$$

(3) there exists a number $\delta_\ell > 0$ such that for every $x \in \Lambda_\iota^\ell$ and $y \in \Lambda_\iota^\ell \cap B(x, \delta_\ell)$ the intersection $V^s(x) \cap V^u(y)$ is nonempty and consists of a single point which depends continuously on x and y;

(4) local stable and unstable distributions depend Hölder continuously on $x \in \Lambda_\iota^\ell$; more precisely, for every $\ell \geq 1$, $x \in \Lambda_\iota^\ell$, and points $z_1, z_2 \in V^s(x)$ or $z_1, z_2 \in V^u(x)$ we have

$$\begin{aligned} d(T_{z_1}V^s(x), T_{z_2}V^s(x)) &\leq C\rho(z_1, z_2)^\alpha, \\ d(T_{z_1}V^u(x), T_{z_2}V^u(x)) &\leq C\rho(z_1, z_2)^\alpha, \end{aligned} \tag{7.41}$$

where $C > 0$ is a constant and d is the distance in the Grassmannian bundle of TM generated by the Riemannian metric.

If f is an Anosov diffeomorphism (i.e., f is uniformly hyperbolic; see Section 1.1), then $\Lambda_\iota^\ell = M$ for some ℓ, and hence,

$$r(x) \geq r_\ell \quad \text{for every } x \in M.$$

7.3.2. Smoothness of local manifolds. One can obtain more refined information about smoothness of local stable manifolds. More precisely, assume that the diffeomorphism f is of class $C^{r+\alpha}$, with $r \geq 1$ and $0 < \alpha \leq 1$ (i.e., the differential $d^r f$ is Hölder continuous with Hölder exponent α). Then $V^s(x)$ is of class C^r; in particular, if f is of class C^r for some $r \geq 2$, then $V^s(x)$ is of class C^{r-1}. These results are immediate consequences of the following version of Theorem 7.4.

Theorem 7.10. *Assume that the conditions of Theorem 7.4 hold. In addition, assume that for $i = 1, 2$:*

(1) *the maps g_m^i are of class $C^{r+\alpha}$ for some $r \geq 2$;*
(2) *there exists $K > 0$ such that for $\ell = 1, \ldots, r$,*
$$\sup_{z \in B(r_0)} \|d_z^\ell g_m^i\| \leq K\gamma^{-m};$$
(3) *if $z_1, z_2 \in B(r_0)$, then*
$$\|d_{z_1}^r g_m^i - d_{z_2}^r g_m^i\| \leq K\gamma^{-m} \|z_1 - z_2\|^\alpha,$$
for some $\alpha \in (0, 1)$.

If ψ is the map constructed in Theorem 7.4, then there exists a number $N > 0$, which depends only on the numbers λ, μ, γ, α, κ, and K such that:

(i) *ψ is of class $C^{r+\alpha}$;*
(ii) *$\sup_{u \in B^1(r)} \|d_u^\ell \psi\| \leq N$ for $\ell = 1, \ldots, r$.*

Sketch of the proof. It is sufficient to show that the map Φ_κ is of class C^r. Indeed, a simple modification of the argument in the proof of Theorem 7.4 allows one to prove that for $2 \leq \ell \leq r$,
$$d_y^\ell \Phi_\kappa(y, z) = (0, 0)$$
and for every $m \geq 1$,
$$d_z^\ell \Phi_\kappa(y, z)(m)$$
$$= \left(\sum_{n=0}^{m-1} \left(\prod_{i=n+1}^{m-1} A_i^1 \right) d_{z(n)}^\ell g_n^1, -\sum_{n=0}^{\infty} \left(\prod_{i=0}^{n} (A_{i+m}^2)^{-1} \right) d_{z(m+n)}^\ell g_{m+n}^2 \right).$$
□

Exercise 7.11. Complete the proof of Theorem 7.10.

It follows that if f is of class C^r, then $V^s(x)$ is of class $C^{r-1+\alpha}$ for every $x \in \Lambda_\iota$ and $0 < \alpha \leq 1$. One can in fact show that $V^s(x)$ is of class C^r; see [**95**].

7.3. Basic properties of stable and unstable manifolds

7.3.3. Graph transform property. There is another proof of the Stable Manifold Theorem (see Theorem 7.1), which is based on a version of the graph transform property—a statement that is well known in the uniform hyperbolicity theory (and is usually referred to as the Inclination Lemma or the λ-Lemma). This approach is essentially an elaboration of the Hadamard method.

Choose a point $x \in \Lambda_\iota$ and positive numbers r_0, b_0, c_0, and d_0. For every $m \geq 0$ set
$$r_m = r_0 e^{-\varepsilon m}, \quad b_m = b_0 e^{-\varepsilon m}, \quad c_m = c_0 e^{-\varepsilon m}, \quad d_m = d_0 (\lambda')^m e^{\varepsilon m}.$$
Denote by $B^s(r_m)$ (respectively, $B^u(r_m)$) the ball in $E^s(f^{-m}(x))$ (respectively, in $E^u(f^m(x))$) centered at 0 of radius r_m.

Consider the set Ψ of $C^{1+\alpha}$ functions $\psi = (\psi^m)_{m \geq 0}$ which are defined on the set $\{(m, v) : m \in \mathbb{N}, v \in B^s(r_m)\}$ and satisfy the following conditions:

(1) for every $m \geq 0$ and $v \in B^s(r_m)$ we have $\psi^m(v) \in E^u(f^{-m}(x))$;
(2) for every $m \geq 0$,
$$\|\psi^m(0)\| \leq b_m, \quad \max_{v \in B^s(r_m)} \|d_v \psi^m\| \leq c_m; \tag{7.42}$$
(3) if $v_1, v_2 \in B^s(r_m)$, then
$$\|d_{v_1} \psi^m - d_{v_2} \psi^m\| \leq d_m \|v_1 - v_2\|^\alpha. \tag{7.43}$$

Theorem 7.12 (Graph Transform Property). *For every $\ell \geq 1$ there are positive constants r_0, b_0, c_0, and d_0, which depend only on ℓ, such that for every $x \in \Lambda_\iota^\ell$ and every function $\psi = (\psi^m)_{m \geq 0} \in \Psi$ there exists a function $\tilde{\psi} = (\tilde{\psi}^m)_{m \geq 0} \in \Psi$ with the property that for all $m \geq 0$,*
$$F_m^{-1}(\{(v, \psi^m(v)) : v \in B^s(r_m)\}) \supset \{(v, \tilde{\psi}^{m+1}(v)) : v \in B^s(r_{m+1})\}, \tag{7.44}$$
where the maps F_m are given by (7.9).

Outline of the proof. Setting $(v^{m+1}, w^{m+1}) = F_m^{-1}(v, \psi^m(v))$, we obtain that
$$(v^{m+1}, w^{m+1}) = ((A_m^1)^{-1} v + \tilde{g}_m^1(v, \psi^m(v)), (A_m^2)^{-1} \psi^m(v) + \tilde{g}_m^2(v, \psi^m(v))),$$
where for $i = 1, 2$ (compare with Section 7.1.3)
$$\tilde{g}_m^i(0) = 0, \quad d_0 \tilde{g}_m^i = 0,$$
$$\|d_{(v_1, w_1)} \tilde{g}_m^i - d_{(v_2, w_2)} \tilde{g}_m^i\| \leq C \gamma^{-m} (\|v_1 - v_2\| + \|w_1 - w_2\|)^\alpha$$
for some constants $C > 0$ and γ satisfying (7.13). For some $c > 0$ we have that
$$\|v_1^{m+1} - v_2^{m+1}\| \geq (\lambda')^{-1} \|v_1 - v_2\| - \|\tilde{g}_m^1(v_1, \psi^m(v_1)) - \tilde{g}_m^1(v_2, \psi^m(v_2))\|$$
$$= [(\lambda')^{-1} - c(1 + c_m)] \|v_1 - v_2\|.$$

Therefore, choosing $r_0, b_0, c_0 > 0$ sufficiently small, one can define a function $v \mapsto \tilde{\psi}^{m+1}(v)$ on $B^s(r_{m+1})$ by

$$\tilde{\psi}^{m+1}(v^{m+1}) = w^{m+1}. \tag{7.45}$$

This function satisfies (7.44). Furthermore, we have

$$\|w_1^{m+1} - w_2^{m+1}\| \le (\mu')^{-1} c_m \|v_1 - v_2\| + c(1 + c_m)\|v_1 - v_2\|$$
$$\le \frac{(\mu')^{-1} c_m + c(1 + c_m)}{(\lambda')^{-1} - c(1 + c_m)} \|v_1^{m+1} - v_2^{m+1}\|.$$

By eventually choosing a smaller $c_0 > 0$, this implies that

$$\|\tilde{\psi}^{m+1}(0)\| \le b_{m+1}, \quad \max_{v \in B^s(r_{m+1})} \|d_v \tilde{\psi}^{m+1}\| \le c_{m+1}.$$

Taking derivatives in (7.45) and using (7.43), one can show that if $v_1, v_2 \in B^s(r_{m+1})$, then

$$\|d_{v_1} \tilde{\psi}_{m+1} - d_{v_2} \tilde{\psi}_{m+1}\| \le d_{m+1} \|v_1 - v_2\|^\alpha.$$

This completes the proof of the theorem. $\square$

Let us now consider the set Φ of $C^{1+\alpha}$ functions $\varphi = (\varphi^m)_{m \ge 0}$ on $\{(m, v) : m \in \mathbb{N}, v \in B^u(r_m)\}$ for which

$$\varphi^m(v) \in E^s(f^m(x)) \quad \text{for every } m \ge 0 \text{ and } v \in B^u(r_m)$$

and such that conditions (7.42) and (7.43) hold with ψ^m replaced by φ^m and with d_m in (7.43) to be $d_0(\mu')^{-1} e^{\varepsilon m}$. Using arguments similar to those in the proof of Theorem 7.12, one can prove the following result.

Theorem 7.13. *For every $x \in \Lambda_\iota^\ell$ and any function $\varphi = (\varphi^m)_{m \ge 0} \in \Phi$ there is a function $\tilde{\varphi} = (\tilde{\varphi}^m)_{m \ge 0} \in \Phi$ with the property that for all $m \ge 0$,*

$$F_m(\{(v, \varphi^m(v)) : v \in B^u(r_m)\}) \supset \{(v, \tilde{\varphi}^{m+1}(v)) : v \in B^u(r_{m+1})\}. \tag{7.46}$$

As an immediate corollary of this theorem we obtain the following result. Let $\chi \colon B^u(r_0) \to E^s(x)$ be a $C^{1+\alpha}$ function satisfying

$$\|\chi(0)\| \le b_0, \quad \max_{v \in B^u(r_0)} \|d_v \chi\| \le c_0,$$

and for $v_1, v_2 \in B^u(r_0)$,

$$\|d_{v_1} \chi - d_{v_2} \chi\| \le d_0 \|v_1 - v_2\|^\alpha.$$

Corollary 7.14. *There is a function $\tilde{\chi} = (\chi^m)_{m \ge 0} \in \Phi$ such that $\chi^0 = \chi$ and for every $m \ge 0$,*

$$F_m(\{(v, \chi^m(v)) : v \in B^u(r_m)\}) \supset \{(v, \chi^{m+1}(v)) : v \in B^u(r_{m+1})\}. \tag{7.47}$$

7.3. Basic properties of stable and unstable manifolds

Proof. Let $\varphi = (\varphi^m)_{m \geq 0}$ be any function in Φ such that $\varphi^0 = \chi = \chi^0$. By Theorem 7.13, there is a function $\tilde{\varphi} = (\tilde{\varphi}^m)_{m \geq 0} \in \Phi$ such that equation (7.46) holds for any $m \geq 0$. In particular, taking $m = 0$ yields
$$F_0(\{(v, \chi^0(v)) : v \in B^u(r_0)\}) \supset \{(v, \tilde{\varphi}^1(v)) : v \in B^u(r_1)\},$$
where $r_1 = r_0 e^{-\varepsilon}$. Set $\chi^1 = \tilde{\varphi}^1$. Proceeding by induction, assume that there are functions χ^m, $0 \leq m \leq n$, for which (7.47) holds. Choose any function $\varphi = (\varphi^m)_{m \geq 0} \in \Phi$ such that $\varphi^m = \chi^m$ for all $0 \leq m \leq n$ and let $\tilde{\varphi} = (\tilde{\varphi}^m)_{m \geq 0} \in \Phi$ be a function given by Theorem 7.13 for which (7.46) holds for all $m \geq 0$. In particular, for $m = n$ we have
$$F_n(\{(v, \chi^n(v)) : v \in B^u(r_n)\}) \supset \{(v, \tilde{\varphi}^{n+1}(v)) : v \in B^u(r_{n+1})\}.$$
Setting $\chi^{n+1} = \tilde{\varphi}^{n+1}$ and using induction we obtain the desired function $\tilde{\chi} = (\chi^m)_{m \geq 0}$. $\square$

One can obtain a different proof of the corollary by utilizing the Lyapunov charts constructed in Section 5.3.3 and using the fact that the action of the diffeomorphism f on any such chart is uniformly hyperbolic (see Theorem 5.15). In fact, one can prove a more general version of the corollary.

For $x \in \Lambda_\iota$ let $r(x)$ be the size of the local unstable manifold at x; see Section 7.3.1. Given $y \in V^s(x)$, we call a smooth submanifold V r-admissible, $0 < r \leq r(x)$, if there is a smooth function $\psi \colon B^u(0, r) \to E^s(x)$ (here $B^u(0, r)$ is the ball in $E^u(x)$ centered at zero of radius r) such that:

(1) $V = \exp_x\{(v, \psi(v)) : v \in B^u(0, r)\}$;

(2) $y = \exp_x(0, \psi(0))$;

(3) there are $a > 0$ and $b > 0$ such that
$$\max_{v \in B^u(0,r)} \|\psi(v)\| \leq a, \quad \max_{v \in B^u(0,r)} \|d_v \psi\| \leq b.$$

We leave the proof of the following result to the reader.

Theorem 7.15. *For any $\varepsilon, \delta > 0$ and any r-admissible submanifold V there is $N > 0$ such that for any $n > N$ the manifold $f^n(V)$ contains a $(1 - \delta)r(f^n(x))$-admissible submanifold V' such that $f^n(y) \in V'$ and there is a submanifold $V'' \subset V^u(f^n(x))$ such that $d_{C^1}(V', V'') \leq \varepsilon$.*

7.3.4. Global manifolds. For every $x \in \Lambda_\iota$ we define the *global stable* and *global unstable manifolds* by
$$W^s(x) = \bigcup_{n=0}^{\infty} f^{-n}(V^s(f^n(x))), \quad W^u(x) = \bigcup_{n=0}^{\infty} f^n(V^u(f^{-n}(x))). \tag{7.48}$$
They are finite-dimensional immersed smooth submanifolds which are invariant under f and are of class $C^{1+\alpha}$ if f is of class $C^{1+\alpha}$ (respectively, of

class $C^{r+\alpha}$ if f is of class $C^{r+\alpha}$ for $r > 1$). They have the following properties which are immediate consequences of the Stable Manifold Theorem (see Theorem 7.1).

Theorem 7.16. *If $x, y \in \Lambda_\iota$, then:*

(1) $W^s(x) \cap W^s(y) = \varnothing$ *if* $y \notin W^s(x)$; $W^u(x) \cap W^u(y) = \varnothing$ *if* $y \notin W^u(x)$;

(2) $W^s(x) = W^s(y)$ *if* $y \in W^s(x)$; $W^u(x) = W^u(y)$ *if* $y \in W^u(x)$;

(3) *for every $y \in W^s(x)$ (or $y \in W^u(x)$) we have $\rho(f^n(x), f^n(y)) \to 0$ as $n \to +\infty$ (respectively, $n \to -\infty$) with an exponential rate; more precisely, for every $x \in \Lambda_\iota$ (where $\iota = \{\lambda, \mu, \varepsilon, j\}$) and every $R > 0$ there exists $C = C(x, R) > 0$ such that for any $n > 0$*
$$\rho(f^n(x), f^n(y)) \leq C\lambda^n e^{\varepsilon n} \rho(x, y), \quad y \in B^s(x, R),$$
$$\rho(f^{-n}(x), f^{-n}(y)) \leq C\mu^{-n} e^{\varepsilon n} \rho(x, y), \quad y \in B^u(x, R),$$
where $B^s(x, R)$ (respectively, $B^u(x, R)$) is the ball in $W^s(x)$ (respectively, $W^u(x)$) centered at x of radius R.

Exercise 7.17. Prove Theorem 7.16. *Hint:* To prove statement (3) note that by the definition of $W^s(x)$, for every $y \in B^s(x, R)$ there is $m = m(y) > 0$ such that $f^m(y) \in V^s(f^m(x))$. Derive from there that there is a small ball in $B^s(x, R)$ centered at y whose image under f^m lies in $V^s(f^m(x))$. Then cover $B^s(x, R)$ by such balls and choose a finite subcover.

Although global stable and unstable manifolds are smooth and pairwise disjoint, they may not form a foliation.[4] Moreover, these manifolds may be of *finite* size in some directions in the manifold; see Chapter 14 for more details.

7.3.5. Stable and unstable manifolds for nonuniformly partially hyperbolic sets. Let $f \colon M \to M$ be a $C^{1+\alpha}$ diffeomorphism of a compact smooth Riemannian manifold M of dimension p and let Λ be a nonuniformly *partially* hyperbolic set for f (see Section 5.4). Consider the corresponding collections of invariant level sets $\{\Lambda_\iota\}$ given by (5.33), where the index set ι consists of numbers $\{\lambda_1, \lambda_2, \mu_1, \mu_2, \varepsilon, j_1, j_2\}$ satisfying (5.32).

Fix $x \in \Lambda_\iota$. Consider the positive semitrajectory $\{f^m(x)\}_{m \geq 0}$, the family of maps $\{\tilde{F}_m = \tilde{f}_{f^m(x)}\}_{m \geq 0}$ generated by the map $\tilde{f}_x$ given by (7.4), and the corresponding sequence of maps $\{F_m\}_{m \in \mathbb{Z}}$ given by (7.9). Observe that Theorem 7.4 applies to this sequence of maps due to the fact that in (7.17) the number μ is allowed to be ≤ 1. As a result, for each $x \in \Lambda_\iota$ we obtain a local stable manifold $V^s(x)$ which has all the properties stated in Theorem 7.1. Similarly, one can construct a local unstable manifold $V^u(x)$.

---

[4]Global manifolds form what can be called a *measurable lamination* of the set Λ_ι.

7.3. Basic properties of stable and unstable manifolds

7.3.6. Stable manifold theorem for flows. Let φ_t be a smooth flow on a compact smooth Riemannian manifold M and let Λ be a nonuniformly hyperbolic set for φ_t. Also let $\{\Lambda_\iota\}$ be the corresponding collection of level sets (see (5.28)). In what follows, we fix numbers λ, μ, ε, and $1 \leq j < p$ and assume that the level set Λ_ι is not empty. The following statement is an analog of Theorem 7.1 for flows.

Theorem 7.18. *For every $x \in \Lambda_\iota$ there exists a local stable manifold $V^s(x)$ such that $x \in V^s(x)$, $T_x V^s(x) = E^s(x)$, and for every $y \in V^s(x)$ and $t > 0$,*
$$\rho(\varphi_t(x), \varphi_t(y)) \leq T(x)\lambda^t \rho(x, y),$$
where $0 < \lambda < 1$ is a constant and $T \colon \Lambda_\iota \to (0, \infty)$ is a Borel function such that for every $\tau \in \mathbb{R}$,
$$T(\varphi_\tau(x)) \leq T(x) e^{10\varepsilon|\tau|}.$$
Furthermore, $\varphi_\tau(V^s(x)) \subset V^s(\varphi_\tau(x))$ for every $\tau \in \mathbb{R}$.

Proof. Consider the time-one map $f = \varphi_1$. It is nonuniformly partially hyperbolic (see Section 5.4) and the desired result follows from Section 7.3.5. □

We call $V^s(x)$ a *local stable manifold* at x. In a similar fashion, by reversing the time, one can show that there exists a *local unstable manifold* $V^u(x)$ at x such that $T_x V^u(x) = E^u(x)$.

For every $x \in \Lambda_\iota$ we define the *global stable* and *unstable manifolds* at x by
$$W^s(x) = \bigcup_{t > 0} \varphi_{-t}(V^s(\varphi_t(x))), \quad W^u(x) = \bigcup_{t > 0} \varphi_t(V^u(\varphi_{-t}(x))). \tag{7.49}$$

These are finite-dimensional immersed smooth submanifolds (of class $C^{r+\alpha}$ if φ_t is of class $C^{r+\alpha}$). They satisfy properties (1) and (2) in Theorem 7.16. Furthermore, for every $y \in W^s(x)$ (or $y \in W^u(x)$) we have $\rho(\varphi_t(x), \varphi_t(y)) \to 0$ as $t \to +\infty$ (respectively, $t \to -\infty$) with an exponential rate.

We also define the *global weakly stable* and *unstable manifolds* at x by
$$W^{s0}(x) = \bigcup_{t \in \mathbb{R}} W^s(\varphi_t(x)), \quad W^{u0}(x) = \bigcup_{t \in \mathbb{R}} W^u(\varphi_t(x)).$$
It follows from (7.49) that
$$W^{s0}(x) = \bigcup_{t \in \mathbb{R}} \varphi_t(W^s(x)), \quad W^{u0}(x) = \bigcup_{t \in \mathbb{R}} \varphi_t(W^u(x)).$$
We remark that each of these two families of invariant immersed manifolds forms a partition of the set Λ_ι.

Chapter 8

Absolute Continuity of Local Manifolds

Let f be a $C^{1+\alpha}$ diffeomorphism of a compact smooth Riemannian manifold M and let W be a continuous foliation of M with smooth leaves (see Section 1.1). Recall that given $x \in M$ and a small number r, the local manifold $V(x) = V_r(x)$ is the connected component of the intersection of the global manifold $W(x)$ with the ball $B(x,r)$ at x of radius r.

For $x \in M$ we choose, in a canonical way, a local transversal to the foliation W. More precisely, denote by $F(x)$ a subspace in T_xM which is transverse to $E(x) = V_r(x)$, so that

$$T_xM = E(x) \oplus F(x).$$

Using the Riemannian metric on TM, without loss of generality, we may assume that $F(x)$ is chosen to be orthogonal to $E(x)$. We denote by $T(x,r)$ the *r-local transversal through x* given by

$$T(x,r) = \exp_x(B_F(0,r))$$

where $B_F(0,r)$ is the ball in $F(x)$ centered at zero of radius r.

We call the set

$$\mathcal{F}(x,r) = \bigcup_{y \in T(x,r)} V(y)$$

the *r-foliation block* at the point x.

Since $V(y)$ is a smooth submanifold, the restriction of the Riemannian metric to $V(y)$ generates a Riemannian volume on $V(y)$ that we call the *leaf-volume*. We denote by m the Riemannian volume on M and by $m_{V(y)}$ the leaf-volume on $V(y)$.

8. Absolute Continuity of Local Manifolds

We describe one of the most crucial properties of foliations (and of their local manifolds) known as *absolute continuity*. It addresses the following question:

> If $E \subset \mathcal{F}(x,r)$ is a Borel set of positive volume, can the intersection $E \cap V(y)$ have zero leaf-volume for almost every $y \in E$?

Absolute continuity is a property of the foliation with respect to volume and does not require the presence of the dynamics.[1] This property can be understood in a variety of ways. The one that addresses the above question in the most straightforward way is the following: the foliation W is *absolutely continuous (in the weak sense)* if for almost every $x \in M$ (with respect to volume), any $r > 0$, any Borel subset $X \subset \mathcal{F}(x,r)$ of positive volume, and almost every $y \in X$ the intersection $X \cap V(y)$ has positive leaf-volume.

If the foliation W is smooth, then by Fubini's theorem, the intersection $E \cap V(y)$ has positive leaf-volume for almost all $y \in \mathcal{F}(x,r)$. If the foliation is only continuous, the absolute continuity property may not hold. A simple example, which illustrates this paradoxical phenomenon, is presented in Section 8.4. In the example a set of full volume meets almost every leaf of the foliation at a single point—the phenomenon known as "Fubini's nightmare" or "Fubini foiled". Such pathological foliations are in a sense "typical" objects in the stable ergodicity theory (see Section 13.3).

We describe a stronger version of the absolute continuity property. Fix $x \in M$ and consider the partition ξ of the foliation block $\mathcal{F}(x,r)$ by local manifolds $V(y)$, $y \in T(x,r)$. Since the partition ξ is measurable,[2] there exist the factor-measure $\tilde{m}$ in the factor-space $\mathcal{F}(x,r)/\xi$ and the system of conditional measures m_y, $y \in \mathcal{F}(x,r)/\xi$, such that for any Borel subset $E \subset \mathcal{F}(x,r)$ of positive volume,

$$m(E) = \int_{\mathcal{F}(x,r)/\xi} \int_{V(y)} \chi_E(y,z) \, dm_y(z) \, d\tilde{m}(y), \tag{8.1}$$

where χ_E is the characteristic function of the set E. Observe that the factor-space $\mathcal{F}(x,r)/\xi$ can be naturally identified with the local transversal $T = T(x,r)$. We say that the foliation W is *absolutely continuous (in the strong sense)* if for almost every $x \in M$, any sufficiently small $r > 0$, any $y \in T(x,r)$, and $z \in V(y)$,

$$d\tilde{m}(y) = h(y) \, dm_T(y), \quad dm_y(z) = g(y,z) \, dm_{V(y)}(z),$$

---

[1] In particular, neither the foliation W nor the volume m are assumed to be invariant under a dynamical system.

[2] Recall that a partition ξ of a measure space (X,μ) is measurable if there is a sequence of finite or countable partitions $(\xi_n)_{n \in \mathbb{N}}$ whose elements are measurable sets of positive measure such that for almost every $x \in M$ we have $\bigcap_{n \in \mathbb{N}} C_{\xi_n}(x) = C_\xi(x)$ where $C_{\xi_n}(x)$ and $C_\xi(x)$ are the elements of the partitions ξ_n and ξ, respectively, containing x.

where m_T is the leaf-volume on T and $h(y)$ and $g(y,z)$ are positive bounded Borel functions. In other words, the conditional measures on the elements $V(y)$ of the partition ξ are absolutely continuous with respect to the leaf-volume on $V(y)$ and the factor-measure is absolutely continuous with respect to the leaf-volume on the transversal T.[3] In particular, this implies that for any Borel subset $E \subset \mathcal{F}(x,r)$ of positive volume, the relation (8.1) can be rewritten as follows:

$$m(E) = \int_T h(y) \int_{V(y)} \chi_E(y,z) g(y,z) \, dm_{V(y)}(z) \, dm_T(y).$$

A celebrated result by Anosov claims that the stable and unstable invariant foliations for Anosov diffeomorphisms are absolutely continuous. We stress that generically (in the set of Anosov diffeomorphisms) these foliations are not smooth and, therefore, the absolute continuity property is not at all trivial and requires a deep study of the structure of these foliations.

An approach to establishing absolute continuity utilizes the holonomy maps associated with the foliation. To explain this, consider a foliation W. Given a point x, choose two transversals T^1 and T^2 to the family of local manifolds $V(y)$, $y \in T(x,r)$. For any $z \in T^1$ there is a unique $y \in T(x,r)$ such that $z \in V(y)$. The holonomy map associates to a point $z \in T^1$ the point $w = V(y) \cap T^2$. If it is absolutely continuous (see the definition below) for all points x and any choice of transversals T^1 and T^2, then the absolute continuity property follows.

For nonuniformly hyperbolic diffeomorphisms the study of absolute continuity is technically much more involved due to the fact that the global stable and unstable manifolds may not form foliations (they may not even exist for some points in M) and the sizes of local manifolds may vary wildly from point to point. In order to overcome this difficulty, one should define and study the holonomy maps associated with local stable (or unstable) manifolds on regular sets.

8.1. Absolute continuity of the holonomy map

Let Λ be an f-invariant nonuniformly completely hyperbolic set for f. Consider the corresponding collection $\{\Lambda_\iota\}$ of level sets (see (5.28)). Here $\iota = \{\lambda, \mu, \varepsilon, j\}$ is the index set where the numbers λ, μ, ε satisfy (5.24) and $1 \leq j < p$ (recall that $j = \dim E^s(x)$ for every $x \in \Lambda_\iota$ and $p = \dim M$). Further, given ι, we consider the corresponding collection of regular sets $\{\Lambda_\iota^\ell : \ell \geq 1\}$ (see (5.29)). Note that the regular sets are compact. In what follows, we fix numbers λ, μ, ε, j, and ℓ.

---

[3]Recall that if ν and μ are two measures on a measure space X, then ν is said to be absolutely continuous with respect to μ if for every $\varepsilon > 0$ there exists $\delta > 0$ such that $\nu(E) < \varepsilon$ for every measurable set E with $\mu(E) < \delta$.

The stable and unstable subspaces, $E^s(x)$ and $E^u(x)$, as well as the local stable and unstable manifolds, $V^s(x)$ and $V^u(x)$, depend continuously on $x \in \Lambda_\iota^\ell$ and their sizes are bounded away from zero by a number r_ℓ (see (7.40)).

8.1.1. Foliation blocks. Fix $x \in \Lambda_\iota^\ell$ and sufficiently small numbers r, δ such that $0 < r \le r_\ell$ and $0 < \delta < r$.

We denote by $T^u(x)$ the *u-canonical local transversal* through x given by
$$T^u(x) = \exp_x(B^u(0, r)) \tag{8.2}$$
where $B^u(0, r)$ is the ball in $E^u(x)$ centered at zero of radius r.

If the number $\delta = \delta(\ell, r)$ is sufficiently small, then for every $w \in \Lambda_\iota^\ell \cap B(x, \delta)$ the intersection $T^u(x) \cap V^s(w)$ is nonempty and consists of a single point $z = z(w)$.[4] Moreover, there is a "sufficiently large piece" of $V^s(w)$ which contains z. More precisely, there are a number c_ℓ satisfying $0 < c_\ell \le 1$ and a $C^{1+\alpha}$ function
$$\psi_z^s \colon B^s(0, c_\ell r) \to T_x M$$
such that $\exp_x(0, \psi_z^s(0)) = z$ and
$$V_z^s := \exp_x\{(u, \psi_z^s(u)) \colon u \in B^s(0, c_\ell r)\} \subset V^s(w).$$

We call V_z^s the *s-manifold* at z.[5] Denote
$$A^u(x) = \{z \in T^u(x) \colon \text{there is } w \in \Lambda_\iota^\ell \cap B(x, \delta)$$
$$\text{such that } z = T^u(x) \cap V^s(w)\}.$$

We consider the family of *s*-manifolds
$$\mathcal{L}^s(x) = \{V_z^s \colon z \in A^u(x)\} \tag{8.3}$$
and the set
$$\mathcal{F}_\ell^s(x) = \bigcup_{z \in A^u(x)} V_z^s, \tag{8.4}$$
called the *s-foliation block* at x (also known as a *Pesin block*).

8.1.2. The holonomy map and the Absolute Continuity Theorem. A *local transversal* T to the family $\mathcal{L}^s(x)$ of *s*-local manifolds is a local smooth submanifold which is uniformly transverse to every *s*-manifold V_z^s in the family, i.e., the angle between T and V_z^s is uniformly away from zero for every $z \in A^u(x)$.

---

[4] Note that the point z may not lie in Λ_ι^ℓ nor in $B(x, \delta)$.

[5] While the point z may not lie in Λ_ι^ℓ (in fact, not even in Λ), V_z^s can be viewed as a local stable manifold at z (of size $c_\ell r$) with the contraction rate controlled by those at the point w. In particular, there is a stable subspace $E^s(z) = T_z V_z^s$. However, unless z lies in Λ, there is no canonical unstable subspace at z.

8.1. Absolute continuity of the holonomy map

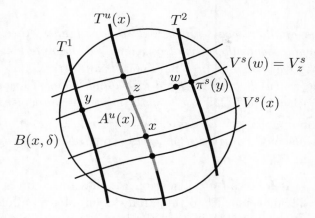

Figure 8.1. A family $\mathcal{L}^s(x)$ of s-manifolds, the u-canonical transversal $T^u(x)$ and the set $A^u(x)$ (in gray), the s-foliation block $\mathcal{F}_\ell^s(x)$, and the holonomy map π^s associated to the transversals T^1 and T^2.

Let T^1 and T^2 be two local transversals to the family $\mathcal{L}^s(x)$ of s-local manifolds. We define the *holonomy map*
$$\pi^s \colon \mathcal{F}_\ell^s(x) \cap T^1 \to \mathcal{F}_\ell^s(x) \cap T^2$$
by setting
$$\pi^s(z_1) = T^2 \cap V_z^s \quad \text{for } z_1 = T^1 \cap V_z^s \text{ and } z \in A^u(x) \tag{8.5}$$
(see Figure 8.1). The map π^s depends upon the choice of the numbers ℓ, r, δ, the point $x \in \Lambda_\iota^\ell$, and the transversals T^1 and T^2.

Exercise 8.1. Show that the holonomy map π^s is a homeomorphism between $\mathcal{F}_\ell^s(x) \cap T^1$ and $\mathcal{F}_\ell^s(x) \cap T^2$.

In a manner similar to that in (8.2)–(8.5) one can define the *s-canonical local transversal* $T^s(x)$ through x, the *family of u-manifolds* $\mathcal{L}^u(x)$, the *u-foliation block* $\mathcal{F}_\ell^u(x)$, and the *holonomy map* π^u.

Recall that a measurable invertible transformation $S \colon X \to Y$ between measure spaces (X, ν) and (Y, μ) is said to be *absolutely continuous* if the measure μ is absolutely continuous with respect to the measure $S_*\nu$. In this case one defines the *Jacobian* $\operatorname{Jac}(S)(x)$ of S at a point $x \in X$ (specified by the measures ν and μ) to be the Radon–Nikodým derivative $d\mu/d(S_*\nu)$. If X is a metric space, then for ν-almost every $x \in X$ one has
$$\operatorname{Jac}(S)(x) = \lim_{r \to 0} \frac{\mu(S(B(x,r)))}{\nu(B(x,r))}. \tag{8.6}$$

Theorem 8.2 (Absolute Continuity). *Given numbers $\ell \geq 1$, $0 < r < r_\ell$, and a point $x \in \Lambda_\iota^\ell$, there is $\delta = \delta(\ell, r)$ such that if T^1 is a local transversal to the family $\mathcal{L}^s(x)$ of s-manifolds (see (8.3)) satisfying*
$$m_{T^1}(\mathcal{F}_\ell^s(x) \cap T^1) > 0 \tag{8.7}$$

and if T^2 is another local transversal to this family, then the holonomy map π^s associated with these two transversals is absolutely continuous (with respect to the leaf-volumes $m_{T^1}|\mathcal{F}^s_\ell(x) \cap T^1$ and $m_{T^2}|\mathcal{F}^s_\ell(x) \cap T^2$) and the Jacobian $\text{Jac}(\pi^s)(y)$ is bounded from above and bounded away from zero for almost every point $y \in \mathcal{F}^s_\ell(x) \cap T^1$ (with respect to the leaf-volume m_{T^1}).

The holonomy map π^u generated by the family $\mathcal{L}^u(x)$ of u-manifolds also satisfies the absolute continuity property provided the condition (8.7) holds with respect to $\mathcal{F}^u_\ell(x)$.

Remark 8.3. It follows from Theorem 8.2 that once condition (8.7) holds for some local transversal T^1 to the family $\mathcal{L}^s(x)$ of s-manifolds, then it also holds for *any* transversal T to this family. This means that this condition is a property of the foliation block itself (with respect to the volume) and does not depend on the choice of the transversal.

8.1.3. The strong absolute continuity property. The proof of Theorem 8.2 will be given in Section 8.2. An important consequence of this theorem is the fact that the families of s-manifolds and u-manifolds are absolutely continuous (in the strong sense). To see this, consider the set Λ of nonuniformly hyperbolic trajectories and assume that it has positive (not necessarily full) volume, i.e., $m(\Lambda) > 0$. Let us stress that here the volume m need not be invariant under f. We can choose an index set ι and a number $\ell \geq 1$ so large that the regular set Λ^ℓ_ι has positive volume. Fix a point $x \in \Lambda^\ell_\iota$ and consider the family $\mathcal{L}^s(x)$ of s-manifolds V^s_z, $z \in A^u(x)$.

There exists a family of u-local transversals $\{T^u(y) : y \in V^s(x)\}$ such that:

(1) every $T^u(y)$ is a smooth submanifold which is transverse to $\mathcal{L}^s(x)$ uniformly in y, i.e., the angle between $T^u(y)$ and V^s_z is uniformly away from zero for every $y \in V^s(x)$ and $z \in A^u(x)$;
(2) $T^u(x)$ is the u-canonical local transversal;
(3) $T^u(y_1) \cap T^u(y_2) = \varnothing$ for any $y_1, y_2 \in V^s(x)$ such that $y_1 \neq y_2$;
(4) $\bigcup_{y \in V^s(x)} T^u(y) \supset \mathcal{F}^s_\ell(x)$;
(5) $T^u(y)$ depends smoothly on $y \in V^s(x)$.

The collection of smooth submanifolds $T^u(y)$ generates a partition of the set $T^u = \bigcup_{y \in V^s(x)} T^u(y)$. We denote this partition by η and also denote by $m_{T^u(y)}$ the leaf-volume on $T^u(y)$.

For $y \in V^s(x)$ denote
$$A^u(y) = \mathcal{F}^s_\ell(x) \cap T^u(y). \tag{8.8}$$

Note that for $q \in T^u(y)$ there is a unique $z \in A^u(x)$ such that $q = T^u(y) \cap V^s_z$ (see Figure 8.2). We write $q = (y, z)$.

8.1. Absolute continuity of the holonomy map

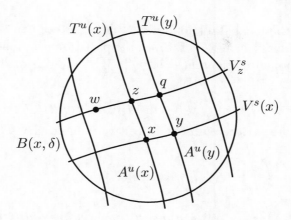

Figure 8.2. The family of transversals $T^u(y)$ and the associated sets $A^u(y)$.

Let ξ^s be the measurable partition of $\mathcal{F}_\ell^s(x)$ into s-manifolds V_z^s, $z \in A^u(x)$ and $m^s(z)$ the conditional measure on V_z^s generated by this partition and volume m. The factor-space $\mathcal{F}_\ell^s(x)/\xi^s$ can be identified with the set $A(y)$ for every $y \in V^s(x)$. Denote by $\hat{m}^s$ the factor-measure generated by the partition ξ^s.

Theorem 8.4. *Assume that for some $y \in V^s(x)$*
$$m_{T^1}(\mathcal{F}_\ell^s(x) \cap T^u(y)) > 0.$$
Then for m-almost every $q = (y, z) \in \mathcal{F}_\ell^s(x)$ the following statements hold:

(1) *the measures $m^s(z)$ and $m_{V_z^s}$ are equivalent;*

(2) *the factor-measure $\hat{m}^s$ is equivalent to the measure $m_{T^u(y)}|A^u(y)$.*

Remark 8.5. Similar results hold for the u-manifolds V_z^u in the foliation block $\mathcal{F}_\ell^u(x)$. More precisely, let $T^s(y)$ be a family of local transversals to the family of u-manifolds and let $A^s(y)$ be given by (8.8) (where the s-foliation block should be replaced by the u-foliation block). Assume that for some $y \in V^u(x)$
$$m_{T^1}(\mathcal{F}_\ell^u(x) \cap T^s(y)) > 0.$$
Then for m-almost every
$$q = (y, z) \in \mathcal{F}_\ell^u(x) \quad (\text{with } y \in V^u(x) \text{ and } z \in A^u(x))$$
the conditional measure $m^u(z)$ on V_z^u is equivalent to the leaf-volume $m_{V_z^u}$ and the factor-measure $\hat{m}^u$ is equivalent to the measure $m_{T^s(y)}|A^s(y)$.

Proof of Theorem 8.4. We follow the approach suggested in [6]. Consider the partition η. For $y \in V^s(x)$, denote by $\{\mu(y)\}$ the system of conditional measures on $T^u(y)$ and denote by $\hat{\mu}$ the factor-measure on T^u/η generated by the partition η and volume m. The factor-space T^u/η can be

identified with the local stable manifold $V^s(x)$ and we can view the factor-measure $\hat{\mu}$ as a measure on $V^s(x)$. Since the partition η is smooth, there exist *positive* continuous functions $h(y)$ for $y \in V^s(x)$ and $g(y,q)$ for $q \in A^u(y)$ such that
$$d\hat{\mu}(y) = h(y)\, dm_{V^s(x)}(y), \quad d\mu(y)(q) = g(y,q)\, dm_{T^u(y)}(q).^6$$
Let $E \subset \mathcal{F}_\ell^s(x)$ be a Borel subset of positive volume. By Theorem 8.2, we have
$$m(E) = \int_{T^u/\eta} \left(\int_{T^u(y)} \chi_E(y,q)\, d\mu(y)(q) \right) d\hat{\mu}(y)$$
$$= \int_{T^u/\eta} \left(\int_{T^u(y)} \chi_E(y,q) g(y,q)\, dm_{T^u(y)}(q) \right) d\hat{\mu}(y)$$
$$= \int_{T^u/\eta} \left(\int_{T^u(x)} \chi_E(y,q) g(y,q) \operatorname{Jac}(\pi_{xy}^s)(z)\, dm_{T^u(x)}(z) \right) d\hat{\mu}(y),$$
where

(1) χ_E is the characteristic function of the set E;
(2) π_{xy}^s is the holonomy map between $A^u(x)$ and $A^u(y)$;
(3) $q \in A^u(y)$, $z \in A^u(x)$, and $q = \pi_{xy}^s(z)$;
(4) $\operatorname{Jac}(\pi_{xy}^s)(z)$ is the Jacobian of the map π_{xy}^s at z.

Applying Fubini's theorem, we obtain
$$m(E) = \int_{T^u(x)} \left(\int_{T^u/\eta} \chi_E(y,q) g(y,q) \operatorname{Jac}(\pi_{xy}^s)(z) d\hat{\mu}(y) \right) dm_{T^u(x)}(z)$$
$$= \int_{T^u(x)} \left(\int_{V^s(x)} \chi_E(y,q) g(y,q) \operatorname{Jac}(\pi_{xy}^s)(z) h(y) dm_{V^s(x)}(y) \right) dm_{T^u(x)}(z).$$
This implies that the conditional measure $m^s(z)$ on V_z^s is equivalent to the leaf-volume $m_{V_z^s}$ with the density function
$$\kappa^s(y,q) = \operatorname{Jac}(\pi_{xy}^s)((\pi_{xy}^s)^{-1}(q)) g(y,q) h(y) > 0. \tag{8.9}$$
The desired result now follows. $\square$

Exercise 8.6. Show that the family $\mathcal{L}^u(x)$ of u-manifolds has the strong absolute continuity property and derive the formula for the density function.

One can use Theorem 8.4 to prove the following result. We leave the details to the reader (see [**13**, Corollaries 8.6.9–8.6.12]).

---

[6] Indeed, there is a local diffeomorphism $G\colon T^u \to \mathbb{R}^p = \mathbb{R}^k \times \mathbb{R}^{p-k}$ (where $p = \dim M$ and $k = \dim T^u(y)$, $y \in V^s(x)$) that *rectifies* the partition η, i.e., it maps it into a partition of the open set $B^k \times B^{p-k}$ (where $B^k \subset \mathbb{R}^k$ and $B^{p-k} \subset \mathbb{R}^{p-k}$ are open unit balls) by subspaces parallel to $\mathbb{R}^k$ and it maps $V^s(x)$ onto B^{p-k}. It remains to apply Fubini's theorem to the measure G_*m, which is equivalent to the volume in $\mathbb{R}^p$.

Theorem 8.7. *The following statements hold:*

(1) *if $A \subset \Lambda$ is a set of positive volume and for each $x \in A$ there is a set $B_x \subset V^s(x)$ of positive leaf-volume, then the set $B = \bigcup_{x \in A} B_x$ has positive volume;*

(2) *if $N \subset M$ is a set of zero volume, then for m-almost every $x \in \Lambda$*
$$m_{V^s(x)}(V^s(x) \cap N) = 0;$$

(3) *for m-almost every $x \in \Lambda_\iota^\ell$, any local transversal T to the family $\mathcal{L}^s(x)$ of s-manifolds, and any set $N \subset \mathcal{F}_\ell^s(x) \cap T$ of zero leaf-volume in T, we have $m(\bigcup_{z \in N} V_z^s) = 0$;*

(4) *if $m(\Lambda) = 1$, then $m_{V^s(x)}(V^s(x) \setminus \Lambda) = 0$ for m-almost every $x \in \Lambda$;*

(5) *similar statements hold with respect to local unstable manifolds $V^u(x)$, $x \in \Lambda$.*

Remark 8.8. In view of its construction the density function $\kappa^s(y, q)$, given by (8.9), depends on the choice of the family of transversals $T^u(y)$. It turns out that this is not the case if the map f preserves volume as can be seen from the following theorem.

Let f be a $C^{1+\alpha}$ volume-preserving diffeomorphism possessing a nonuniformly hyperbolic set Λ of positive volume. Choose a regular set Λ_ι^ℓ of positive volume and a point $x \in \Lambda_\iota^\ell$ and consider the family $\mathcal{L}^s(x)$ of s-manifolds V_z^s, $z \in A^u(x)$. For any $n \geq 0$, $z \in A^u(x)$, and $q \in V_z^s$ set

$$\rho_n^s(z, q) = \prod_{i=0}^{n} \frac{\text{Jac}(d_{f^i(z)} f | E^s(f^i(z)))}{\text{Jac}(d_{f^i(q)} f | T_q V^s(f^i(q)))}.$$

Theorem 8.9. *The following statements hold:*

(1) *the infinite product*

$$\rho^s(z, q) = \lim_{n \to \infty} \rho_n^s(z, q) = \prod_{i=0}^{\infty} \frac{\text{Jac}(d_{f^i(z)} f | E^s(f^i(z)))}{\text{Jac}(d_{f^i(q)} f | T_q V^s(f^i(q)))} \qquad (8.10)$$

converges uniformly in z and q;

(2) *the function $\rho^s(z, q)$ is Hölder continuous in $z \in A^u(x)$ and is smooth in $q \in V^s(z)$;*

(3) *the conditional measure $m^s(z)$ on V_z^s is equivalent to the leaf-volume $m_{V_z^s}$ with the density function $\rho^s(z)^{-1} \rho^s(z, q)$ where*

$$\rho^s(z) = \int_{V^s(z)} \rho^s(z, q) \, dm^s(z)(q) \qquad (8.11)$$

is the normalizing factor.

For the proof of this theorem we refer the reader to [**13**, Theorem 9.3.4] where the result is shown for a larger class of invariant smooth measures.

Remark 8.10. Statement (4) of Theorem 8.7 may not be true if we only require $m(\Lambda) > 0$. In fact, it fails for *dissipative* Anosov diffeomorphisms for which there is no invariant measure that is absolutely continuous with respect to volume. To see this let f be such a diffeomorphism. By a result in [**51**], there is a set A of positive volume such that $f^i(A) \cap f^j(A) = \varnothing$ for any $i, j \in \mathbb{Z}$, $i \neq j$, and $\bigcup_{i \in \mathbb{Z}} f^i(A) = M$ up to a set of zero volume. Clearly, there is a subset $B \subset A$ of positive volume whose complement in A also has positive volume. Then the set $\Lambda = \bigcup_{i \in \mathbb{Z}} f^i(B)$ is of positive volume and is invariant, and so is its complement $\Lambda^c \subset M$. It is easy to see that there is a Lebesgue density point $x \in \Lambda$ with the property that for every $r > 0$ the intersections of the ball $B(x, r)$ with Λ and Λ^c both have positive volume. This contradicts statement (4).

We complete this remark by observing that in view of Theorem 9.6, if volume is invariant, statement (4) is true even if we assume that $m(\Lambda) > 0$.

Remark 8.11. Consider the set $\mathcal{E} \subset M$ of LP-regular points with nonzero Lyapunov exponents, given by (5.3). Observe that for a dissipative Anosov diffeomorphism, while Lyapunov exponents at every point are nonzero, the (noninvariant) volume $m(\mathcal{E}) = 0$.[7] Therefore, it is natural to ask whether Theorems 8.4 and 8.7 remain true if we replace the set Λ with $\mathcal{E}$ and require that $m(\mathcal{E}) = 1$. In this regard we quote the following result from [**1**].

We call a subset $\mathcal{P} \subset M$ *positively recurrent with respect to* $\mathcal{E}$ if $\mathcal{P} \subset \Lambda_\iota^\ell \cap \mathcal{E}$ for some $\iota = \{\lambda, \mu, \varepsilon\}$ and $\ell \geq 1$ and every $x \in \mathcal{P}$ returns to $\Lambda_\iota^\ell \cap \mathcal{E}$ infinitely often in forward and backward times with positive frequency.

Theorem 8.12. *Let $f \colon M \to M$ be a $C^{1+\alpha}$ diffeomorphism of a compact Riemannian manifold M. Assume that there is a set $\mathcal{P}$ of positive volume which is positively recurrent with respect to $\mathcal{E}$. Then f preserves a smooth measure.*

A proof of the absolute continuity property for Anosov diffeomorphisms was given by Anosov in [**5**] and by Anosov and Sinai in [**6**] where they used the approach based on the holonomy map. For nonuniformly hyperbolic systems the absolute continuity property was established by Pesin in [**88**]. The proof presented in the next section is an elaboration of his proof.

---

[7]This is because any invariant measure for this diffeomorphism must be singular with respect to volume.

8.2. A proof of the Absolute Continuity Theorem

Fix $x \in \Lambda_\iota^\ell$ and numbers $0 < r \le r_\ell$, $0 < \delta < r$. Consider the family $\mathcal{L}^s(x)$ of s-manifolds and the s-foliation block $\mathcal{F}_\ell^s(x)$ (see (8.3) and (8.4)). Let T^1 and T^2 be local transversals to the family $\mathcal{L}^s(x)$. Fix $w \in \Lambda_\iota^\ell \cup B(x, \delta)$ and let $y_1 \in T^1 \cap V^s(w)$. Given a number $\tau > 0$, we let

$$B^1(\tau) = \mathcal{F}_\ell^s(x) \cap B^1(y_1, \tau), \quad B^2(\tau) = \pi^s(\mathcal{F}_\ell^s(x) \cap B^1(y_1, \tau)),$$

where $B^1(y_1, \tau) \subset T^1$ is the ball of radius τ centered at y_1. In view of (8.7) we may assume that y_1 is a Lebesgue density point of the set $B^1(\tau)$. It suffices to show that there is a constant $K > 0$ that is independent of the choice of the point x, the transversals T^1, T^2, and the point y such that for every $\tau > 0$,

$$\frac{1}{K} \le \frac{m_{T^2}(B^2(\tau))}{m_{T^1}(B^1(\tau))} \le K.$$

Indeed, letting $\tau \to 0$, we conclude from (8.6) that

$$\frac{1}{K} \le \operatorname{Jac}(\pi^s)(y_1) \le K.$$

In what follows, we fix the number τ. We split the proof into five steps.

Step 1. Let $y_2 = T^2 \cap V^s(w)$. Choose $n \ge 0$, $q > 0$ and for $i = 1, 2$ set

$$T_n^i = f^n(T^i), \quad w_n = f^n(w), \quad y_n^i = f^n(y^i), \quad q_n = qe^{-\varepsilon n}. \tag{8.12}$$

Note that $T_0^i = T^i$, $w_0 = w$, $y_0^i = y^i$, and $q_0 = q$.

In view of Corollary 7.14 for $i = 1, 2$, there exists an open neighborhood $T_n^i(w, q)$ of the point y_n^i such that

$$T_n^i(w, q) = \exp_{w_n}\{(v, \varphi_n^i(v)) \colon v \in B^u(q_n)\} \subset T_n^i, \tag{8.13}$$

where $B^u(q_n) \subset E^u(w_n)$ is the open ball centered at zero of radius q_n and $\varphi_n^i \colon B^u(q_n) \to E^s(w_n)$ is a $C^{1+\alpha}$ map satisfying (compare to (7.42) and (7.43))

$$\|\varphi_n^i(0)\| \le b_0 e^{-\varepsilon n}, \quad \max_{v \in B^u(q_n)} \|d_v \varphi_n^i\| \le c_0 e^{-\varepsilon n},$$

and for $v_1, v_2 \in B^u(q_n)$

$$\|d_{v_1}\varphi_n^i - d_{v_2}\varphi_n^i\| \le d_0 (\mu^{-1})^n e^{\varepsilon n} \|v_1 - v_2\|^\alpha.$$

Here b_0, c_0, and d_0 are some positive constants. By Corollary 7.14, for any $k = 1, \ldots, n$ we have

$$f^{-1}(T_k^i(w, q)) \subset T_{k-1}^i(w, q), \quad i = 1, 2. \tag{8.14}$$

Note also that $y_n^i = \exp_{w_n}(0, \varphi_n^i(0)) \in T_n^i(w, q)$.

We wish to compare the measures $m_{T_n^1}|T_n^1(w, q)$ and $m_{T_n^2}|T_n^2(w, q)$ for sufficiently large n.

Lemma 8.13. *There exist $K_1 > 0$ and $q = q(\ell)$ such that for each $n > 0$,*

$$\frac{1}{K_1} \leq \frac{m_{T_n^1}(T_n^1(w,q))}{m_{T_n^2}(T_n^2(w,q))} \leq K_1.$$

Proof of the lemma. Writing

$$\frac{m_{T_n^1}(T_n^1(w,q))}{m_{T_n^2}(T_n^2(w,q))} = \frac{m_{T_n^1}(T_n^1(w,q))}{m(B^u(q_n))} \times \frac{m(B^u(q_n))}{m_{T_n^2}(T_n^2(w,q))},$$

by Corollary 7.14, each term in the product on the right-hand side of the formula is bounded below and above by constants independent of n and the choice of the points $x \in \Lambda_\iota^\ell$ and $w \in \Lambda_\iota^\ell \cup B(x,\delta)$. The desired result follows. □

Step 2. We continue with the following covering lemma.

Lemma 8.14. *For any sufficiently large $n > 0$ there are points $w_j \in \Lambda_\iota^\ell \cap B(x,\delta)$, $j = 1,\ldots, t = t(n)$, such that for $i = 1, 2$ the sets $T_n^i(w_j, q)$ form an open cover of the set $f^n(B^i(\tau))$. These covers have finite multiplicity which depends only on the dimension of T^1.*

Proof of the lemma. We recall the Besicovich Covering Lemma. It states that for each $Z \subset \mathbb{R}^k$, if $r \colon Z \to \mathbb{R}^+$ is a bounded function, then the cover $\{B(x, r(x)) : x \in Z\}$ of Z contains a countable (Besicovich) subcover of finite multiplicity depending only on k. Clearly, this statement extends to subsets of smooth manifolds.

It follows from the definition of the sets $T_n^i(w,q)$ (see (8.12)–(8.13)) that there is $N > 0$ and for all $n > N$ a number $R = R(\ell, n) > 0$ such that for every $w \in \Lambda_\iota^\ell \cap B(x,\delta)$,

$$B_n^i(w, \tfrac{1}{2}R) \subset T_n^i(w, q) \subset B_n^i(w, R) \subset T_n^i(w, 10q),$$

where $B_n^i(w, R)$ is the ball in T_n^i centered at y_n^i of radius R. This allows us to apply the Besicovich Covering Lemma to the cover of the set $Z^i = f^n(B^i(\tau))$ by balls

$$\left\{ B_n^i(w, \tfrac{1}{2}R) : w \in \Lambda_\iota^\ell \cap B(x,\delta) \right\}.$$

Thus, we obtain a Besicovich subcover $\{B_n^i(w_j, \tfrac{1}{2}R) : j = 1,\ldots, t\}$, $t = t(n)$, of the set Z^i of finite multiplicity M_i (which does not depend on n). The sets $T_n^i(w_j, q)$ also cover Z^i and the multiplicity L_i of this cover does not depend on n. Indeed, L_i does not exceed the multiplicity K_i of the cover of Z^i by balls $B_n^i(w_j, R)$. Note that every ball $B_n^i(w_j, R)$ can be covered by at most CM_i balls $B_n^i(w, \tfrac{1}{2}R)$ with some $w \in \Lambda_\iota^\ell$ where C depends only on the dimension of T^i. Furthermore, given w, we have $w_j \in B_n^i(w, \tfrac{1}{2}R)$ for

8.2. A proof of the Absolute Continuity Theorem

at most M_i points w_j (otherwise at least $M_i + 1$ balls $B_n^i(w_j, \frac{1}{2}R)$ would contain w). Therefore, $w_j \in B_n^i(w, R)$ for at most CM_i points w_j. This implies that $K_i \leq CM_i$. □

Step 3. We now compute the measures that are the pullbacks under f^{-n} of the measures $m_{T_n^i}|T_n^i(w,q)$ for $i = 1, 2$. Note that for every $w \in \Lambda_\iota^\ell \cap B(x, \delta)$ the points y_n^1 and y_n^2 (see (8.12)) lie on the stable manifold $V^s(w_n)$. Choose $z \in f^{-n}(T_n^i(w,q)) \subset T_0^i(w,q)$ (with $i = 1$ or $i = 2$; see (8.14)). Set $z_n = f^n(z)$ and
$$H^i(z, n) = \operatorname{Jac}(d_{z_n} f^{-n} | T_{z_n} T_n^i(w, q))$$
(recall that $\operatorname{Jac}(S)$ denotes the Jacobian of the transformation S). We need the following lemma.

Lemma 8.15. *There exist $K_2 > 0$ and $n_1(\ell) > 0$ such that for every $w \in \Lambda_\iota^\ell \cap B(x, \delta)$ and $n \geq n_1(\ell)$ we have*
$$\left| \frac{H^2(y_n^2, n)}{H^1(y_n^1, n)} \right| \leq K_2, \quad \left| \frac{H^1(z_n, n)}{H^1(y_n^1, n)} \right| \leq K_2.$$

Proof of the lemma. For any $0 < k \leq n$ denote by $V(k, z)$ the tangent space $T_{z_k} T_k^1(w, q)$ to the submanifold $T_k^1(w, q)$ at the point z_k. We use the parallel translation of the subspace $V(k, z) \subset T_{z_k} M$ along the geodesic connecting the points z_k and y_k^1 (this geodesic is uniquely defined since these points are sufficiently close to each other) to obtain a new subspace $\tilde{V}(k, z) \subset T_{y_k^1} M$. We have
$$\left| \operatorname{Jac}(d_{z_k} f^{-1} | V(k, z)) - \operatorname{Jac}(d_{y_k^1} f^{-1} | V(k, y_1)) \right|$$
$$\leq \left| \operatorname{Jac}(d_{z_k} f^{-1} | V(k, z)) - \operatorname{Jac}(d_{y_k^1} f^{-1} | \tilde{V}(k, z)) \right|$$
$$+ \left| \operatorname{Jac}(d_{y_k^1} f^{-1} | \tilde{V}(k, z)) - \operatorname{Jac}(d_{y_k^1} f^{-1} | V(k, y_1)) \right|.$$

Since the diffeomorphism f is of class $C^{1+\alpha}$, in view of (8.14) and Exercise 5.22, we obtain
$$\left| \operatorname{Jac}(d_{z_k} f^{-1} | V(k, z)) - \operatorname{Jac}(d_{y_k^1} f^{-1} | V(k, y_1)) \right|$$
$$\leq C_1 \rho(z_k, y_k^1)^\alpha + C_2 d(\tilde{V}(k, z), V(k, y_1)),$$
where $C_1 > 0$ and $C_2 > 0$ are constants (recall that ρ is the distance in M and d is the distance in the Grassmannian bundle of TM generated by the Riemannian metric). It follows from (7.41) that
$$d(\tilde{V}(k, z), V(k, y_1)) \leq C_3 \rho(z_k, y_k^1)^\alpha,$$
where $C_3 > 0$ is a constant. Since $z_k, y_k^1 \in T_k^1(w, q)$, we obtain that
$$\rho(z_k, y_k^1) \leq C_4 q_k,$$

where $C_4 > 0$ is a constant. This implies that

$$\left|\mathrm{Jac}(d_{z_k} f^{-1}|V(k,z)) - \mathrm{Jac}(d_{y_k^1} f^{-1}|V(k,y_1))\right| \leq C_3 C_4 (e^{-\varepsilon k} q)^\alpha.$$

Note that for any $0 < k \leq n$ and $z \in T_0^1(w,q)$ we have

$$C_5^{-1} \leq |\mathrm{Jac}(d_{z_k} f^{-1}|V(k,z))| \leq C_5,$$

where $C_5 > 0$ is a constant. We obtain

$$\begin{aligned}
\frac{H^1(z_n, n)}{H^1(y_n^1, n)} &= \prod_{k=1}^n \frac{\mathrm{Jac}(d_{z_k} f^{-1}|V(k,z))}{\mathrm{Jac}(d_{y_k^1} f^{-1}|V(k,y_1))} \\
&= \exp \sum_{k=1}^n \log \frac{\mathrm{Jac}(d_{z_k} f^{-1}|V(k,z))}{\mathrm{Jac}(d_{y_k^1} f^{-1}|V(k,y_1))} \\
&\leq \exp \sum_{k=1}^n \left(\frac{\mathrm{Jac}(d_{z_k} f^{-1}|V(k,z))}{\mathrm{Jac}(d_{y_k^1} f^{-1}|V(k,y_1))} - 1 \right) \\
&\leq \exp \left(\sum_{k=1}^n C_3 C_4 C_5 (e^{-\varepsilon k} q)^\alpha \right) \\
&\leq \exp \left(\frac{C_3 C_4 C_5 q^\alpha}{1 - e^{-\varepsilon \alpha}} \right) < \infty.
\end{aligned}$$

This proves the second inequality. The first inequality can be proven in a similar fashion. $\square$

Lemma 8.15 allows one to compare the measures of the preimages under f^{-n} of $T_n^1(w,q)$ and $T_n^2(w,q)$. More precisely, the following statement holds.

Lemma 8.16. *There exist $K_3 > 0$ and $n_2(\ell) > 0$ such that for any $w \in \Lambda_\iota^\ell \cap B(x,\delta)$ and $n \geq n_2(\ell)$ we have*

$$\frac{m_{T^1}(f^{-n}(T_n^1(w,q)))}{m_{T^2}(f^{-n}(T_n^2(w,q)))} \leq K_3.$$

Proof of the lemma. For $i = 1, 2$ we have

$$\begin{aligned}
m_{T^i}(f^{-n}(T_n^i(w,q))) &= \int_{T_n^i(w,q)} H^i(z,n) \, dm_{T_n^i}(z) \\
&= H^i(z_n^i, n) m_{T_n^i}(T_n^i(w,q)),
\end{aligned}$$

where $z_n^i \in T_n^i(w,q)$ are some points. It follows from the assumptions of the lemma, (8.14), and Lemmas 8.13 and 8.15 that for sufficiently large n and

8.2. A proof of the Absolute Continuity Theorem

sufficiently small q,

$$\frac{m_{T^1}(f^{-n}(T_n^1(w,q)))}{m_{T^2}(f^{-n}(T_n^2(w,q)))}$$
$$\leq \left|\frac{H^1(z_n^1,n)}{H^2(z_n^2,n)}\right| \times \frac{\nu_{T_n^1}(T_n^1(w,q))}{m_{T_n^2}(T_n^1(w,q))}$$
$$\leq \left|\frac{H^1(z_n^1,n)}{H^1(y_n^1,n)} \cdot \frac{H^2(y_n^2,n)}{H^2(z_n^2,n)} \cdot \frac{H^1(y_n^1,n)}{H^2(y_n^2,n)}\right|$$
$$\leq K_2^3 + K_1 \leq K_3$$

for some $K_3 > 0$. The lemma follows. $\square$

Step 4. Given $n > 0$, choose points $w_j \in \Lambda_\iota^\ell \cap B(x,r)$ as in Lemma 8.14. Consider the sets

$$\hat{T}_n^1 = \bigcup_{j=1}^t T_n^1(w_j,q), \quad \hat{T}_n^2 = \bigcup_{j=1}^t T_n^2(w_j,q).$$

Note that $\hat{T}_n^1 \supset f^n(B^1(\tau))$ and $\hat{T}_n^2 \supset f^n(B^1(\tau))$ (see Figure 8.3).

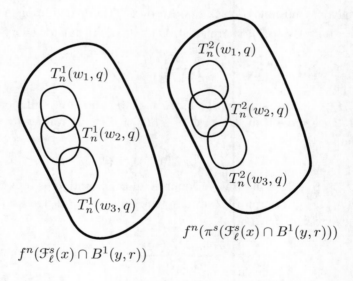

Figure 8.3. Sets $f^n(\mathcal{F}_\ell^s(x) \cap T^i)$ and their covers by $T_n^i(w_j,q)$.

We wish to compare the measures $m_{T^1}|f^{-n}(\hat{T}_n^1)$ and $m_{T^2}|f^{-n}(\hat{T}_n^2)$. For $i = 1, 2$ let L_i be the multiplicity of the cover of the set $B^i(\tau)$ by the sets

$\{f^{-n}(T_n^i(w_j, q))\}$ (see Lemma 8.14) and let $L = \max\{L_1, L_2\}$. We have

$$\frac{1}{L}\sum_{j=1}^{t} m_{T^i}(f^{-n}(T_n^i(w_j, q))) \leq m_{T^i}(f^{-n}(\hat{T}_n^i))$$

$$\leq \sum_{j=1}^{t} m_{T^i}(f^{-n}(T_n^i(w_j, q))).$$

It follows from (8.7) and Lemma 8.16 that

$$\frac{m_{T^2}(f^{-n}(\hat{T}_n^2))}{m_{T^1}(f^{-n}(\hat{T}_n^1))} \leq L \frac{\sum_{j=1}^{t(n)} m_{T^2}(f^{-n}(T_n^2(w_j, q)))}{\sum_{j=1}^{t(n)} m_{T^1}(f^{-n}(T_n^1(w_j, q)))}$$

$$\leq L \max\left\{\frac{m_{T^2}(f^{-n}(T_n^2(w_j, q)))}{m_{T^1}(f^{-n}(T_n^1(w_j, q)))} : j = 1, \ldots, t(n)\right\}$$

$$\leq LK_3,$$

with a similar bound for the inverse ratio. We conclude that

$$K_4^{-1} \leq \frac{m_{T^2}(f^{-n}(\hat{T}_n^2))}{m_{T^1}(f^{-n}(\hat{T}_n^1))} \leq K_4, \tag{8.15}$$

where $K_4 > 0$ is a constant independent of n.

Step 5. Given a number $\beta > 0$, denote by U_β^i the β-neighborhood of the set $B^i(\tau)$, $i = 1, 2$. By (8.7), there exists $\beta_0 > 0$ such that for every $0 < \beta \leq \beta_0$,

$$\frac{m_{T^1}(B^1(\tau))}{m_{T^1}(U_\beta^1)} \geq \frac{1}{2}.$$

For any $\beta > 0$, any sufficiently large $n > 0$, and any sufficiently small q, for $i = 1, 2$ we have that $B^i(\tau) \subset f^{-n}(\hat{T}_n^i) \subset U_\beta^i$. It follows from (8.7) and (8.15) that

$$m_{T^2}(f^{-n}(\hat{T}_n^2)) > C$$

where $C > 0$ is a constant independent of n and q. This implies that $m_{T^2}(B^2(\tau)) > 0$. Therefore, reducing the number β_0 if necessary, we obtain that for every $0 < \beta \leq \beta_0$,

$$\frac{m_{T^2}(B^2(\tau))}{m_{T^2}(U_\beta^2)} \geq \frac{1}{2}.$$

It follows from (8.15) that

$$\frac{1}{2K_4} \leq \frac{m_{T^1}(B^1(\tau))}{m_{T^2}(B^2(\tau))} \leq 2K_4.$$

This completes the proof of Theorem 8.2.

Remark 8.17. One can obtain a sharper estimate on the Jacobian of the holonomy map. Consider the family of local stable manifolds $\mathcal{L}^s(x)$ (see (8.3)). Let T^1 and T^2 be two transversals to this family. We can choose them such that for $i = 1, 2$ the set $\exp_x^{-1} T^i$ is the graph of a smooth map $\psi^i \colon B^u(q) \subset E^u(x) \to E^s(x)$ (for some $q > 0$) with sufficiently small C^1 norm. Set
$$\Delta = \Delta(T^1, T^2) = \|\psi^1 - \psi^2\|_{C^1}.$$

Theorem 8.18. *Under the assumption of Theorem 8.2 there exists a constant $K = K(\ell) > 0$ such that for almost every $y \in \mathcal{F}_\ell^s(x) \cap T^1$,*
$$|\mathrm{Jac}(\pi^s)(y) - 1| \leq K\Delta.$$

Remark 8.19. Let Λ be a nonuniformly *partially* hyperbolic set for a $C^{1+\alpha}$ diffeomorphism of a compact Riemannian manifold M. In particular, for every $x \in \Lambda$ we have a splitting
$$T_x M = E^s(x) \oplus E^0(x) \oplus E^u(x)$$
into invariant stable $E^s(x)$, center $E^0(x)$, and unstable $E^u(x)$ subspaces. Consider the corresponding collections of invariant level sets $\{\Lambda_\iota\}$ and regular sets $\{\Lambda_\iota^\ell\}$ given by (5.33) and (5.34), where the index set ι consists of numbers $\{\lambda_1, \lambda_2, \mu_1, \mu_2, \varepsilon, j_1, j_2\}$ satisfying (5.32) and $\ell \geq 1$. For every level set Λ_ι and every point $x \in \Lambda_\iota$ one can construct local stable $V^s(x)$ and unstable $V^u(x)$ manifolds (see Section 7.3.5). Their sizes may vary with x but are uniformly bounded away from zero on every regular set Λ_ι^ℓ.

In a manner similar to that in Sections 8.1.1 and 8.1.2 one can define the $(u0)$-canonical local transversal $T^{(u0)}(x)$ through x, the family of s-manifolds $\mathcal{L}^s(x)$, the s-foliation block $\mathcal{F}_\ell^s(x)$, and the holonomy map π^s (see (8.2)–(8.5)). One can then extend Theorems 8.2 and 8.4 to the case of nonuniformly partially hyperbolic maps as well as prove their analogs to the family of u-manifolds $\mathcal{L}^u(x)$.

8.3. Computing the Jacobian of the holonomy map

In this section we strengthen the Absolute Continuity Theorem (see Theorem 8.2) and obtain an explicit formula for the Jacobian of the holonomy map.

Theorem 8.20. *Under the assumption of Theorem 8.2, for any transversals T^1, T^2 and almost every $y \in \mathcal{F}_\ell^s(x) \cap T^1$ (with respect to the leaf-volume m_{T^1}),*
$$\mathrm{Jac}(\pi^s)(y) = \prod_{k=0}^{\infty} \frac{\mathrm{Jac}(d_{f^k(\pi^s(y))} f^{-1} | T_{f^k(\pi^s(y))} f^k(T^2))}{\mathrm{Jac}(d_{f^k(y)} f^{-1} | T_{f^k(y)} f^k(T^1))} \tag{8.16}$$

(in particular, the infinite product on the right-hand side converges).

Proof. Fix a point $y \in \mathcal{F}_\ell^s(x) \cap T^1$ and for $k \geq 0$ set $z_k = f^k(\pi^s(y))$ and $y_k = f^k(y)$. Repeating the argument in the proof of Lemma 8.15, it is straightforward to verify that the infinite product

$$\prod_{k=0}^{\infty} \frac{\mathrm{Jac}(d_{z_k} f^{-1} | T_{z_k} f^k(T^2))}{\mathrm{Jac}(d_{y_k} f^{-1} | T_{y_k} f^k(T^1))} =: J$$

converges. Therefore, given $\varepsilon > 0$, there exists $n_1 > 0$ such that for every $n \geq n_1$ we have

$$\left| J - \prod_{k=0}^{n-1} \frac{\mathrm{Jac}(d_{z_k} f^{-1} | T_{z_k} f^k(T^2))}{\mathrm{Jac}(d_{y_k} f^{-1} | T_{y_k} f^k(T^1))} \right| \leq \varepsilon. \tag{8.17}$$

Fix $n > 0$. One can choose small neighborhoods $U_n^1 \subset T^1$ and $U_n^2 \subset T^2$ of the points y and $\pi^s(y)$, respectively, and define the holonomy map

$$\pi_n^s \colon f^n(U_n^1) \to f^n(U_n^2) \quad \text{by} \quad \pi_n^s = f^n \circ \pi^s \circ f^{-n}.$$

We have

$$\mathrm{Jac}(\pi^s)(y) = \prod_{k=0}^{n-1} \frac{\mathrm{Jac}(d_{z_k} f^{-1} | T_{z_k} f^k(T^2))}{\mathrm{Jac}(d_{y_k} f^{-1} | T_{y_k} f^k(T^1))} \mathrm{Jac}(\pi_n^s)(f^n(y)). \tag{8.18}$$

We now choose $n > n_1$ so large and neighborhoods U_n^1 and U_n^2 so small that

$$\Delta(f^n(U_n^1), f^n(U_n^2)) \leq \varepsilon$$

(see Remark 8.17). Theorem 8.18, applied to the holonomy map π_n^s, yields that for almost every $y \in \mathcal{F}_\ell^s(x) \cap T^1$,

$$|\mathrm{Jac}(\pi_n^s)(f^n(y)) - 1| \leq K\varepsilon. \tag{8.19}$$

It follows from (8.17), (8.18), and (8.19) that

$$|\mathrm{Jac}(\pi^s)(y) - J| \leq \varepsilon + (J + \varepsilon)K\varepsilon.$$

Since ε is arbitrary, this completes the proof of the theorem. $\square$

Remark 8.21. A formula similar to (8.16) holds for the unstable holonomy map π^u. More precisely, for any transversals T^1, T^2 and almost every $y \in \mathcal{F}_\ell^u(x) \cap T^1$ (with respect to the leaf-volume m_{T^1}),

$$\mathrm{Jac}(\pi^u)(y) = \prod_{k=0}^{\infty} \frac{\mathrm{Jac}(d_{f^k(\pi^s(y))} f | T_{f^k(\pi^s(y))} f^k(T^2))}{\mathrm{Jac}(d_{f^k(y)} f | T_{f^k(y)} f^k(T^1))}.$$

8.4. An invariant foliation that is not absolutely continuous

We describe an example due to Katok of a continuous foliation with smooth leaves that is not absolutely continuous (another version of this example can be found in [**82**]). We will see below that the Lyapunov exponent along this foliation is zero at every point in the manifold. In Section 13.3 we present an elaborate discussion of foliations with smooth leaves that are not absolutely continuous such that the Lyapunov exponents along the foliation are all negative.

Consider a hyperbolic automorphism T of the torus $\mathbb{T}^2$ generated by the matrix $A = \begin{pmatrix} 2 & 1 \\ 1 & 1 \end{pmatrix}$ (see Section 1.1). Given $\varepsilon > 0$, let $\{\psi_t : t \in [0,1]\}$ be a one-parameter family of real-valued C^∞ functions on $[0,1]$ satisfying:

(a) $0 < 1 - \psi_t(u) \leq \varepsilon$ for $u \in [0,1]$;

(b) $\psi_t(u) = 1$ for $u \geq r_0$ for some $0 < r_0 < 1$;

(c) $\psi_t'(u) \geq 0$ for every $0 \leq u < r_0$;

(d) ψ_t depends smoothly on t and $\psi_0(u) = \psi_1(u) = 1$ for $u \in [0,1]$.

For $t \in [0,1]$ let $f_t = G_{\mathbb{T}^2}(\psi_t)$ be the Katok map constructed via the function ψ_t (see the end of Section 1.5). The map f_t preserves the area m and $f_0 = f_1 = T$. Hence, f_t is a loop through T in the space of C^1 diffeomorphisms of $\mathbb{T}^2$.

Exercise 8.22. Show that for a sufficiently small $\varepsilon > 0$ and any family of functions $\{\psi_t : t \in [0,1]\}$ satisfying conditions (a)–(d) above, the family $\{f_t : t \in S^1\}$ has the following properties:

(1) f_t is a small perturbation of T for every $t \in S^1$; in particular, f_t is an Anosov diffeomorphism (see also Exercise 1.30);

(2) f_t depends smoothly on t.

Since for any $t \in [0,1]$ the map f_t is Anosov and preserves area m, by Theorem 1.1 it is ergodic with respect to m. We denote by χ_t^1 and χ_t^2 the Lyapunov exponents of m for f_t, so that $\chi_t^1 < 0 < \chi_t^2$. By the entropy formula (9.15), the metric entropy $h_m(f_t)$ of f_t with respect to m is $h_m(f_t) = \chi_t^2$ (see Section 9.3.1 for the definition of the metric entropy and relevant results). Also by results in Section 1.5.3, for all sufficiently small $t > 0$ we have $\chi_t^2 < \log \frac{1+\sqrt{5}}{2} = \chi_0^2$. In particular, for those t we obtain that

$$h_m(f_t) < h_m(f_0). \tag{8.20}$$

We introduce a diffeomorphism $F \colon \mathbb{T}^2 \times S^1 \to \mathbb{T}^2 \times S^1$ by

$$F(x,t) = (f_t(x), t).$$

Since f_t is sufficiently close to T, they are conjugate via a Hölder homeomorphism g_t, i.e.,
$$f_t = g_t \circ T \circ g_t^{-1}$$
(see Section 1.1). Given $x \in \mathbb{T}^2$, consider the set
$$H(x) = \{(g_t(x), t) : t \in S^1\}.$$
It is diffeomorphic to the circle S^1 and the collection of these sets forms a foliation H of $\mathbb{T}^2 \times S^1 = \mathbb{T}^3$ which is invariant under F, i.e., $F(H(x)) = H(T(x))$. Note that $H(x)$ depends Hölder continuously on x. However, the holonomy maps associated with the foliation H are not absolutely continuous. To see this, consider the holonomy map
$$\pi_{t_1, t_2} \colon \mathbb{T}^2 \times \{t_1\} \to \mathbb{T}^2 \times \{t_2\}.$$
We have that
$$\pi_{0,t}(x, 0) = (g_t(x), t)$$
and $F(\pi_{0,t}(x,0)) = \pi_{0,t}(T(x), 0)$. If the map $\pi_{0,t}$ (with t being fixed) were absolutely continuous, the measure $(\pi_{0,t})_* m$ would be absolutely continuous with respect to m. Since f_t is ergodic, m is the only absolutely continuous f_t-invariant probability measure. This implies that $(\pi_{0,t})_* m = m$.

The metric entropy of the map f_t is preserved by the conjugacy. Hence, $h_m(f_t) = h_m(f_0)$ for any sufficiently small t. This contradicts (8.20), and hence, the holonomy maps associated with the foliation H are not absolutely continuous.

Chapter 9

Ergodic Properties of Smooth Hyperbolic Measures

In this chapter we move to the core of smooth ergodic theory and consider smooth dynamical systems preserving smooth hyperbolic invariant measures (i.e., invariant measures which are equivalent to the Riemannian volume and have nonzero Lyapunov exponents almost everywhere). A sufficiently complete description of their ergodic properties is one of the main manifestations of the above results on local instability (see Sections 7.1 and 7.3) and absolute continuity (see Chapter 8). It turns out that smooth hyperbolic invariant measures have an abundance of ergodic properties. This makes smooth ergodic theory a deep and well-developed part of the general theory of smooth dynamical systems.

9.1. Ergodicity of smooth hyperbolic measures

One of the main manifestations of absolute continuity is a description of the ergodic properties of diffeomorphisms preserving smooth hyperbolic measures.

Let f be a $C^{1+\alpha}$ diffeomorphism of a smooth compact Riemannian manifold M without boundary. Throughout this chapter we assume that f preserves a *smooth measure* ν, i.e., a measure that is equivalent to the Riemannian volume with density bounded from above and bounded away from zero. We also assume that ν is hyperbolic, i.e., that the set $\mathcal{E}$ defined in (5.3) has full measure.

As shown in Section 5.2.1, $\mathcal{E}$ is a nonuniformly completely hyperbolic set which can be extended to a larger nonuniformly completely hyperbolic set Λ of full measure. We denote by $\{\Lambda_\iota\}_{\iota=\{\lambda,\mu,\varepsilon,j\}}$ the corresponding collection of level sets (see (5.28)) and by $\{\Lambda_\iota^\ell\}_{\ell\geq 1}$ the corresponding collection of regular sets (see (5.29)).

9.1.1. Conditional measures on local stable and unstable manifolds.
For every point $x \in \Lambda$ there are an index set ι and a number $\ell \geq 1$ such that $x \in \Lambda_\iota^\ell$. Without loss of generality we may assume that $\nu(\Lambda_\iota^\ell) > 0$. Consider the s-foliation block $\mathcal{F}_\ell^s(x)$ and its partition by s-manifolds V_z^s, $z \in A^u(x)$; see Section 8.1.1. We denote by $\nu^s(z)$ the conditional measure generated by ν on V_z^s. Note that if $y \in \Lambda_\iota^\ell$ is such that for some $z_1 \in A^u(x)$ and $z_2 \in A^u(y)$ the intersection $U = V_{z_1}^s \cap V_{z_2}^s$ is not empty, then $\nu^s(z_1)|U = \nu^s(z_2)|U$ up to a normalization factor. It follows that for ν-almost every $x \in \Lambda$ there is a well-defined *conditional measure* $\nu^s(x)$ on the local stable manifold $V^s(x)$. Similarly, for almost every $x \in \Lambda$ one can define the notion of the *conditional measure* $\nu^u(x)$ on the local unstable manifold $V^u(x)$.

Exercise 9.1. Show that Theorem 8.4 holds if the volume is replaced by the measure ν. Use this to show that the conditional measures $\nu^s(x)$ and $\nu^u(x)$ on local stable and unstable manifolds, respectively, are equivalent to the leaf-volumes on these manifolds. Show also that Theorem 8.7 extends to smooth measures.

9.1.2. Ergodic decomposition.
We begin by describing the decomposition of the measure ν into its ergodic components. Recall that a measurable partition χ is a *decomposition into ergodic components* if for almost every element C of χ the map $S|C$ is ergodic with respect to the conditional measure ν_C induced by ν.

The following theorem established in [**89**] is one of the main results of smooth ergodic theory. It shows that every ergodic component of ν has positive measure. The proof exploits a simple yet deep argument due to Hopf [**56**] (see the proof of Lemma 9.4 below) and is known as *Hopf's argument*.[1]

---

[1] Hopf studied ergodicity of geodesic flows on surfaces of negative curvature and exploited the fact that the stable and unstable foliations of the flow are smooth. However, this may no longer be the case for flows on manifolds of negative curvature in higher dimension. Anosov [**5**] (and also Anosov and Sinai [**6**]) realized that Hopf's argument works well if the stable and unstable foliations are only absolutely continuous. Later Pesin [**89**] has modified Hopf's argument, so that it can be used to study ergodicity of dynamical systems with nonzero Lyapunov exponents.

9.1. Ergodicity of smooth hyperbolic measures

Theorem 9.2. *There exist invariant sets* $\Lambda_0, \Lambda_1, \ldots$ *such that:*

(1) $\bigcup_{i \geq 0} \Lambda_i = \Lambda$ *and* $\Lambda_i \cap \Lambda_j = \varnothing$ *whenever* $i \neq j$;

(2) $\nu(\Lambda_0) = 0$ *and* $\nu(\Lambda_i) > 0$ *for each* $i \geq 1$;

(3) $f|\Lambda_i$ *is ergodic for each* $i \geq 1$.

Proof. We begin with the following statement.

Lemma 9.3. *For any f-invariant Borel function $\varphi \in L^1(M, \nu)$ there exists a set $N_\varphi \subset M$ of measure zero such that for any $x \in \Lambda$ we have:*

(1) *if* $y, z \in V^s(x) \setminus N_\varphi$, *then* $\varphi(y) = \varphi(z)$;

(2) *if* $y, z \in V^u(x) \setminus N_\varphi$, *then* $\varphi(y) = \varphi(z)$.

Proof of the lemma. Choose a continuous function ψ and define

$$\overline{\psi}(w) = \lim_{n \to \infty} \frac{1}{2n+1} \sum_{k=-n}^{n} \psi(f^k(w)),$$

$$\psi^+(w) = \lim_{n \to \infty} \frac{1}{n} \sum_{k=1}^{n} \psi(f^k(w)), \quad \psi^-(w) = \lim_{n \to \infty} \frac{1}{n} \sum_{k=1}^{n} \psi(f^{-k}(w)).$$
(9.1)

By Birkhoff's Ergodic Theorem, there is a set $N_\psi \subset M$ such that $\nu(N_\psi) = 0$ and for every $w \in M \setminus N_\psi$ the above limits exist and $\overline{\psi}(w) = \psi^+(w) = \psi^-(w)$. Moreover, the functions $\overline{\psi}(w)$, $\psi^+(w)$, and $\psi^-(w)$ are integrable and invariant under f.

Fix $x \in \Lambda$. Since the function ψ is continuous, we obtain that:

(a) if $y, z \in V^s(x)$, then $\rho(f^n(y), f^n(z)) \to 0$ as $n \to +\infty$, and hence, $\psi^+(y) = \psi^+(z)$;

(b) if $y, z \in V^u(x)$, then $\rho(f^n(y), f^n(z)) \to 0$ as $n \to -\infty$, and hence, $\psi^-(y) = \psi^-(z)$.

Since the points y and z do not lie in N_ψ, we conclude that

$$\overline{\psi}(y) = \psi^+(y) = \psi^+(z) = \overline{\psi}(z) \quad \text{for } y, z \in V^s(x);$$

$$\overline{\psi}(y) = \psi^-(z) = \psi^-(z) = \overline{\psi}(z) \quad \text{for } y, z \in V^u(x).$$

Given an f-invariant Borel function $\varphi \in L^1(M, \nu)$, we can find a sequence $(\psi_k)_{k \geq 1}$ of continuous functions such that $\|\varphi - \psi_k\|_{L^1} < \frac{1}{k}$ where $\|\cdot\|_{L^1}$ is the standard L^1 norm in $L^1(M, \nu)$. Fix $n \geq 0$. Since both the measure ν and the function φ are invariant, for any $0 \leq i < n$ we have that

$$\|\varphi - \psi_k \circ f^i\|_{L^1} = \|\varphi \circ f^i - \psi_k \circ f^i\|_{L^1} = \|\varphi - \psi_k\|_{L^1} < \frac{1}{k},$$

and hence,
$$\left\|\varphi - \frac{1}{n}\sum_{i=1}^{n}\psi_k \circ f^i\right\|_{L^1} < \frac{1}{k}.$$
Applying the Lebesgue Dominated Convergence Theorem, we obtain
$$\|\varphi - \psi_k^+\|_{L^1} < \frac{1}{k}.$$
Hence, we can find a subsequence ψ_{n_k} such that $\varphi(x) = \lim_{k\to\infty}\psi_{n_k}(x)$ for every $x \in M$ outside a set $\tilde{N}$ of measure zero. Setting $N_\varphi = \tilde{N} \cup \bigcup_{k \geq 1} N_{\psi_{n_k}}$ completes the proof of the lemma. □

We proceed with the proof of the theorem. Fix a regular set Λ_ι^ℓ and let $x \in \Lambda_\iota^\ell$ be a Lebesgue density point.[2] For each $r > 0$ set
$$P_\ell^s(x,r) = \bigcup_{y \in \Lambda_\iota^\ell \cap B(x,r)} V^s(y). \tag{9.2}$$

Lemma 9.4. *There exists $r = r(\ell) > 0$ such that the map f is ergodic on the set*
$$P^s(x) = \bigcup_{n \in \mathbb{Z}} f^n(P_\ell^s(x,r)). \tag{9.3}$$

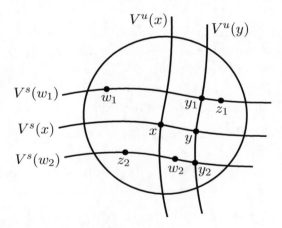

Figure 9.1. Illustration to the proof of Lemma 9.4; $y_1, y_2 \notin R^s$, $z_1, z_2 \notin N_\varphi$.

Proof of the lemma. Let φ be an f-invariant function. We wish to prove that φ is constant almost everywhere and to achieve this it suffices to show this for the restriction $\varphi|P_\ell^s(x,r)$. Let N_φ be the set of measure zero constructed in Lemma 9.3. Choose $0 < r < \min\{r_\ell, \delta_\ell\}$ (see (7.40)), a regular set Λ_ι^ℓ of positive measure, and a point $x \in \Lambda_\iota^\ell$. Consider the collection

---
[2] Recall that given a Borel subset $A \subset M$ of positive measure, a point $x \in A$ is called a Lebesgue density point if $\lim_{r \to 0} \nu(A \cap B(x,r))/\nu(B(x,r)) = 1$.

of local unstable manifolds $\{V^u(y) : y \in \Lambda_\iota^\ell \cap B(x,r)\}$. By statement (5) of Theorem 8.7 and Exercise 9.1, there exists $y \in \Lambda_\iota^\ell \cap B(x,r)$ such that $m^u(y)(V^u(y) \cap N_\varphi) = 0$ (indeed, almost every $y \in \Lambda_\iota^\ell \cap B(x,r)$ has this property). Let
$$R^s = \bigcup V^s(w),$$
where the union is taken over all points $w \in \Lambda_\iota^\ell \cap B(x,r)$ for which $V^s(w) \cap V^u(y) \in N_\varphi$. Using again Theorem 8.7 and Exercise 9.1, we obtain that $\nu(R^s) = 0$.

To prove the lemma we need to show that $\varphi(z_1) = \varphi(z_2)$ for any $z_1, z_2 \in P_\ell^s(x,r) \setminus (R^s \cup N_\varphi)$. There are points $w_i \in \Lambda_\iota^\ell \cap B(x,r)$ such that $z_i \in V^s(w_i)$ for $i = 1, 2$. Note that the intersection $V^s(w_i) \cap V^u(y)$ is nonempty and consists of a single point y_i for $i = 1, 2$ (see Figure 9.1) and that $z_i, y_i \notin N_\varphi$. By Lemma 9.3, $\varphi(z_1) = \varphi(y_1) = \varphi(y_2) = \varphi(z_2)$. This completes the proof of the lemma. □

Lemma 9.5. *We have $\nu(P^s(x)) > 0$.*

Proof of the lemma. Since $P^s(x) \supset P_\ell^s(x,r) \supset \Lambda_\iota^\ell \cap B(x,r)$, we have that $\nu(P^s(x)) \geq \nu(\Lambda_\iota^\ell \cap B(x,r)) > 0$ and the lemma follows. □

We say that the index set $\iota = \{\lambda, \mu, \varepsilon, j\}$ is *rational* if λ, μ, and ε are rational numbers. Observe that the set Λ can be covered by a countable collection of level sets Λ_ι with rational index sets. Therefore, to prove the theorem it suffices to do this for a level set Λ_ι of positive measure. Since almost every point $x \in \Lambda_\iota$ is a Lebesgue density point of Λ_ι^ℓ for some $\ell \geq 1$, the invariant sets $P^s(x)$ (defined by (9.3) for different values of ℓ) cover the set Λ_ι (mod 0). By Lemma 9.5, there are at most countably many such sets. We denote them by $P_1, P_2, \ldots$. We have $\nu(P_i) > 0$ for each $i \geq 1$, and the set $P_0 = \Lambda_\iota \setminus \bigcup_{i \geq 1} P_i$ has measure zero. By Lemma 9.4, for each $i \geq 1$ the map $f|P_i$ is ergodic, and hence, $P_i \cap P_j = \varnothing$ (mod 0) whenever $i \neq j$. This completes the proof of the theorem. □

As an immediate consequence of Theorem 9.2 we obtain the following result. Let ν_x^s and ν_x^u be, respectively, the conditional measure on $V^s(x)$ and $V^u(x)$ generated by the measure ν.

Theorem 9.6. *For ν-almost every $x \in \Lambda$ we have*
$$\nu^s(x)(V^s(x) \setminus \Lambda) = 0, \quad \nu^u(x)(V^u(x) \setminus \Lambda) = 0.$$

In other words, for almost every $x \in \Lambda$ the local stable and unstable manifolds at x lie in Λ up to a set of zero conditional measure (compare to Remark 8.10).

Proof. Let Λ_ι^ℓ be a regular set of positive measure for some index set ι and $\ell \geq 1$. It suffices to prove the theorem for every Lebesgue density point $x \in \Lambda_\iota^\ell$. Consider the sets $P_\ell(x,r)$ and $P(x)$ given by (9.5) below. We have $\nu(P_\ell(x,r) \cap \Lambda_\iota^\ell) > 0$. Let $Q(x) = \bigcup_{n \in \mathbb{Z}} f^n(P_\ell(x,r) \cap \Lambda_\iota^\ell)$. This is an invariant set of positive measure in $P(x)$ and since $f|P(x)$ is ergodic (see Remark 9.7 below), we obtain that $Q(x) = P(x) \pmod 0$. Since $P(x) \subset \Lambda \pmod 0$, the desired result now follows from Theorem 8.7 and Exercise 9.1. □

Remark 9.7. Set

$$P_\ell^u(x,r) = \bigcup_{y \in \Lambda_\iota^\ell \cap B(x,r)} V^u(y), \quad P^u(x) = \bigcup_{n \in \mathbb{Z}} f^n(P_\ell^u(x,r)) \qquad (9.4)$$

and

$$P_\ell(x,r) = \bigcup_{y \in \Lambda_\iota^\ell \cap B(x,r)} (V^s(y) \cup V^u(y)), \quad P(x) = \bigcup_{n \in \mathbb{Z}} f^n(P_\ell(x,r)). \qquad (9.5)$$

The sets $P_\ell^s(x,r)$, $P_\ell^u(x,r)$, and $P_\ell(x,r)$ in (9.2), (9.4), and (9.5), respectively, are known as *Pesin blocks*. Repeating the argument in the proof of Lemma 9.4, one can show that $f|P^u(x)$ and $f|P(x)$ are ergodic, and hence, $P^s(x) = P^u(x) = P(x) \pmod 0$. It follows that every ergodic component Λ_m of positive measure is of the form

$$\Lambda_m = \bigcup_{n \in \mathbb{Z}} f^n(P_\ell(x,r)) = \bigcup_{n \in \mathbb{Z}} f^n(P_\ell^s(x,r)) = \bigcup_{n \in \mathbb{Z}} f^n(P_\ell^u(x,r)),$$

where x is a Lebesgue density point of Λ_ι^ℓ for some index set ι and $\ell \geq 1$.

Remark 9.8. Given a hyperbolic periodic point $p \in \Lambda$, set

$$\Lambda^s(p) = \{x \in \mathcal{R} : W^s(x) \text{ intersects transversally } W^u(O(p))\},$$
$$\Lambda^u(p) = \{x \in \mathcal{R} : W^u(x) \text{ intersects transversally } W^s(O(p))\}.$$

Here $\mathcal{R}$ is the set of LP-regular points, $O(p)$ is the orbit of p, and $W^s(x)$, $W^u(x)$ are, respectively, global stable and unstable manifolds at x. The sets $\Lambda^s(p)$ and $\Lambda^u(p)$ are called, respectively, *s-saturated* and *u-saturated*, and the set $\Lambda(p) = \Lambda^s(p) \cap \Lambda^u(p)$ is called the *ergodic homoclinic class* of p (see Remark 1.16 on homoclinic classes).

Let Λ_m be an ergodic component of ν of positive measure (see Theorem 9.2). There are an index set ι, $\ell \geq 1$, and a point $x \in \Lambda_\iota^\ell$ such that $\Lambda_m = P(x)$ where $P(x)$ is given by (9.5) (in fact, this is true for almost every $x \in \Lambda_m$). By Theorem 11.10, we can find a periodic point $p \in \Lambda_\iota^{\ell'}$ for some $\ell' \geq \ell$ which is arbitrarily close to x.

9.1. Ergodicity of smooth hyperbolic measures

Exercise 9.9. Show that the set $\Lambda(p)$ has positive measure (and hence, the sets $\Lambda^s(p)$ and $\Lambda^u(p)$ have positive measure too).

It follows that $\Lambda_m = \Lambda(p)$ (mod 0) yielding a somewhat more subtle description of ergodic components: *every ergodic component of positive measure is an ergodic homoclinic class*.

We now assume that the map f possesses a nonuniformly *partially* hyperbolic set Λ of full measure, i.e., that the measure ν is *partially* hyperbolic. In this case ergodic homoclinic classes may a priori contain points with some zero Lyapunov exponents and have zero measure. The following theorem shows that under some additional assumptions, an ergodic homoclinic class $\Lambda(p)$ is an ergodic component of ν of positive measure; moreover, $\nu|\Lambda(p)$ is a hyperbolic measure.

Theorem 9.10 ([97]). *Assume that $\nu(\Lambda^s(p)) > 0$ and $\nu(\Lambda^u(p)) > 0$ for some periodic point p. Then:*

(1) $\Lambda(p) = \Lambda^s(p) = \Lambda^u(p)$ (mod 0);

(2) $f|\Lambda(p)$ *is ergodic and has nonzero Lyapunov exponents almost everywhere.*

We outline a proof of statement (2) assuming that statement (1) is proven, and hence, in particular, $\nu(\Lambda(p)) > 0$. We follow the line of argument in [97]. Observe that for any $\delta > 0$ there is an index set $\iota = \{\lambda_1, \lambda_2, \mu_1, \mu_2, \varepsilon, j_1, j_2\}$ (see (5.32)) such that the invariant set $\Lambda_\iota(p) = \Lambda_\iota \cap \Lambda(p)$ has measure at least $1 - \delta$, and hence, it suffices to show that $f|\Lambda_\iota(p)$ is ergodic.

Without loss of generality we assume that p is a fixed point. Choose a continuous function $\psi \colon \Lambda_\iota(p) \to \mathbb{R}$ and consider the functions ψ^+ and ψ^- given by (9.1). Since ψ is continuous, for every $z \in \Lambda(p)$ the function ψ^+ is constant on $W^s(z)$ and the function ψ^- is constant on $W^u(z)$. By Birkhoff's Ergodic Theorem and Theorem 9.6, there is an invariant set $A_\iota(p) \subset \Lambda_\iota(p)$ with the following properties: $A_\iota(p) = \Lambda_\iota(p)$ (mod 0) and for any $x \in A_\iota(p)$:

(1) x is LP-regular and $\psi^+(x) = \psi^-(x)$;

(2) $W^s(x)$ intersects transversally $W^u(p)$ and $W^u(x)$ intersects transversally $W^s(p)$;

(3) $\nu^s(x)(W^s(x) \setminus A_\iota(p)) = 0$ and $\nu^u(x)(W^u(x) \setminus A_\iota(p)) = 0$, where $\nu^s(x)$ and $\nu^u(x)$ are the conditional measures on $W^s(x)$ and $W^u(x)$, respectively.

By property (1), for $x \in A_\iota(p)$ we have that $j_1 = \dim W^s(x) = \dim W^s(p)$ and $j_2 = \dim W^u(x) = \dim W^u(p)$, and hence, $j_1 + j_2 = \dim M$. This implies that almost every point $x \in \Lambda_\iota(p)$ has nonzero Lyapunov exponents.

Choose $\ell \geq 1$ such that the set $A_\iota^\ell(p) = A_\iota(p) \cap \Lambda_\iota^\ell$ has positive measure. There is a set $B_\iota^\ell(p) \subset A_\iota^\ell(p)$ with the following properties: $B_\iota^\ell(p) = A_\iota^\ell(p)$ (mod 0) and every point $x \in B_\iota^\ell(p)$:

(4) is a Lebesgue density point;

(5) returns to $B_\iota^\ell(p)$ infinitely often in the future and the past.

Points in $B_\iota^\ell(p)$ have properties (1)–(5) above. Moreover, there is $r_\ell > 0$ such that for every $x \in B_\iota^\ell(p)$ the size of the local stable manifold $V^s(x)$ is at least r_ℓ (see Section 7.3.1).

Since the number ℓ can be chosen arbitrarily large, we can prove statement (2) by showing that $\psi^+(x) = \psi^+(y)$ for every $x, y \in B_\iota^\ell(p)$. Thus, we fix $x, y \in B_\iota^\ell(p)$. By property (5), there is a large enough $n > 0$ such that $x_n = f^n(x) \in B_\iota^\ell(p)$ and $\rho(x_n, W^u(p)) < \frac{r_\ell}{4}$.

By property (4), x_n is a Lebesgue density point of $B_\iota^\ell(p)$, and hence, we can find a point $q \in B_\iota^\ell(p)$ near x_n such that for any $r > 0$ if $B^u(q, r) \subset V^u(q)$ is the ball centered at q of radius r, then the set $E(r) = B^u(q, r) \cap B_\iota^\ell(p)$ has positive $\nu^u(q)$-measure in $V^u(q)$. Therefore, the stable holonomy $\pi_{q,x_n}^s : E \to V^u(x_n)$ is well-defined and by Theorem 8.2, the set $F(r) = \pi_{q,x_n}^s(E(r))$ has positive $\nu^u(x_n)$-measure in $V^u(x_n)$.

Since $y \in B_\iota^\ell(p)$, by property (2), its global unstable manifold $W^u(y)$ intersects transversally the global stable manifold $W^s(p)$ at a point which we denote by t. As a consequence of the λ-Lemma (see Section 7.3.3), we can find $k > 0$ so large that if $t_k = f^k(t)$, then $W^u(t_k)$ intersects transversally $V^s(x_n)$ at a point z for which $\rho(x_n, z) < \frac{r_\ell}{2}$. For sufficiently small $r > 0$ the holonomy map $\pi_{x_n,z}$ is well-defined on the set $F(r)$ and by Theorem 8.2, the set $G(r) = \pi_{x_n,z}(F(r))$ has $\nu^u(z)$-positive measure in $V^u(z)$.

By properties (1) and (3), the function ψ^+ is constant on stable and unstable global manifolds of points in $B_\iota^\ell(p)$. Therefore ψ^+ is constant on the set $F(r)$ and takes value $\psi^+(x)$. Hence, ψ^+ is constant on the set $G(r)$ and takes the same value $\psi^+(x)$. On the other hand, since $G(r) \subset W^u(y_k)$, the function ψ^+ is constant on $G(r)$ and takes value $\psi^+(y)$ implying that $\psi^+(x) = \psi^+(y)$. This completes the proof of ergodicity.

9.1.3. An example of a diffeomorphism with nonzero Lyapunov exponents and more than one ergodic components. We describe an example of a diffeomorphism with nonzero Lyapunov exponents that has more than one ergodic component. Consider Katok's diffeomorphism $G_{\mathbb{T}^2}$ of the torus $\mathbb{T}^2$ constructed in Section 1.6. The map $G_{\mathbb{T}^2}$ preserves area and is ergodic. The punctured torus $\mathbb{T}^2 \setminus \{0\}$ is C^∞-diffeomorphic to the manifold $\mathbb{T}^2 \setminus \overline{U}$ with boundary ∂U, where U is a small open disk around 0 and $\overline{U}$ denotes its closure. Therefore, we obtain a C^∞ diffeomorphism $F_{\mathbb{T}^2}$ of the manifold $\mathbb{T}^2 \setminus U$ with $F_{\mathbb{T}^2}|\partial U = \text{Id}$. The map $F_{\mathbb{T}^2}$ preserves a smooth measure, has nonzero Lyapunov exponents, and is ergodic.

Let $(\tilde{M}, \tilde{F}_{\mathbb{T}^2})$ be a copy of $(M, F_{\mathbb{T}^2})$. By gluing the manifolds M and $\tilde{M}$ along ∂U, we obtain a smooth compact manifold $\mathcal{M}$ without boundary and a diffeomorphism $\mathcal{F}$ of $\mathcal{M}$.

Exercise 9.11. Show that the map $\mathcal{F}$ preserves a smooth measure and has nonzero Lyapunov exponents almost everywhere.

Note that the map $\mathcal{F}$ is not ergodic and has two ergodic components of positive measure (M and $\tilde{M}$). Similarly, one can obtain a diffeomorphism with nonzero Lyapunov exponents with n ergodic components of positive measure for an arbitrary n. However, this construction cannot be used to obtain a diffeomorphism with nonzero Lyapunov exponents which has countably many ergodic components of positive measure. Such an example is constructed in Section 14.1 and it illustrates that Theorem 9.2 cannot be improved.

9.1.4. Other ergodic properties of smooth hyperbolic measures. In ergodic theory there is a hierarchy of ergodic properties of which ergodicity (or the description of ergodic components) is the weakest one. Among the stronger properties let us mention, without going into detail, mixing and K-property (including the description of the π-partition). The strongest property that implies all others is the Bernoulli property or the description of Bernoulli components. In connection with smooth ergodic measures the latter is known as the Spectral Decomposition Theorem. It was established in [89].

Theorem 9.12. *For each $i \geq 1$ the following properties hold:*

(1) *the set Λ_i is a disjoint union of sets Λ_i^j, for $j = 1, \ldots, n_i$, which are cyclically permuted by f, i.e., $f(\Lambda_i^j) = \Lambda_i^{j+1}$ for $j = 1, \ldots, n_i - 1$ and $f(\Lambda_i^{n_i}) = \Lambda_i^1$;*

(2) $f^{n_i}|\Lambda_i^j$ *is a Bernoulli automorphism for each j.*

We illustrate the statement of this theorem by considering the diffeomorphism $\mathcal{G} = \mathcal{F} \circ I$, where $\mathcal{F}$ is the diffeomorphism constructed in the previous section and $I \colon \mathcal{M} \to \mathcal{M}$ is the radial symmetry along the boundary of U. This diffeomorphism preserves a smooth measure, has nonzero Lyapunov exponents, and is ergodic. However, the map $\mathcal{G}^2$ is not ergodic and has two ergodic components of positive measure (M and $\tilde{M}$) which are cyclically permuted by $\mathcal{G}$.

9.1.5. The continuous-time case. We now consider the case of a smooth flow φ_t on a compact manifold M. We assume that ν is a smooth measure which is φ_t-invariant. This means that $\nu(\varphi_t A) = \nu(A)$ for any Borel set $A \subset M$ and $t \in \mathbb{R}$.

Note that the Lyapunov exponent along the flow direction is zero. We assume that all other values of the Lyapunov exponent are nonzero for ν-almost every point, and hence, φ_t is a nonuniformly hyperbolic flow (see Section 5.2). We also assume that ν vanishes on the set of fixed points of φ_t.

Since the time-one map of the flow is nonuniformly partially hyperbolic, we conclude that the families of stable and unstable local manifolds possess the absolute continuity property (see Remark 8.19). This allows one to study ergodic properties of nonuniformly hyperbolic flows. The following result from [91] (see also [13, Section 9.6]) describes the decomposition into ergodic components and establishes the Bernoulli property for flows with nonzero Lyapunov exponents with respect to smooth invariant measures.

Theorem 9.13. *There exist invariant sets* $\Lambda_0, \Lambda_1, \ldots$ *such that:*

(1) $\bigcup_{i \geq 0} \Lambda_i = \Lambda$ *and* $\Lambda_i \cap \Lambda_j = \varnothing$ *whenever* $i \neq j$;

(2) $\nu(\Lambda_0) = 0$ *and* $\nu(\Lambda_i) > 0$ *for each* $i \geq 1$;

(3) $\varphi_t | \Lambda_i$ *is ergodic for each* $i \geq 1$;

(4) *if the flow* $\varphi_t | \Lambda_i$ *is weakly mixing,*[3] *then it is a Bernoulli flow.*

Using the flow described in Section 1.7, one can construct an example of a flow with nonzero Lyapunov exponents having an arbitrary finite number of ergodic components.

9.2. Local ergodicity

According to Theorem 9.2, the ergodic components of a hyperbolic smooth measure invariant under a $C^{1+\alpha}$ diffeomorphism of a compact smooth Riemannian manifold M are Borel sets of positive measure. In this section we discuss a problem known as the *local ergodicity problem* which asks for additional conditions that guarantee that every ergodic component is open (up to a set of measure zero).

The following are main obstacles for local ergodicity:

(1) the stable and unstable distributions are measurable but not necessarily continuous;

---

[3] Recall that a flow φ_t preserving a finite measure μ is called *weakly mixing* if for any measurable sets A and B

$$\lim_{t \to \infty} \frac{1}{t} \int_0^t |\mu(A \cap \varphi_\tau(B)) - \mu(A)\mu(B)| \, d\tau = 0.$$

(2) the global stable (or unstable) manifolds may not form a foliation;

(3) the unstable manifolds may not expand under the action of f^n (we remind the reader that they were defined as being exponentially *contracting* under f^{-n}); the same is true for stable manifolds with respect to the action of f^{-n}.

9.2.1. Locally continuous foliations. We describe the approach to local ergodicity developed in [89]. Roughly speaking, it requires that the stable (or unstable) leaves form a foliation of a measurable subset in M of positive measure. First, we extend the notion of continuous foliation of M with smooth leaves introduced in Section 1.1 to the foliation of a measurable subset.

Given a subset $X \subset M$, an integer $k > 0$, and a continuous function $q\colon X \to (0, \infty)$, we call a partition ξ of X a *q-foliation of X with smooth leaves* or simply a *q-foliation of X* if for each $x \in X$:

(1) there exists a smooth immersed k-dimensional submanifold $W(x)$ containing x for which $\xi(x) = W(x) \cap X$ where $\xi(x)$ is the element of the partition ξ containing x; the manifold $W(x)$ is called the *global leaf* of the q-foliation at x; for $y, z \in X$ either $W(y) \cap W(z) = \varnothing$ or $W(y) = W(z)$;

(2) there exists a continuous map $\varphi_x\colon B(x, q(x)) \to C^1(D, M)$ (where $D \subset \mathbb{R}^k$ is the open unit ball) such that setting $V(y) = \varphi_x(y)(D)$ we have: (a) either $V(y) \cap V(z) = \varnothing$ or $V(y) = V(z)$ for $y, z \in B(x, q(x))$, and (b) $V(y) \subset W(y)$ for $y \in X$; the manifold $V(y)$ is called the *local leaf* of the q-foliation at y.

The function $\varphi_x(y, z) = \varphi_x(y)(z)$, $x \in X$, is called the *foliation coordinate chart*. This function is continuous in $y \in B(x, q(x))$ and is differentiable in $z \in D$; moreover, the derivative $(\partial \varphi_x/\partial z)(y, z)$ is continuous in y.

9.2.2. Main theorem. We now present a theorem that provides sufficient conditions for local ergodicity.

Theorem 9.14. *Let f be a $C^{1+\alpha}$ diffeomorphism of a compact smooth Riemannian manifold M preserving a smooth measure ν and let Λ be a nonuniformly hyperbolic set for f of positive measure. Assume that there exists a q-foliation W of Λ such that $W(x) = W^s(x)$ for every $x \in \Lambda$ (where $W^s(x)$ is the global stable manifold at x; see Section 7.3). Then every ergodic component of f of positive measure is open (mod 0) in Λ (with respect to the induced topology). In particular, the set Λ is open (mod 0).*

Proof. By Lemma 9.4, for every ergodic component P of positive measure there exist an index set $\iota = \{\lambda, \mu, \varepsilon, j\}$, a number $\ell > 0$, and a

Lebesgue density point $x \in \Lambda_\iota^\ell$ such that $P = P(x)$ (mod 0) (where $P(x)$ is given by (9.5)). Since $P(x) \subset \Lambda_\iota$, without loss of generality we may restrict ourselves to only consider the set Λ_ι. We wish to show that $P(x)$ is open (mod 0). Reducing, if necessary, the size of the local unstable manifold $V^u(x)$, we may assume without loss of generality that $V^u(x) \subset B(x, q(x))$.[4] By Theorem 9.6, $\nu^u(x)$-almost every point $y \in V^u(x)$ belongs to Λ_ι. Since W is a q-foliation of Λ_ι, there is $\delta = \delta(x) > 0$ such that for any $y \in V^u(x)$ we have $B_{V(y)}(y, \delta) \subset V(y)$ (where $V(y)$ is the local leaf of the foliation W at y and $B_{V(y)}(y, \delta)$ is the ball in $V(y)$ centered at y of radius δ).

For a $\nu^u(x)$-measurable set $Y \subset V^u(x)$ and $0 < r \leq \delta$ let

$$R(x, r, Y) = \bigcup_{y \in Y} B_{V(y)}(y, r)$$

be a "fence" through the set Y (see Figure 9.2). We also set

$$R(r) = R(x, r, V^u(x)), \quad \tilde{R}(r) = R(x, r, V^u(x) \cap \Lambda_\iota),$$

and for $m \geq 1$

$$R^m(r) = R(x, r, V^u(x) \cap \Lambda_\iota^m).$$

Clearly, $R^m(r) \subset \tilde{R}(r) \subset R(r)$.

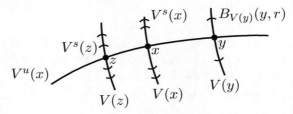

Figure 9.2. The "fence" $R(r)$; $z \in \Lambda_\iota^m$, $x \in \Lambda_\iota^\ell$, $m > \ell$.

By Theorem 7.4 and condition (7.40), for every $m \geq 1$ there exists $0 < r_m \leq \delta$ such that $B_{V(y)}(y, r_m) \subset V^s(y)$ for any $y \in V^u(x) \cap \Lambda_\iota^m$.

Fix $r \in (0, r_m]$. Given $y \in R^m(r/2)$, we denote by $n_i(y)$ the successive return times of the positive semitrajectory of y to the set $R^m(r/2)$. We also denote by $z_i \in V^u(x) \cap \Lambda_\iota^m$ a point for which $f^{n_i(y)}(y) \in B_{V(y)}(z_i, r/2)$. In view of the Poincaré Recurrence Theorem, one can find a subset $N \subset M$ of zero measure for which the sequence $n_i(y)$ is well-defined provided that $y \in R^m(r/2) \setminus N$.

Lemma 9.15. *For every $y \in R^m(r/2) \setminus N$ we have*

$$W^s(y) = \bigcup_{i=1}^\infty f^{-n_i(y)}(B_{V(y)}(z_i, r)). \tag{9.6}$$

---

[4]Note that decreasing the size of $V^u(x)$ will require choosing a smaller number r in (9.2) but the corresponding set $P(x)$ in (9.5) is still the same ergodic component of ν.

9.2. Local ergodicity

Proof of the lemma. Let $y \in R^m(r/2) \setminus N$ and $w \in W^s(y)$. By Theorem 7.16, we have $\rho(f^{n_i(y)}(y), f^{n_i(y)}(w)) \leq r/2$ for all sufficiently large i. Therefore,
$$f^{n_i(y)}(w) \in B_{V(y)}(f^{n_i(y)}(y), r/2) \subset B_{V(y)}(z_i, r)$$
and the lemma follows. □

Denote by $\xi^m(r)$ the partition of the set $R^m(r)$ into the sets $B_{V(y)}(y, r)$, $y \in V^u(x) \cap \Lambda_\iota^m$.

Lemma 9.16. *The partition $\xi^m(r/2)$ is measurable and has the following properties:*

(1) *the conditional measure on the element $B_{V(y)}(y, r/2)$ of this partition is absolutely continuous with respect to the measure $\nu^s(y)$;*

(2) *the factor-measure on the factor-space $R^m(r/2)/\xi^m(r/2)$ is absolutely continuous with respect to the measure $\nu^u(x)|V^u(x) \cap \Lambda_\iota^m$.*

Proof of the lemma. By Theorem 8.7, Exercise 9.1, and Lemma 9.15, for $\nu^u(x)$-almost every point $w \in V^u(x) \cap \Lambda_\iota^m$ one can find a point $y(w) \in B_{V(y)}(w, r/2) \subset R^m(r/2)$ for which (9.6) holds. Moreover, we can choose $y(w)$ in such a way that the map
$$w \in V^u(x) \cap \Lambda_\iota^m \mapsto y(w) \in R^m(r/2)$$
is measurable. For each $n > 0$ define
$$R_n = \bigcup_{w \in V^u(x) \cap \Lambda_\iota^m} \bigcup_{n_i(y) \leq n} \left(f^{-n_i(y(w))}(B_{V(y)}(z_i, r)) \cap R^m(3r/4) \right).$$

Observe that $R^m(3r/4) = \bigcup_{n \in \mathbb{N}} R_n$, and hence, given $\varepsilon > 0$, there exists $p > 0$ and a set $Y \subset V^u(x) \cap \Lambda_\iota^m$ such that
$$\nu_x^u((V^u(x) \cap \Lambda_\iota^m) \setminus Y) \leq \varepsilon \quad \text{and} \quad R(r/2, Y) \subset \bigcup_{n \leq p} R_n. \quad (9.7)$$

It follows from Theorem 8.4 that the partition $\xi^m(r/2)|R_n$ satisfies statements (1) and (2) of the lemma for each $n > 0$. Since ε is arbitrary, the desired result follows. □

We proceed with the proof of the theorem. Denote by $\xi(r)$ the partition of the set $R(r)$ into the sets $B_{V(y)}(y, r)$ and denote by $\tilde{\xi}(r)$ the partition of the set $\tilde{R}(r)$ into the sets $B_{V(y)}(y, r)$. Since W is a q-foliation of Λ_ι, the factor-space $R(r)/\xi(r)$ can be identified with $V^u(x)$ and the factor-space $\tilde{R}(r)/\tilde{\xi}(r)$ can be identified with $V^u(x) \cap \Lambda_\iota$.

Letting $m \to \infty$ in Lemma 9.16, we conclude that the partition $\tilde{\xi}(r/2)$ also satisfies statements (1) and (2) of the lemma. By Lemma 9.4, we have $P \supset \tilde{R}(r/2)$. This implies that $P = \bigcup_{n \in \mathbb{Z}} f^n(\tilde{R}(r/2))$.

Note that the set $R(r/2)$ is open. We will show that the set

$$A = \Lambda_\iota \cap (R(r/2) \setminus \tilde{R}(r/2)) \tag{9.8}$$

has measure zero. Assuming the contrary, we have that the set

$$A^m = \Lambda_\iota^m \cap (R(r/2) \setminus \tilde{R}(r/2)) \tag{9.9}$$

has positive measure for all sufficiently large m. Therefore, for ν-almost every $z \in A^m$ we obtain that $\nu^u(z)(V^u(z) \cap A^m) > 0$. Consider the set $R(z, r/2, A^m)$. Clearly,

$$R(z, r/2, A^m) \subset R(r/2) \setminus \tilde{R}(r/2).$$

Let $B^m = R(z, r/2, A^m) \cap V^u(x)$. This set is nonempty. It follows from (9.9) that $B^m \subset V^u(x) \setminus \Lambda_\iota$, and thus, $\nu^u(x)(B^m) = 0$ (see Theorem 9.6). Moreover, repeating arguments in the proof of Lemma 9.16 (see (9.7)), we find that the holonomy map π, which moves $V^u(z) \cap A^m$ onto B^m, is absolutely continuous. Hence, $\nu^u(x)(B^m) > 0$. This contradiction implies that the set A in (9.8) has measure zero and completes the proof of Theorem 9.14. □

Theorem 9.14 provides a way to establish ergodicity of the map $f|\Lambda$. Recall that a map f is said to be *topologically transitive* if given two nonempty open sets A and B, there exists a positive integer n such that $f^n(A) \cap B \neq \varnothing$.

Theorem 9.17. *Under the assumptions of Theorem 9.14, if $f|\Lambda$ is topologically transitive, then $f|\Lambda$ is ergodic.*

Proof. Let $C, D \subset \Lambda$ (mod 0) be two distinct ergodic components of positive measure. We have that $\nu(f^n(C) \cap D) = 0$ for any integer n. By Theorem 9.14, the sets C and D are open (mod 0). Since f is topologically transitive, we have $f^m(C) \cap D \neq \varnothing$ for some m. Furthermore, the set $f^m(C) \cap D$ is also open (mod 0). Since ν is equivalent to the Riemannian volume, we conclude that $\nu(f^m(C) \cap D) > 0$, leading to a contradiction. □

For a general diffeomorphism preserving a smooth hyperbolic measure, one should not expect the unstable (and stable) leaves to form a q-foliation for some function $q(x)$. To explain why this may be the case, consider a local unstable manifold $V^u(x)$ passing through a point $x \in \Lambda$. For some index set ι, a typical $x \in \Lambda_\iota$, and a sufficiently large ℓ, the set $V^u(x) \cap \Lambda_\iota^\ell$ has positive leaf-volume but, in general, is a Cantor-like set. When the local manifold is moved forward a given time n, one should expect a sufficiently small neighborhood of the set $V^u(x) \cap \Lambda_\iota^\ell$ to expand. Other pieces of the local manifold (corresponding to bigger values of ℓ) will also expand but with smaller rates. This implies that the global leaf $W^u(x)$ (defined by (7.48))

9.2. Local ergodicity

may bend "uncontrollably". As a result, the map $x \mapsto \varphi_x$ in the definition of q-foliations may not be, indeed, continuous.

Furthermore, the global manifold $W^u(x)$ may turn out to be "bounded", i.e., it may not admit an embedding of an arbitrarily large ball in $\mathbb{R}^k$ (where $k = \dim W^u(x)$) (see Section 14.1).

The local continuity of the global unstable leaves often comes up in the following setting. Using some additional information on the system, one can build an invariant foliation whose leaves contain local unstable leaves. This alone may not yet guarantee that global unstable leaves form a foliation. However, one often may find that the local unstable leaves expand in a "controllable" and somewhat uniform way when they are moved forward. As we see below, this guarantees the desired properties of unstable leaves. Such a situation occurs, for example, for geodesic flows on compact Riemannian manifolds of nonpositive curvature (see Section 10.4).

We now state a formal criterion for local ergodicity. Let f be a $C^{1+\alpha}$ diffeomorphism of a compact smooth Riemannian manifold M, preserving a smooth measure ν, and let Λ be a nonuniformly hyperbolic set of positive measure.

Theorem 9.18. *Let W be a q-foliation of Λ satisfying:*

(1) $W(x) \supset V^s(x)$ *for every* $x \in \Lambda$;

(2) *there exists a number $\delta > 0$ and a measurable function $n(x)$ on Λ such that for almost every $x \in \Lambda$ and any $n \geq n(x)$,*

$$f^{-n}(V^s(x)) \supset B_{V(y)}(f^{-n}(x), \delta).$$

Then every ergodic component of f of positive measure is open (mod 0).

Proof. Let x be a Lebesgue density point of a regular set Λ_ι^ℓ of positive measure for some index set ι and sufficiently large $\ell > 0$. Set $A(r) = \Lambda_\iota^\ell \cap B(x, r)$. Applying Lemma 9.15 to the set $A(r)$ for sufficiently small r and using the conditions of the theorem, we find that $W^s(y) \supset B_{V(y)}(y, \delta_0)$ for almost every $y \in A(r)$. One obtains the desired result by repeating arguments in the proof of Theorem 9.14. $\square$

9.2.3. The one-dimensional case. In the case of one-dimensional (δ, q)-foliations the second condition of Theorem 9.18 holds automatically and, hence, can be omitted.

Theorem 9.19. *Let W be a one-dimensional q-foliation of Λ such that $W(x) \supset V^s(x)$ for every $x \in \Lambda$. Then every ergodic component of f of positive measure is open* (mod 0). *Moreover, $W^s(x) = W(x)$ for almost every $x \in \Lambda$.*

Proof. Fix an index set ι and a number $\ell \geq 1$ such that the regular set Λ_ι^ℓ has positive measure. For almost every $x \in \Lambda_\iota^\ell$, the intersection $A(x) = V^s(x) \cap \Lambda_\iota^\ell$ has positive Lebesgue measure in $V^s(x)$. For every $y \in A(x)$, let $s(y)$ be the distance between x and y measured along $V^s(x)$. Then there exists a differentiable curve $\gamma \colon [0, s(y)] \to V^s(x)$ with $\gamma(0) = x$ and $\gamma(s(y)) = y$, satisfying for all sufficiently large n

$$\rho_{W(f^{-n}(x))}(f^{-n}(x), f^{-n}(y)) = \int_0^{s(y)} \|d_{\gamma(t)} f^{-n} \gamma'(t)\| \, dt$$
$$\geq \int_0^{s(y)} \|d_{f^n(\gamma(t))} f^n \gamma'(t)\|^{-1} \, dt$$
$$\geq \ell^{-1} e^{-\varepsilon n} \lambda^{-n} \geq \delta_0$$

(see Section 5.2), where δ_0 is a positive constant. Therefore, the second condition of Theorem 9.18 holds and the desired result follows. □

One can readily extend Theorems 9.18 and 9.19 to dynamical systems with continuous time.

Theorem 9.20. *Let φ_t be a smooth flow of a compact smooth Riemannian manifold preserving a smooth measure ν and let Λ be a nonuniformly hyperbolic set of positive measure. Let also W be a q-foliation of Λ such that:*

(1) *$W(x) \supset V^s(x)$ for every $x \in \Lambda$;*
(2) *there exists a number $\delta > 0$ and a measurable function $t(x)$ on Λ such that for almost every $x \in \Lambda$ and any $t \geq t(x)$,*

$$\varphi_{-t}(V^s(x)) \supset B_{V(y)}(\varphi_{-t}(x), \delta).$$

Then every ergodic component of $\varphi_t|\Lambda$ of positive measure is open (mod 0). Furthermore, if the foliation $W(x)$ is one-dimensional, then the second requirement can be dropped.

9.3. The entropy formula

One of the main concepts of smooth ergodic theory is that sufficient instability of trajectories yields rich ergodic properties of the system. The entropy formula is in a sense a "quantitative manifestation" of this idea and is yet another pearl of the theory. It expresses the metric entropy $h_\nu(f)$[5] of a diffeomorphism with respect to a smooth hyperbolic measure, in terms of the values of the Lyapunov exponent.

---
[5] Other commonly used terms are *measure-theoretic entropy* and *Kolmogorov–Sinai entropy*.

9.3. The entropy formula

9.3.1. The metric entropy of a diffeomorphism. We briefly describe some relevant notions from the theory of measurable partitions and entropy theory.

Let $(X, \mathcal{B}, \mu)$ be a Lebesgue measure space and let ξ, η be two finite or countable partitions of X whose elements are measurable sets of positive measure. We say that η is a *refinement* of ξ and we write $\xi \subset \eta$ if every element of η is contained (mod 0) in an element of ξ. We say that η is *equivalent* to ξ and we write $\xi = \eta$ if these partitions are refinements of each other. The *common refinement* is the partition $\xi \vee \eta$ with elements $C \cap C'$ where $C \in \xi$ and $C' \in \eta$. Finally, we say that ξ is *independent* of η if $\mu(C \cap C') = \mu(C)\mu(C')$ for all $C \in \xi$ and $C' \in \eta$.

The *entropy* of the measurable partition ξ (with respect to μ) is given by

$$H_\mu(\xi) = -\sum_{C \in \xi} \mu(C) \log \mu(C),$$

with the convention that $0 \log 0 = 0$.

Exercise 9.21. Let ξ and η be two finite or countable partitions of X. Show that:

(1) $H_\mu(\xi) \geq 0$ and $H_\mu(\xi) = 0$ if and only if ξ is the trivial partition;

(2) if $\xi \subset \eta$, then $H_\mu(\xi) \leq H_\mu(\eta)$ and equality holds if and only if $\xi = \eta$;

(3) if ξ has n elements, then $H_\mu(\xi) \leq \log n$ and equality holds if and only if each element has measure $1/n$;

(4) $H_\mu(\xi \vee \eta) \leq H_\mu(\xi) + H_\mu(\eta)$ and equality holds if and only if ξ is independent of η.

Hint: Use the fact that the function $x \log x$ is continuous and strictly concave.

Given two measurable partitions ξ and ζ, we also define the *conditional entropy of ξ with respect to ζ* by

$$H_\mu(\xi|\zeta) = -\sum_{C \in \xi} \sum_{D \in \zeta} \mu(C \cap D) \log \frac{\mu(C \cap D)}{\mu(D)}.$$

One can show that the conditional entropy has the following properties. Let ξ, ζ, and η be three finite or countable partitions of X. Then:

(1) $H_\mu(\xi|\zeta) \geq 0$ and $H_\mu(\xi|\zeta) = 0$ if and only if $\xi \subset \zeta$;

(2) if $\zeta \subset \eta$, then $H_\mu(\xi|\zeta) \geq H_\mu(\xi|\eta)$ and equality holds if and only if $\zeta = \eta$;

(3) if $\zeta \subset \eta$, then $H_\mu(\xi \vee \zeta|\eta) = H_\mu(\xi|\eta)$;

(4) if $\xi \subset \eta$, then $H_\mu(\xi|\zeta) \leq H_\mu(\eta|\zeta)$ and equality holds if and only if $\xi \vee \zeta = \eta \vee \zeta$;

(5) $H_\mu(\xi \vee \zeta|\eta) = H_\mu(\xi|\eta) + H_\mu(\zeta|\chi \vee \eta)$ and $H_\mu(\xi \vee \zeta) = H_\mu(\xi) + H_\mu(\zeta|\xi)$;

(6) $H_\mu(\xi|\zeta \vee \eta) \leq H_\mu(\xi|\eta)$;

(7) $H_\mu(\xi|\zeta) = \sum_{D \in \zeta} \nu(D) H(\xi|D)$ where $H(\xi|D)$ is the entropy of ξ with respect to the conditional measure on D induced by μ;

(8) $H_\mu(\xi|\zeta) \leq H_\mu(\xi)$ and equality holds if and only if ξ is independent of ζ.

Now consider a measurable transformation $T: X \to X$. Given a partition ξ and a number $k \geq 0$, denote by $T^{-k}\xi$ the partition of X by the sets $T^{-k}(C)$ with $C \in \xi$.

We define the *entropy* of T with respect to the partition ξ by the formula

$$h_\mu(T, \xi) = \inf_n \frac{1}{n} H_\mu \left(\bigvee_{k=0}^{n-1} T^{-k}\xi \right). \tag{9.10}$$

Using the properties of the conditional entropy described above, one can show that

$$\begin{aligned} h_\mu(T, \xi) &= \lim_{n \to \infty} \frac{1}{n} H_\mu \left(\bigvee_{k=0}^{n-1} T^{-k}\xi \right) \\ &= \lim_{n \to \infty} H_\mu \left(\xi \Big| \bigvee_{k=1}^{n} T^{-k}\xi \right) = H_\mu(\xi|\xi^-), \end{aligned} \tag{9.11}$$

where $\xi^- = \bigvee_{k=0}^\infty T^{-k}\xi$. One can show that the entropy of T with respect to the partition ξ has the following properties:

(1) $h_\mu(T, \xi) \leq H_\mu(\xi)$;

(2) $h_\mu(T, T^{-1}\xi) = h_\mu(T, \xi)$;

(3) $h_\mu(T, \xi) = h_\mu(T, \bigvee_{i=0}^n T^{-i}\xi)$.

We define the *metric entropy* of T with respect to μ by

$$h_\mu(T) = \sup_\xi h_\mu(T, \xi),$$

where the supremum is taken over all measurable partitions ξ with finite entropy (which actually coincides with the supremum over all finite measurable partitions). Using (9.11), one can show that:

(1) $h_\mu(T^m) = m h_\mu(T)$ for every $m \in \mathbb{N}$;
(2) if T is invertible with measurable inverse, then $h_\mu(T^m) = |m| h_\mu(T)$ for every $m \in \mathbb{Z}$;[6]
(3) let $(X, \mathcal{B}, \mu)$ and $(Y, \mathcal{A}, \nu)$ be two Lebesgue measure spaces, let $T \colon X \to X$ and $S \colon Y \to Y$ be two measurable transformations, and let $P \colon X \to Y$ be a measurable transformation with measurable inverse such that $P \circ T = S \circ P$ and $P_* \mu = \nu$ (in other words, P is an isomorphism of measure spaces that conjugates T and S); then $h_\mu(T) = h_\nu(S)$.

9.3.2. Upper bound for the metric entropy. Let f be a C^1 diffeomorphism of a smooth compact Riemannian manifold M and let ν be an invariant Borel probability measure on M. Consider the Lyapunov exponent $\chi(x, \cdot)$ and its Lyapunov spectrum

$$\operatorname{Sp} \chi(x) = \{(\chi_i(x), k_i(x)) : 1 \le i \le s(x)\};$$

see Section 5.1.

Theorem 9.22. *The entropy of f admits the upper bound*

$$h_\nu(f) \le \int_M \sum_{i : \chi_i(x) > 0} k_i(x) \chi_i(x) \, d\nu(x). \tag{9.12}$$

This upper bound was established by Ruelle in [**99**]. Independently, Margulis obtained this estimate in the case of volume-preserving diffeomorphisms (unpublished). Inequality (9.12) is often referred to as the Margulis–Ruelle inequality. The proof presented below is due to Katok and Mendoza [**65**].

Proof of the theorem. By decomposing ν into its ergodic components, we may assume without loss of generality that ν is ergodic. Then $s(x) = s$, $k_i(x) = k_i$, and $\chi_i(x) = \chi_i$ are constant ν-almost everywhere for each $1 \le i \le s$. Fix $m > 0$. Since the manifold M is compact, there exists a number $r_m > 0$ such that for every $0 < r \le r_m$, $y \in M$, and $x \in B(y, r)$ we have

$$\frac{1}{2} d_x f^m \left(\exp_x^{-1} B(y, r) \right) \subset \exp_{f^m(x)}^{-1} f^m(B(y, r))$$
$$\subset 2 d_x f^m \left(\exp_x^{-1} B(y, r) \right), \tag{9.13}$$

where for a set $A \subset T_z M$ and $z \in M$, we write $\alpha A = \{\alpha v : v \in A\}$.

We begin by constructing a special partition ξ of the manifold M with some "nice" properties that will be used to estimate the metric entropy of f

---

[6] To see this, observe that one can replace $T^{-k} \xi$ by $T^k \xi$ in (9.10) and (9.11) and conclude that $h_\mu(T^{-1}, \xi) = h_\mu(T, \xi)$ and hence that $h_\mu(T^{-1}) = h_\mu(T)$.

from above. First, each element C of the partition is roughly a ball of a small radius (which is independent of the element of the partition) such that the image of the element C under f^m is also approximated by the image of the ball. Second, the metric entropy of f^m with respect to ξ is close to the metric entropy of f^m. More precisely, the following statement holds.

Lemma 9.23. *Given $\varepsilon > 0$, there exists a partition ξ of M satisfying the following conditions:*

(1) $h_\nu(f^m, \xi) \geq h_\nu(f^m) - \varepsilon$;
(2) *there are numbers $r < 2r' \leq r_m/20$ such that $B(x, r') \subset C \subset B(x, r)$ for every element $C \in \xi$ and every $x \in C$;*
(3) *there exists $0 < r < r_m/20$ such that*
   (a) *if $C \in \xi$, then $C \subset B(y, r)$ for some $y \in M$ and*
   (b) *for every $x \in C$,*

$$\frac{1}{2} d_x f^m \left(\exp_x^{-1} B(y, r) \right) \subset \exp_{f^m(x)}^{-1} f^m(C)$$
$$\subset 2 d_x f^m \left(\exp_x^{-1} B(y, r) \right).$$

Proof of the lemma. Given $\alpha > 0$, we call a finite set of points $E \subset M$ α-separated if $\rho(x, y) > \alpha$ whenever $x, y \in E$. Among all α-separated sets there is a maximal one which we denote by Γ. For each $x \in \Gamma$ define

$$\mathcal{D}_\Gamma(x) = \{y \in M : \rho(y, x) \leq \rho(y, z) \text{ for all } z \in \Gamma \setminus \{x\}\}.$$

Obviously, $B(x, \alpha/2) \subset \mathcal{D}_\Gamma(x) \subset B(x, \alpha)$. Note that the sets $\mathcal{D}_\Gamma(x)$ corresponding to different points $x \in \Gamma$ intersect only along their boundaries, i.e., at a finite number of submanifolds of codimension greater than zero. Since ν is a Borel measure, if necessary, we can move the boundaries slightly so that they have measure zero. Thus, we obtain a partition ξ with $\operatorname{diam} \xi \leq \alpha$. Moreover, we can choose a partition ξ such that

$$h_\nu(f^m, \xi) > h_\nu(f^m) - \varepsilon \quad \text{and} \quad \operatorname{diam} \xi < r_m/10.$$

This implies statements (1) and (2). Statement (3) follows from (9.13). □

We proceed with the proof of the theorem. Using the properties of conditional entropy, we obtain that

$$h_\nu(f^m, \xi) = \lim_{k \to \infty} H_\nu(\xi | f^m \xi \vee \cdots \vee f^{km} \xi)$$
$$\leq H_\nu(\xi | f^m \xi) = \sum_{D \in f^m \xi} \nu(D) H(\xi | D) \quad (9.14)$$
$$\leq \sum_{D \in f^m \xi} \nu(D) \log \left(\operatorname{card} \{ C \in \xi : C \cap D \neq \varnothing \} \right),$$

9.3. The entropy formula

where card denotes the cardinality of the set. In view of this estimate our goal now is first to obtain a uniform exponential estimate for the number of those elements $C \in \xi$ which have nonempty intersection with a given element $D \in f^m \xi$ and then to establish an exponential bound on the number of those sets $D \in f^m \xi$ that contain LP-regular orbits $\{f^k(x)\}$ along which for $v \in T_x M$ the length $\|d_x f^k v\|$ admits effective exponential estimates in terms of the Lyapunov exponents $\chi(x, v)$.

Lemma 9.24. *There exists a constant $K_1 > 0$ such that for $D \in f^m \xi$,*
$$\mathrm{card}\{C \in \xi : D \cap C = \varnothing\} \leq K_1 \sup\{\|d_x f\|^{mp} : x \in M\},$$
where $p = \dim M$.

Proof. By the Mean Value Theorem,
$$\mathrm{diam}(f^m(C)) \leq \sup\{\|d_x f\|^m : x \in M\} \, \mathrm{diam}\, C.$$
Thus, if $C \cap D \neq \varnothing$, then C is contained in the $4r'$-neighborhood of D. Therefore,
$$\mathrm{diam} \bigcup_{C \cap D \neq \varnothing} C \leq (\sup\{\|d_x f\|^m : x \in M\} + 2) 4r',$$
and hence, the volume of $\bigcup_{C \cap D \neq \varnothing} C$ is bounded from above by
$$K r^p \sup\{\|d_x f\|^{mp} : x \in M\},$$
where $K > 0$ is a constant. On the other hand, by property (2) of the partition ξ, the set C contains a ball $B(x, r')$, and hence, the volume of C is at least $K'(r')^p$, where K' is a positive constant. This implies the desired result. □

Fix $\varepsilon > 0$ and let $R_m = R_m(\varepsilon)$ be the set of LP-regular points $x \in M$ satisfying the following condition: for $k > m$ and $v \in T_x M$,
$$e^{k(\chi(x,v)-\varepsilon)} \|v\| \leq \|d_x f^k v\| \leq e^{k(\chi(x,v)+\varepsilon)} \|v\|.$$

Lemma 9.25. *If $D \in f^m \xi$ has nonempty intersection with R_m, then there exists a constant $K_2 > 0$ such that*
$$\mathrm{card}\{C \in \xi : D \cap C \neq \varnothing\} \leq K_2 e^{\varepsilon m} \prod_{i:\chi_i > 0} e^{m(\chi_i + \varepsilon) k_i}.$$

Proof. Let $C' \in \xi$ be such that $C' \cap R_m \neq \varnothing$ and $f^m(C') = D$. Pick a point $x \in C' \cap f^{-m}(R_m)$ and let $B = B(x, 2\,\mathrm{diam}\, C')$. The set
$$\tilde{B}_0 = d_x f^m(\exp_x^{-1} B) \subset T_{f^m x} M$$
is an ellipsoid and $D \subset B_0 = \exp_{f^m(x)}(\tilde{B}_0)$. If a set $C \in \xi$ has nonempty intersection with B_0, then it lies in the set
$$B_1 = \{y \in M : \rho(y, B_0) < \mathrm{diam}\, \xi\}.$$

Therefore,
$$\operatorname{card}\{C \in \xi : D \cap C \neq \varnothing\} \leq \operatorname{vol}(B_1)(\operatorname{diam} \xi)^{-p},$$

where $\operatorname{vol}(B_1)$ denotes the volume of B_1. Up to a bounded factor, $\operatorname{vol}(B_1)$ is bounded by the product of the lengths of the axes of the ellipsoid $\tilde{B}_0$. Those of them corresponding to nonpositive exponents are at most subexponentially large. The remaining ones are of size at most $e^{m(\chi_i+\varepsilon)}$, up to a bounded factor, for all sufficiently large m. Thus,

$$\operatorname{vol}(B_1) \leq K e^{m\varepsilon}(\operatorname{diam} B)^p \prod_{i:\chi_i>0} e^{m(\chi_i+\varepsilon)k_i}$$
$$\leq K e^{m\varepsilon}(2\operatorname{diam} \xi)^p \prod_{i:\chi_i>0} e^{m(\chi_i+\varepsilon)k_i},$$

for some constant $K > 0$ and the lemma follows. $\square$

We proceed with the proof of the theorem and estimate the metric entropy of f^m with respect to the partition ξ. Namely, by (9.14), we obtain that

$$h_\nu(f^m, \xi) \leq \sum_{D \cap R_m \neq \varnothing} \nu(D) \log \operatorname{card}\{C \in \xi : C \cap D \neq \varnothing\}$$
$$+ \sum_{D \cap R_m = \varnothing} \nu(D) \log \operatorname{card}\{C \in \xi : C \cap D \neq \varnothing\}$$
$$= \sum_1 + \sum_2.$$

By Lemma 9.25, the first sum can be estimated as

$$\sum_1 \leq \left(\log K_2 + \varepsilon m + m \sum_{i:\chi_i>0}(\chi_i + \varepsilon)k_i\right)\nu(R_m)$$

and by Lemma 9.24, the second sum can be estimated as

$$\sum_2 \leq \left(\log K_1 + pm \log \sup\{\|d_x f\| : x \in M\}\right)\nu(M \setminus R_m).$$

Note that by the Multiplicative Ergodic Theorem (see Theorem 5.3), for every sufficiently small ε,

$$\bigcup_{m \geq 0} R_m(\varepsilon) = M \quad (\operatorname{mod} 0),$$

9.3. The entropy formula

and hence, $\nu(M \setminus R_m) \to 0$ as $m \to \infty$. We are now ready to obtain the desired upper bound. By Lemma 9.23, we have that

$$mh_\nu(f) - \varepsilon = h_\nu(f^m) - \varepsilon \le h_\nu(f^m, \xi)$$
$$\le \log K_2 + \varepsilon m + m \sum_{i:\chi_i>0} (\chi_i + \varepsilon)k_i$$
$$+ \big(\log K_1 + pm \log \sup\{\|d_x f\| : x \in M\}\big)\nu(M \setminus R_m).$$

Dividing by m and letting $m \to \infty$, we find that

$$h_\nu(f) \le 2\varepsilon + \sum_{i:\chi_i>0}(\chi_i + \varepsilon)k_i.$$

Letting $\varepsilon \to 0$, we obtain the desired upper bound. $\square$

An important consequence of Theorem 9.22 is that any C^1 diffeomorphism with positive topological entropy has an invariant measure with at least one positive and one negative Lyapunov exponent. In particular, a surface diffeomorphism with positive topological entropy always has a hyperbolic invariant measure.

Exercise 9.26. Show that for an arbitrary invariant measure the inequality (9.12) may be strict. *Hint:* Examine a diffeomorphism with a hyperbolic fixed point and an atomic measure concentrated at this point.

9.3.3. Lower bound for the metric entropy. We shall now prove the lower bound for the metric entropy assuming that the invariant measure is smooth, and hence, we obtain the entropy formula. We stress that while the upper bound for the metric entropy holds for diffeomorphisms of class C^1, the lower bound requires that f be of class $C^{1+\alpha}$.

Theorem 9.27 (Entropy Formula). *If f is of class $C^{1+\alpha}$ and ν is a smooth invariant measure on M, then*

$$h_\nu(f) = \int_M \sum_{i:\chi_i(x)>0} k_i(x)\chi_i(x)\, d\nu(x). \tag{9.15}$$

In the case of Anosov diffeomorphisms, a statement equivalent to the entropy formula was proved by Sinai in [**106**] and for a general diffeomorphism preserving a smooth measure, it was established by Pesin in [**89**] (see also [**90**]). The proof presented below follows the original argument in [**89**] and exploits in an essential way the machinery of stable manifolds and their absolute continuity. There is another proof of the lower bound due to Mañé [**81**] that is more straightforward and does not use this machinery. While the requirement that the map f be of class $C^{1+\alpha}$ for some

$\alpha > 0$ is crucial for both proofs, it is worth mentioning that by a result of Tahzibi [**108**] for a C^1 *generic* surface diffeomorphism the lower bound still holds.

Proof of the theorem. We only need to show that
$$h_\nu(f) \geq \int_M \sum_{i:\chi_i(x)>0} k_i(x)\chi_i(x)\,d\nu(x),$$
or equivalently (by replacing f by f^{-1} and using Theorem 5.3) that
$$h_\nu(f) \geq -\int_M \sum_{i:\chi_i(x)<0} k_i(x)\chi_i(x)\,d\nu(x). \tag{9.16}$$

Consider the invariant set
$$\Gamma = \{x \in M : \chi^+(x,v) < 0 \text{ for some } v \in T_x M\}.$$

Since the invariant measure ν is smooth, for every $x \in \Gamma$ there is $v \in T_x M$ such that $\chi^+(x,v) > 0$, and in particular, the set Γ is nonuniformly partially hyperbolic for f. If $\nu(\Gamma) = 0$, then $\chi^+(x,v) = 0$ for almost every $x \in M$ and the desired result follows immediately from (9.16). We, therefore, assume that $\nu(\Gamma) > 0$. Since Γ is a nonuniformly partially hyperbolic set for f, we can consider the corresponding collection of level sets $\{\Lambda_\iota\}$ and for each $\iota = \{\lambda_1, \lambda_2, \mu_1, \mu_2, \varepsilon, j_1, j_2\}$ (see (5.32)) the collection of regular sets $\{\Lambda_\iota^\ell : \ell \geq 1\}$. It suffices to prove the theorem for each level set Λ_ι of positive measure. Observe that for every $x \in \Lambda_\iota$ the largest negative Lyapunov exponent does not exceed $\log \lambda_2 < 0$ uniformly in x.

Set $\tilde{\Gamma} = \Gamma \cap \mathcal{R}$, where $\mathcal{R}$ is the set of LP-regular points. For each $x \in \tilde{\Gamma} \cap \Lambda_\iota$ set
$$T_n(x) = \operatorname{Jac}(d_x f^n | E^s(x)) \quad \text{and} \quad g(x) = \prod_{i:\chi_i(x)<0} e^{k_i(x)\chi_i(x)}.$$

The following result is an immediate consequence of Corollary 5.2.

Lemma 9.28. *Given $\varepsilon > 0$, there exists a positive Borel function $L(x)$ such that for any $x \in \tilde{\Gamma}$ and $n > 0$,*
$$T_n(x) \leq L(x) g(x)^n e^{\varepsilon n}.$$

Fix $\varepsilon > 0$ and consider the invariant sets
$$\tilde{\Gamma}_n = \{x \in \tilde{\Gamma} : (1+\varepsilon)^{-n} < g(x) \leq (1+\varepsilon)^{-n+1}\}, \quad n > 0.$$

We shall evaluate the metric entropy of the restriction $f|\tilde{\Gamma}_n$ with respect to the measure $\nu_n = \nu|\tilde{\Gamma}_n$.

9.3. The entropy formula

Fix $n > 0$ and $\ell > 0$, and define the collection of measurable sets
$$\tilde{\Gamma}_n^\ell = \{x \in \tilde{\Gamma}_n \cap \Lambda_\iota^\ell : L(x) \leq \ell\}, \quad \Gamma_n^\ell = \bigcup_{x \in \tilde{\Gamma}_{n,\ell}} V^s(x), \quad \hat{\Gamma}_n = \bigcup_{j \in \mathbb{Z}} f^j(\Gamma_n^\ell).$$

Note that $\hat{\Gamma}_n$ is f-invariant. Fix $\beta > 0$. By choosing a sufficiently large ℓ, we obtain $\nu_n(\tilde{\Gamma}_n^\ell) > 1 - \beta$. Since
$$\tilde{\Gamma}_n^\ell \subset \Gamma_n^\ell \subset \hat{\Gamma}_n \subset \tilde{\Gamma}_n \pmod{0},$$
we also have
$$\nu_n(\hat{\Gamma}_n \setminus \Gamma_n^\ell) \leq \beta \quad \text{and} \quad \nu_n(\tilde{\Gamma}_n \setminus \hat{\Gamma}_n) \leq \beta. \tag{9.17}$$

Let ξ be a finite measurable partition of M constructed in Lemma 9.23; each element $\xi(x)$ of ξ is homeomorphic to a ball, has piecewise smooth boundary, and has diameter at most r_ℓ.

We define a partition η of $\hat{\Gamma}_n$ composed of $\hat{\Gamma}_n \setminus \Gamma_n^\ell$ and of the elements $V^s(y) \cap \xi(x)$ for each $y \in \xi(x) \cap \tilde{\Gamma}_n^\ell$.

Given $x \in \Gamma_n^\ell$ and a sufficiently small $r \leq r_\ell$ (see Chapter 8), we set
$$B_\eta(x, r) = \{y \in \eta(x) : \rho(y, x) < r\}.$$

Lemma 9.29. *There exists $q_\ell > 0$ and a set $A^\ell \subset \Gamma_n^\ell$ with $\nu_n(\Gamma_n^\ell \setminus A^\ell) \leq \beta$ such that $\eta^-(x) \supset B_\eta(x, q_\ell)$ for every $x \in A^\ell$.*

Proof of the lemma. For each $\delta > 0$ set
$$\partial \xi = \bigcup_{y \in M} \partial(\xi(y)) \quad \text{and} \quad \partial \xi_\delta = \{y : \rho(y, \partial \xi) \leq \delta\}.$$

One can easily show that there exists $C_1 > 0$ such that
$$\nu_n(\partial \xi_\delta) \leq C_1 \delta. \tag{9.18}$$

Let $D_q = \{x \in \Gamma_n^\ell : B_\eta(x, q) \setminus \eta^-(x) \neq \varnothing\}$. If $x \in D_q$, then there exist $m \in \mathbb{N}$ and $y \in B_\eta(x, q)$ such that $y \notin (f^{-m}\eta)(x)$. Hence, $\partial \xi \cap f^m(B_\eta(x, q)) \neq \varnothing$. Therefore, by Theorem 7.1, if $x \in D_q$, then $f^m(x) \in \partial \xi_{C_2 \lambda^m e^{\varepsilon m} q}$ for some $C_2 = C_2(\ell) > 0$. Thus, in view of (9.18),
$$\nu_n(D_q) \leq C_1 \sum_{m=0}^\infty C_2 \lambda^m e^{\varepsilon m} q \leq C_3 q$$

for some $C_3 = C_3(\ell) > 0$. The lemma follows by setting $q_\ell = \beta C_3^{-1}$ and $A^\ell = \Gamma_n^\ell \setminus D_{q_\ell}$. $\square$

The following statement is crucial in our proof of the lower bound. It exploits the absolute continuity property of the measure ν in an essential way. Its proof is an immediate consequence of Theorem 8.4, Remark 8.19, and Lemma 9.29.

Lemma 9.30. *There exists $C_5 = C_5(\ell) \geq 1$ such that for $x \in A^\ell \cap V^s(y)$ and $y \in \tilde{\Gamma}_n^\ell$,*
$$C_5^{-1} \leq \frac{d\nu_x^-}{dm_y} \leq C_5,$$
where ν_x^- is the conditional measure on the element $\eta^-(x)$ of the partition η^- and m_y is the leaf-volume on $V^s(y)$.

We also need the following statement.

Lemma 9.31. *There exists $C_4 = C_4(\ell) > 0$ such that for every $x \in \Gamma_n^\ell \cap V^s(y)$ with $y \in \tilde{\Gamma}_n^\ell$ and for every $k > 0$ we have*
$$m_{f^k(y)}(f^k(\eta(x))) \leq C_4 T_k(y).$$

Proof of the lemma. We have
$$m_{f^k(y)}(f^k(\eta(x))) = \int_{\eta(x)} T_k(z)\, dm_y(z).$$

A similar argument to that in the proof of Lemma 8.15 shows that there exists $C' = C'(\ell) > 0$ such that for every $z \in \eta(x)$ and all sufficiently large $k \in \mathbb{N}$,
$$\left|\frac{T_k(z)}{T_k(y)} - 1\right| \leq C'.$$
The desired result follows. □

We now complete the proof of the theorem. Write $\hat{f} = f|\hat{\Gamma}$. For every $k > 0$ we have
$$h_{\nu_n}(f|\tilde{\Gamma}_n) \geq h_{\nu_n}(\hat{f}) = \frac{1}{k}h_{\nu_n}(\hat{f}^k) \geq \frac{1}{k}H_{\nu_n}(\hat{f}^k\eta|\eta^-), \qquad (9.19)$$
where $\eta^- = \bigvee_{k=0}^\infty \hat{f}^{-k}\eta$. To estimate the last expression, we find a lower bound for
$$H_{\nu_n}(\hat{f}^k\eta|\eta^-(x)) = \int_{\eta^-(x)} -\log \nu_x^-(\eta^-(x) \cap (f^k\eta)(y))\, d\nu_x^-(y).$$
Note that for $x \in A^\ell$,
$$\nu_x(B_\eta(x, q_\ell)) \geq C_6 q_\ell^{\dim M}$$
for some $C_6 = C_6(\ell) > 0$. It follows from Lemmas 9.28, 9.30, and 9.31 that for $x \in A^\ell$ and $n \in \mathbb{N}$,
$$H_{\nu_n}(\hat{f}^k\eta|\eta^-(x)) \geq -C_5 \log\left(C_5\nu_x(\eta^-(x) \cap (f^k\eta)(x))\right)\nu_x(B_\eta(x, q_\ell))$$
$$\geq -C_5 \log(C_5 C_4 T_k(x))\nu_x(B_\eta(x, q_\ell))$$
$$\geq -C_5 \log(C_5 C_4 L(x)g(x)^k e^{\varepsilon k})C_6 q_\ell^{\dim M}$$
$$\geq -C_5 \log(C_5 C_4 \ell(1+\varepsilon)^{(-n+1)k} e^{\varepsilon k})C_6 q_\ell^{\dim M}.$$

9.3. The entropy formula

Choosing q_ℓ sufficiently small, we may assume that $C_5 C_6 q_\ell^{\dim M} < 1$. Therefore,

$$H_{\nu_n}(\hat{f}^k \eta | \eta^-(x)) \geq -k(-n+1)\log(1+\varepsilon) - k\varepsilon - C_7$$
$$\geq -k \log g(x) - k(\log(1+\varepsilon) + \varepsilon) - C_7$$

for some $C_7 = C_7(\ell) > 0$. Integrating this inequality over the elements of η^-, we obtain

$$\frac{1}{k} H_{\nu_n}(\hat{f}^k \eta | \eta^-) \geq \int_{A^\ell} \left(-\log g(x) - (\log(1+\varepsilon) + \varepsilon) - \frac{C_7}{k}\right) d\nu_n(x).$$

Given $\delta > 0$, we can choose numbers k and ℓ sufficiently large and ε sufficiently small such that (in view of (9.17), (9.19), and Lemma 9.29)

$$h_{\nu_n}(f | \tilde{\Gamma}_n) \geq -\frac{1}{\nu(\tilde{\Gamma}_n)} \int_{\tilde{\Gamma}_n} \sum_{i:\chi_i(x)<0} k_i(x) \chi_i(x) \, d\nu(x) - \delta.$$

Summing up these inequalities over n (note that the sets $\tilde{\Gamma}_n$ are disjoint and invariant), we obtain

$$h_\nu(f) \geq \sum_{n=1}^\infty \nu(\tilde{\Gamma}_n) h_{\nu_n}(f | \tilde{\Gamma}_n)$$
$$\geq \sum_{n=1}^\infty \left(-\int_{\tilde{\Gamma}_n} \sum_{i:\chi_i(x)<0} k_i(x) \chi_i(x) \, d\nu(x) - \delta \nu(\tilde{\Gamma}_n)\right)$$
$$= \int_M \sum_{i:\chi_i(x)<0} k_i(x) \chi_i(x) \, d\nu(x) - \delta.$$

Since δ is arbitrary, the desired result follows. $\square$

A slight modification of the above proof allows one to establish the entropy formula (9.15) for measures which are absolutely continuous with respect to the Riemannian volume (not necessarily equivalent to it).

9.3.4. The Ledrappier–Young entropy formula. In this section we present a formula for the entropy of a $C^{1+\alpha}$ diffeomorphism of a smooth compact Riemannian manifold M with respect to an invariant Borel probability measure ν. It was obtain by Ledrappier and Young in [**73, 74**].

We will assume that for almost every $x \in M$ there are vectors $v, w \in T_x M$ such that $\chi(x,v) < 0$ and $\chi(x,w) > 0$. The general case can be easily reduced to this one (see [**74**]). Using an ergodic decomposition of ν, we may assume without loss of generality that ν is ergodic. Consider the Lyapunov spectrum

$$\operatorname{Sp}\chi(\nu) = \{(\chi_i, k_i) : 1 \leq i \leq s\};$$

see (5.1) and (5.2). We have that
$$\chi_1 < \chi_2 < \cdots < \chi_s$$
and by our assumption $\chi_1 < 0$ and $\chi_s > 0$. Therefore, there are numbers $1 \leq k < m \leq s$ such that $\chi_k < 0$, $\chi_m > 0$, and $\chi_i = 0$ for $k < i < m$. Thus k is the number of negative Lyapunov exponents with χ_k being the largest and $s - m + 1$ is the number of positive Lyapunov exponents with χ_m being the smallest. Note that ν is a hyperbolic measure if and only if $m = k + 1$.

We denote by X the set of LP-regular points for which
$$\operatorname{Sp}\chi^{\pm}(x) = \operatorname{Sp}\chi(\nu) = \{(\chi_i, k_i), i = 1, \ldots, s\}$$
(see Section 5.1.1 and (5.2)). Consider a point $x \in X$ and the Oseledets decomposition
$$T_x M = \bigoplus_{i=1}^{s} E_i(x).$$
Choose a positive integer l, $m \leq l \leq s$, and write
$$T_x M = F^s(x) \oplus F^0(x) \oplus F_l^u(x), \tag{9.20}$$
where
$$F^s(x) = \bigoplus_{i=1}^{k} E_i(x), \quad F^0(x) = \bigoplus_{i=k+1}^{l-1} E_i(x), \quad F_l^u(x) = \bigoplus_{i=l}^{s} E_i(x).$$

The decomposition (9.20) allows us to view the set X as a nonuniformly partially hyperbolic set for f with parameters of hyperbolicity
$$\lambda_1 = e^{\chi_1}, \quad \lambda_2 = e^{\chi_k}, \quad \mu_2 = e^{\chi_t}, \quad \mu_2 = e^{\chi_s}, \quad j = \dim F_l^u(x)$$
and $\varepsilon > 0$ to be any sufficiently small number. Now using the discussions in Sections 7.3.5 and 9.1.1 and using Remark 8.19, we obtain for every $x \in X$ a local unstable manifold $V_l^u(x)$ and a conditional measure $\nu_l^u(x)$ on $V_l^u(x)$ generated by ν.

To proceed with the entropy formula we need some information from geometric measure theory. Let Y be a complete metric space and let μ be a finite Borel measure on Y. Given $x \in Y$, define the *lower* and *upper pointwise dimensions* of μ at x by
$$\underline{d}_\mu(x) = \liminf_{r \to 0} \frac{\log \mu(B(x,r))}{\log r}, \quad \overline{d}_\mu(x) = \limsup_{r \to 0} \frac{\log \mu(B(x,r))}{\log r}.$$
If $\underline{d}_\mu(x) = \overline{d}_\mu(x) =: d_\mu(x)$, then d_μ is called the *pointwise dimension* of μ at x. Observe that for every $\varepsilon > 0$ there is $r(x, \varepsilon) > 0$ such that for every $0 < r < r(x, \varepsilon)$ we have that
$$r^{d_\mu(x) + \varepsilon} \leq \mu(B(x,r)) \leq r^{d_\mu(x) - \varepsilon}.$$
We consider some examples.

9.3. The entropy formula

Example 9.32. If $Y \subset [0,1]$ is a middle-third Cantor set and μ is a probability measure generated by the probability vector $\{\frac{1}{2}, \frac{1}{2}\}$, then $d_\mu(x) = \frac{\log 2}{\log 3}$ for every $x \in Y$.

Example 9.33. If Y is a smooth Riemannian manifold of dimension p and μ is a smooth measure on Y, then $d_\mu(x) = p$ for every $x \in Y$.

Example 9.34 (see [14] and also [13]). If Y is a compact smooth Riemannian manifold and μ is a finite Borel measure on Y which is invariant under a $C^{1+\alpha}$ diffeomorphism of Y, then $d_\mu(x)$ exists almost everywhere; moreover, $d_\mu(x) = d_\mu^s(x) + d_\mu^u(x)$ where $d_\mu^s(x)$ and $d_\mu^u(x)$ are, respectively, the *stable* and *unstable pointwise dimensions* of μ at x given by

$$d_\mu^s(x) = \lim_{r \to 0} \frac{\log \mu^s(x)(B^s(x,r))}{\log r},$$

$$d_\mu^u(x) = \lim_{r \to 0} \frac{\log \mu^u(x)(B^u(x,r))}{\log r}$$

(here $\mu^s(x)$, $\mu^u(x)$ are conditional measures on local stable and unstable local manifolds and $B^s(x,r)$, $B^u(x,r)$ are balls in these manifolds centered at x of radius r; also the limits exist almost everywhere).

In the particular case when $\dim Y = 2$ for almost every $x \in Y$ one can obtain (see [113]) the following relations between stable and unstable pointwise dimensions, metric entropy $h_\mu(f)$, and the Lyapunov exponents $\chi_1(x) < 0 < \chi_2(x)$:

$$h_\mu(f) = -\chi_1(x) d_\mu^s(x) = \chi_2(x) d_\mu^u(x).$$

Exercise 9.35. Let f be a diffeomorphism of a smooth Riemannian manifold M and let μ be an f-invariant measure. Show that the pointwise dimension of μ is an invariant function, i.e., $d_\mu(f(x)) = d_\mu(x)$. *Hint*: Show that there is $c > 0$ such that

$$B(f(x), c^{-1}r) \subset f(B(x,r)) \subset B(f(x), cr)$$

for any $x \in M$ and $r > 0$. Use this and the invariance of μ to derive the desired result. Prove similar statements for the stable and unstable pointwise dimensions of μ.

We proceed with the discussion of the Ledrappier–Young formula. For each $x \in X$ consider the local unstable manifolds $V_l^u(x)$ and the conditional measures $\nu_l^u(x)$ on $V_l^u(x)$ generated by ν and define the *lower* and *upper leaf pointwise dimension* of ν by

$$\underline{d}_l^u(x) = \liminf_{r \to 0} \frac{\log \nu_l^u(x)(B_l^u(x,r))}{\log r},$$

$$\overline{d}_l^u(x) = \limsup_{r \to 0} \frac{\log \nu_l^u(x)(B_l^u(x,r))}{\log r}$$

(here $B_l^u(x,r) \subset V_l^u(x)$ is the ball in $V_l^u(x)$ centered at x of radius r). Clearly, the lower and upper conditional pointwise dimensions of ν are invariant functions of x and hence are constant almost everywhere, which we denote by $\underline{d}_l^u$ and $\overline{d}_l^u$. One can show that $\underline{d}_l^u = \overline{d}_l^u =: d_l^u$ (see Exercise 9.35, although the proof in this case requires a more elaborate argument).

The following result establishes the Ledrappier–Young formula for the entropy of f with respect to ν.

Theorem 9.36. *We have that*
$$h_\nu(f) = \sum_{m \leq l \leq s} \chi_l(d_l^u - d_{l-1}^u).$$

Note that if ν is a smooth measure, then $d_l^u - d_{l-1}^u = k_l$ is the multiplicity of the Lyapunov exponent χ_l leading to the formula (9.15).

We outline the proof of the theorem. For $x \in M$ and $\varepsilon > 0$ set
$$\underline{h}_l^u(x, \varepsilon) = \liminf_{n \to \infty} -\frac{1}{n} \log \nu_l^u(x)(B_n^l(x, \varepsilon)),$$
$$\overline{h}_l^u(x, \varepsilon) = \limsup_{n \to \infty} -\frac{1}{n} \log \nu_l^u(x)(B_n^l(x, \varepsilon)),$$
where
$$B_n^l(x, \varepsilon) = \left\{ y \in V_l^u(x) : \rho_l(f^k(x), f^k(y)) < \varepsilon \text{ for } 0 \leq k < n \right\}$$
is the (n, ε)-*Bowen's ball* in $V_l^u(x)$ (here ρ_l is the induced distance in $V_l^u(x)$). We also define *lower* and *upper local leaf entropies*
$$\underline{h}_l^u(x) = \lim_{\varepsilon \to 0} \liminf_{n \to \infty} -\frac{1}{n} \log \nu_l^u(x)(B_n^l(x, \varepsilon)),$$
$$\overline{h}_l^u(x) = \lim_{\varepsilon \to 0} \limsup_{n \to \infty} -\frac{1}{n} \log \nu_l^u(x)(B_n^l(x, \varepsilon)).$$

The proof goes by showing that:

(1) $\underline{h}_l^u(f(x)) = \underline{h}_l^u(x)$ and $\overline{h}_l^u(f(x)) = \overline{h}_l^u(x)$ for every $x \in M$;

(2) $h_l^u(x) := \underline{h}_l^u(x) = \overline{h}_l^u(x)$ for ν-almost every $x \in M$.

It follows that $h_l^u(x) =: h_l^u$ is constant almost everywhere. The next step is to obtain the following crucial relation between local leaf entropies, Lyapunov exponents, and leaf pointwise dimensions:
$$h_l^u - h_{l-1}^u = \chi_l(d_l^u - d_{l-1}^u).$$

Summing up, one has that
$$h_s^u = \sum_{l=m}^s \chi_l(d_l^u - d_{l-1}^u).$$

Finally, one can show that $h_s^u = h_\nu(f)$.

Chapter 10

Geodesic Flows on Surfaces of Nonpositive Curvature

For a long time geodesic flows have played an important stimulating role in the development of hyperbolicity theory. In the beginning of the 20th century Hadamard and Morse, while studying the statistics of geodesics on surfaces of negative curvature, made the important observation that the local instability of trajectories gives rise to some global properties of dynamical systems such as ergodicity and topological transitivity. The results obtained during this period were summarized in the survey by Hedlund [53] (see also the article [56] by Hopf).

The subsequent study of geodesic flows has revealed the true nature and importance of instability and has started a remarkable shift from differential-geometric methods to dynamical techniques (see the seminal book by Anosov [5] and the article [6]). Geodesic flows were a source of inspiration (and an excellent model) for introducing concepts of both uniform and nonuniform hyperbolicity.

On the other hand, they always were a touchstone for applying new advanced methods of the general theory of dynamical systems. This, in particular, has led to some new interesting results in differential and Riemannian geometry (such as lower and upper estimates of the number of closed geodesics and their distribution in the phase space). In this chapter we will present some of these results. While describing ergodic properties of geodesic flows in the spirit of the book, we consider the Liouville measure that is invariant under the flow (we allow other invariant measures when discussing an upper bound for the metric entropy), and we refer the reader to the excellent survey [**71**] where measures of maximal entropy are studied. We only consider the case of geodesic flows on surfaces and we refer the reader to [**13**] for the general case.

10.1. Preliminary information from Riemannian geometry

For the reader's convenience we collect here some basic notions and results from Riemannian geometry that will be used throughout this chapter. Consider a compact p-dimensional smooth manifold (of class C^r, $r \geq 3$) without boundary, endowed with a Riemannian metric of class C^r. For $x \in M$ we denote by $\langle \cdot, \cdot \rangle_x$ the inner product in $T_x M$.

10.1.1. The canonical Riemannian metric. We endow the second tangent bundle $T(TM)$ with a special Riemannian metric. Let $\pi \colon TM \to M$ be the natural projection, that is, $\pi(x, v) = x$ for each $x \in M$ and each $v \in T_x M$. The map $d_v \pi \colon T(TM) \to TM$ is linear and its p-dimensional kernel $H(v) \subset T_v TM$ is the space of horizontal vectors.

Consider the connection operator $K \colon T(TM) \to TM$ defined as follows. For $v \in TM$ and $\xi \in T_v TM$ let $Z(t)$ be any curve in TM such that $Z(0) = d\pi \xi$ and $\frac{d}{dt} Z(t)|_{t=0} = \xi$. Set $K\xi = (\nabla Z)(t)|_{t=0}$ where ∇ is the covariant derivative associated with the Riemannian (Levi-Civita) connection.[1] The map K is linear and its kernel is the p-dimensional space of vertical vectors $V(v) \subset T_v TM$. For every $v \in TM$ we have that $T_v TM = H(v) \oplus V(v)$. This allows one to introduce the special *canonical metric* on $T(TM)$ by

$$\langle \xi, \eta \rangle_v = \langle d_v \pi \xi, d_v \pi \eta \rangle_{\pi(v)} + \langle K\xi, K\eta \rangle_{\pi(v)}$$

in which the spaces $H(v)$ and $V(v)$ are orthogonal.

10.1.2. Geodesics. These are curves along which the tangential vector field is parallel. For any $x \in M$ there is a local coordinate chart in which for any $v \in T_x M$ the geodesic $\gamma_v(t) = (\gamma_1(t), \ldots, \gamma_p(t))$ with $\gamma_v(0) = x$ and $\dot\gamma_v(0) = v$ satisfies a certain system of second-order differential equations on functions $\gamma_i(t)$. We shall always assume that the parameter t along the geodesic $\gamma(t)$ is the arc length; this ensures that $\|\dot\gamma(t)\| = 1$ and $(\nabla \dot\gamma)(t) = 0$.

On complete (in particular, compact) manifolds, geodesics are infinitely extendible and for any $x, y \in M$ there is a (perhaps nonunique) geodesic joining x and y. Among such geodesics there is always at least one whose

---

[1] Recall that $\nabla Z(t)|_{t=0} = \frac{d}{dt} \tilde{Z}(t)|_{t=0}$ where the vector $\tilde{Z}(t)$ is the parallel translation of the vector $Z(t)$ along the curve $\gamma(t) = \pi(Z(t))$.

10.1. Preliminary information from Riemannian geometry

length is the distance $\rho(x,y)$ between the points x and y. Furthermore, for any $v \in TM$ there is a unique geodesic $\gamma_v(t)$ such that $\gamma_v(0) = \pi(v)$ and $\dot{\gamma}_v(0) = v$.

10.1.3. Curvature. For $x \in M$ and a plane $P \subset T_xM$ given by vectors $v, w \in T_xM$, the sectional curvature $K_x(P)$ of M at x in the direction of P is given by
$$K_x(P) = \frac{\langle R(v,w)w, v\rangle}{\langle v,v\rangle\langle w,w\rangle - \langle v,w\rangle^2},$$
where $R(v,w)$ is the curvature tensor associated with v and w.[2] It can be shown that $K_x(P)$ does not depend on the choice of v and w specifying the plane P. The Riemannian metric is said to be of negative (respectively, nonpositive) curvature if the sectional curvature at each point and in the direction of every plane P is negative (respectively, nonpositive). If M is a surface, then the curvature at a point $x \in M$ is a function of x only.

10.1.4. Fermi coordinates. Given a geodesic $\gamma(t)$, chose an orthonormal basis $\{e_i\}$, $i = 1, \ldots, p$, in the space $T_{\gamma(0)}M$ such that $e_n = \dot{\gamma}(0)$. Let $\{e_i(t)\}$ be the vector field along $\gamma(t)$ obtained by the parallel translation of e_i along $\gamma(t)$ (so that $e_i(0) = e_i$). The vectors $\{e_i(t)\}$ form an orthonormal basis in $T_{\gamma(t)}M$ and the corresponding coordinates in $T_{\gamma(t)}M$ are the Fermi coordinates.

10.1.5. Jacobi fields. A Jacobi field is a vector field $Y(t)$ along a geodesic $\gamma(t)$ that satisfies the Jacobi equation
$$Y''(t) + R(X,Y)X = 0, \tag{10.1}$$
where $X(t) = \dot{\gamma}(t)$, $R(X,Y)$ is the curvature tensor associated with the vector fields X and Y, and $Y''(t)$ is the second covariant derivative of the vector field $Y(t)$. Using the Fermi coordinates $\{e_i(t)\}$, $i = 1, \ldots, p$, one can rewrite equation (10.1) as a second-order linear differential equation along $\gamma(t)$,
$$\frac{d^2}{dt^2}Y(t) + K(t)Y(t) = 0, \tag{10.2}$$
where $Y(t) = (Y_1(t), \ldots, Y_p(t))$ is a vector and $K(t) = (K_{ij}(t))$ is a matrix whose entry $K_{ij}(t)$ is the sectional curvature at $\gamma(t)$ in the direction of the plane given by vectors $e_i(t)$ and $e_j(t)$. It follows that every Jacobi field $Y(t)$ is uniquely determined by the values $Y(0)$ and $\frac{d}{dt}Y(0)$ and is well-defined for all t.

---

[2] Recall that $R(v,w) = R(V,W)$ where V and W are any vector fields on M such that $V(x) = v$ and $W(x) = w$ and $R(V,W) = \nabla_V\nabla_W - \nabla_W\nabla_V - \nabla_{[V,W]}$. One can show that the value $R(v,w)$ does not depend on the choice of the vector fields V and W.

The Jacobi fields arise in the formula for the second variation of the geodesic, which describes some infinitesimal properties of geodesics. More precisely, let $\gamma(t)$ be a geodesic and let $Z(s)$, $-\varepsilon \leq s \leq \varepsilon$, be a curve in TM such that $Z(0) = \dot\gamma(0)$. Consider the variation $r(t,s)$ of $\gamma(t)$ of the form

$$r(t,s) = \exp(tZ(s)), \quad t \geq 0, \ -\varepsilon \leq s \leq \varepsilon$$

(in other words, for a fixed $s \in [-\varepsilon, \varepsilon]$ the curve $r(t,s)$ is the geodesic through $\pi(Z(s))$ in the direction of $Z(s)$). One can show that the vector field

$$Y(t) = \frac{\partial}{\partial s} r(t,s)|_{s=0}$$

along the geodesic $\gamma(t)$ is the Jacobi field along $\gamma(t)$.

10.2. Definition and local properties of geodesic flows

The *geodesic flow* g_t is a flow on the tangent bundle TM that acts by the formula

$$g_t(v) = \dot\gamma_v(t).$$

See Figure 10.1. The geodesic flow generates a vector field V on TM given by

$$V(v) = \frac{d_v g_t}{dt}\bigg|_{t=0}.$$

Since M is compact, the flow g_t is well-defined for all $t \in \mathbb{R}$ and is a smooth flow of class C^{r-1} where r is the class of smoothness of the Riemannian metric.

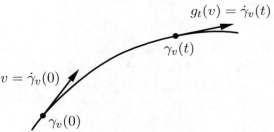

Figure 10.1. The geodesic flow.

We present a different definition of the geodesic flow in terms of the special symplectic structure generated by the Riemannian metric. Consider the cotangent bundle T^*M and define the canonical 1-form ω on T^*M by the formula $\omega(x,q) = q(d\pi^*(x,q))$, where $x \in M$, $q \in T^*M$ is a 1-form, and $\pi^*: T^*M \to M$ is the natural projection (i.e., $\pi^*(x,q) = x$; note that $d\pi^*(x,q) \in T_xM$). The canonical 2-form $\Omega_* = d\omega$ is nondegenerate and induces a symplectic structure on T^*M. The Riemannian metric allows one

10.2. Definition and local properties of geodesic flows

to identify the tangent and cotangent bundles via the map $\mathcal{L} \colon TM \to T^*M$ defined by
$$\mathcal{L}(x, v) = (x, q), \quad x \in M, \quad v \in T_xM,$$
where the 1-form q satisfies $q(w) = \langle v, w \rangle$ for any $w \in T_xM$. This identification produces the canonical 2-form Ω on TM by the formula
$$\Omega(Y, Z) = \Omega_*(\mathcal{L}(Y), \mathcal{L}(Z)), \quad Y, Z \in TM.$$
Consider the function $K \colon TM \to TM$ given by
$$K(x, v) = \frac{1}{2}\langle v, v \rangle = \frac{1}{2}\|v\|^2. \tag{10.3}$$
The vector field V, generating the geodesic flow g_t, is now defined as the vector field corresponding to the 1-form dK relative to the canonical 2-form Ω, that is, $\Omega(V, Z) = dK(Z)$ for any vector field $Z \in TM$.

Geodesic flows serve as mathematical models of classical mechanics, that is, as the phase flows of certain Hamiltonian systems. Here M is viewed as the configuration space and TM as the phase space of the mechanical system; points $x \in M$ are treated as "generalized" coordinates, vectors $v \in TM$ as "generalized" velocities, and 1-forms $q \in T^*M$ as impulses. According to Hamilton's principle, the trajectory of the mechanical system in the configuration space passing through the points x and y is an extremum of the energy functional $E = \frac{1}{2}\int_{t_0}^{t_1}\langle v(t), v(t)\rangle \, dt$,[3] where $\pi(v(t_1)) = x$ and $\pi(v(t_2)) = y$. One can show that these extrema are geodesics connecting x and y; that is, the Euler differential equation for the corresponding variational problem with fixed endpoints is the differential equation for the geodesics.

To describe the trajectories of the mechanical system in the phase space, we observe that the geodesic flow is a Hamiltonian flow with the Hamiltonian function $H(x, v) = K(x, v)$ given by (10.3).[4] Its vector field in local coordinates (x, v) is of the form $\left(\frac{\partial K}{\partial v}, -\frac{\partial K}{\partial x}\right)$.

Since the total energy of the system is a first integral, the hypersurfaces $K(x, v) = \text{const}$ (that is, $\|v\| = \text{const}$) are invariant under the geodesic flow; therefore, we always consider the geodesic flow as acting on the unit tangent bundle $SM = \{v \in TM : \|v\| = 1\}$.

---

[3] We assume that there is no potential energy.

[4] This is a Hamiltonian system of a special type. In general, a Hamiltonian system is defined on a symplectic manifold N (endowed with a symplectic 2-form Ω) by a Hamiltonian function $H \colon N \to \mathbb{R}$ so that the vector field of the Hamiltonian flow is a field corresponding to the 1-form dH relative to the 2-form Ω. In the local coordinate system (x, p) in which $\Omega = \sum dp_i \wedge dx^i$, this vector field has the form $(\partial H/\partial p, -\partial H/\partial x)$. For a conservative mechanical system with a configuration manifold M we have $N = T^*M$ and $H = K - U$ where K is the kinetic energy and U the potential energy of the system. By the Maupertuis–Lagrange–Jacobi principle, for a fixed value E of the Hamiltonian H the motion can be reduced to a geodesic flow on M by introducing a new Riemannian metric $\langle \cdot, \cdot \rangle'_x = (E - U(x))\langle \cdot, \cdot \rangle_x$ and making a certain time change. In our case everything simplifies since $U = 0$ and T^*M is naturally identified with TM.

The representation of a geodesic flow g_t as a Hamiltonian flow allows one to obtain an invariant measure μ for g_t; namely, by Liouville's theorem, this measure is given by $d\mu = d\sigma dm$ where $d\sigma$ is a surface element on the $(p-1)$-dimensional unit sphere and m is the Riemannian volume on M.

In order to study the stability of trajectories of the geodesic flow, one should examine the system of variational equations (6.2). To this end, let us fix $v \in SM$ and $\xi \in T_v SM$ and consider the Jacobi equation (10.2) along the geodesic $\gamma_v(t)$. Let $Y_\xi(t)$ be the unique solution of this equation satisfying the initial conditions:

$$Y_\xi(0) = d_v \pi \xi, \quad \frac{d}{dt} Y_\xi(0) = K\xi. \tag{10.4}$$

One can show that the map $\xi \mapsto Y_\xi(t)$ is an isomorphism for which

$$Y_\xi(t) = d_{g_t(v)} \pi d_v g_t \xi, \quad \frac{d}{dt} Y_\xi(t) = K d_v g_t \xi. \tag{10.5}$$

This map establishes the identification between solutions of the system of variational equations (6.2) and solutions of the Jacobi equation (10.7).

10.3. Hyperbolic properties and Lyapunov exponents

From now on we assume that M is a compact surface endowed with a Riemannian metric of class C^3 and of *nonpositive curvature*, i.e., for any $x \in M$ the Riemannian curvature at x satisfies

$$K(x) \leq 0. \tag{10.6}$$

Take $v \in TM$ and consider the geodesic $\gamma_v(t)$. The Jacobi equation along it is a scalar second-order differential equation of the form

$$\frac{d^2}{dt^2} Y(t) + K(t) Y(t) = 0, \tag{10.7}$$

where $K(t) = K(\gamma_v(t))$. Due to condition (10.6) and compactness of the manifold M, we have for some $k > 0$ and all $t \in \mathbb{R}$ that

$$-k^2 \leq K(t) \leq 0. \tag{10.8}$$

It follows that the boundary value problem for equation (10.7) has a unique solution, i.e., for any s_i and y_i, $i = 1, 2$, there exists a unique solution $Y(t)$ of (10.7) satisfying $Y(s_i) = y_i$ (see [19]).

Proposition 10.1. *Given $s \in \mathbb{R}$, let $Y_s(t)$ be the unique solution of equation (10.7) satisfying the boundary conditions: $Y_s(0) = 1$ and $Y_s(s) = 0$. Then there exists the limit*

$$\lim_{s \to \infty} \frac{d}{dt} Y_s(t) \bigg|_{t=0} = Y^+. \tag{10.9}$$

10.3. Hyperbolic properties and Lyapunov exponents

Proof. Let $A(t)$ be the solution of equation (10.7) satisfying the initial conditions $A(0) = 0$ and $\frac{d}{dt}A(0) = 1$. For $t > 0$ consider the function

$$Z_s(t) = A(t) \int_t^s A^{-2}(u)\, du. \tag{10.10}$$

Exercise 10.2. Show that:

(1) $Z_s(t)$ is the solution of equation (10.7) satisfying the initial conditions $Z_s(s) = 0$ and $\frac{d}{dt}Z_s(s) = -A^{-1}(s)$;

(2) $Z_s(0) = 1$. *Hint:* Use the fact that the Wronskian of any two solutions of equation (10.7) is a constant and apply this to the Wronskian of the solutions $A(t)$ and $Z_s(t)$.

This implies that $Z_s(t) = Y_s(t)$. It is easy to verify that for any numbers $0 < q < s$ and all $t > 0$ we have

$$Y_s(t) - Y_q(t) = A(t) \int_q^s A^{-2}(u)\, du$$

and

$$\frac{d}{dt}Y_s(0) - \frac{d}{dt}Y_q(0) = \int_q^s A^{-2}(u)\, du.$$

Furthermore, one can show that the function $\int_q^s A^{-2}(u)\, du$ is monotonically increasing in t, and hence, the limit

$$\lim_{s \to +\infty} \left(\frac{d}{dt}Y_s(0) - \frac{d}{dt}Y_q(0) \right)$$

exists. It follows that the limit $\lim_{s \to +\infty} \frac{d}{dt}Y_s(0)$ exists as well and the desired result follows. $\square$

We define the *positive limit solution* $Y^+(t)$ of (10.7) as the solution that satisfies the initial conditions:

$$Y^+(0) = 1 \quad \text{and} \quad \frac{d}{dt}Y^+(t)\bigg|_{t=0} = Y^+$$

(see (10.9)). Since solutions of equation (10.7) depend continuously on the initial conditions, by (10.10), we obtain that for every $t > 0$

$$Y^+(t) = \lim_{s \to +\infty} Y_s(t) = A(t) \int_t^\infty A^{-2}(u)\, du. \tag{10.11}$$

It follows that this solution is nondegenerate (i.e., $Y^+(t) \neq 0$ for every $t \in \mathbb{R}$).

Similarly, letting $s \to -\infty$, we can define the *negative limit solution* $Y^-(t)$ of equation (10.7).

For every $v \in SM$ set
$$E^+(v) = \{\xi \in T_v SM : \langle \xi, V(v)\rangle = 0 \text{ and } Y_\xi(t) = Y^+(t)\|d_v\pi\xi\|\}, \\ E^-(v) = \{\xi \in T_v SM : \langle \xi, V(v)\rangle = 0 \text{ and } Y_\xi(t) = Y^-(t)\|d_v\pi\xi\|\}, \quad (10.12)$$
where V is the vector field generated by the geodesic flow and Y_ξ is the solution of equation (10.7) satisfying the initial conditions (10.4).

Using the relations (10.5) and (10.11) (and a similar relation for $Y^-(t)$), one can prove the following properties of the subspaces $E^-(v)$ and $E^+(v)$.

Proposition 10.3. *The following properties hold:*
(1) *$E^-(v)$ and $E^+(v)$ are one-dimensional linear subspaces of $T_v SM$;*
(2) *$d_v \pi E^-(v) = d_v \pi E^+(v) = \{w \in T_{\pi v} M : w \text{ is orthogonal to } v\}$;*
(3) *the subspaces $E^-(v)$ and $E^+(v)$ are invariant under the differential $d_v g_t$, i.e., $d_v g_t E^-(v) = E^-(g_t(v))$ and $d_v g_t E^+(v) = E^+(g_t(v))$;*
(4) *if $\tau : SM \to SM$ is the involution defined by $\tau v = -v$, then*
$$E^+(-v) = d_v \tau E^-(v) \quad \text{and} \quad E^-(-v) = d_v \tau E^+(v);$$
(5) *if $\xi \in E^+(v)$ or $\xi \in E^-(v)$, then $\|K\xi\| \le k\|d_v \pi \xi\|$ where $k > 0$ is given by (10.8);*
(6) *if $\xi \in E^+(v)$ or $\xi \in E^-(v)$, then $Y_\xi(t) \ne 0$ for every $t \in \mathbb{R}$;*
(7) *$\xi \in E^+(v)$ (respectively, $\xi \in E^-(v)$) if and only if*
$$\langle \xi, V(v)\rangle = 0 \quad \text{and} \quad \|d_{g_t(v)}\pi d_v g_t \xi\| \le c$$
for every $t > 0$ (respectively, $t < 0$) and some $c > 0$;
(8) *if $\xi \in E^+(v)$ (respectively, $\xi \in E^-(v)$), then the function $t \mapsto |Y_\xi(t)|$ is nonincreasing (respectively, nondecreasing).*

In view of properties (5) and (7), we have $\xi \in E^+(v)$ (respectively, $\xi \in E^-(v)$) if and only if $\langle \xi, V(v)\rangle = 0$ and $\|d_v g_t \xi\| \le c$ for $t > 0$ (respectively, $t < 0$), for some constant $c > 0$.

The subspaces $E^+(v)$ and $E^-(v)$ are natural candidates for stable and unstable subspaces for the geodesic flows. However, in general, these subspaces may not span the whole second tangent space $T_v SM$, i.e., the intersection $E(v) = E^-(v) \cap E^+(v)$ may be a nontrivial subspace of $T_v SM$. If this is the case, then since the subspaces $E^-(v)$ and $E^+(v)$ are invariant under dg_t, the intersection $E(g_t(v)) = E^-(g_t(v)) \cap E^+(g_t(v))$ coincides with $d_v g_t E(v)$ and is a nontrivial subspace of $T_{g_t(v)}SM$. For every $\xi \in E(v)$ the vector field $Y_\xi(t)$ is parallel along the geodesic $\gamma_v(t)$.[5] Hence, for every $\xi \in E(v)$ the Lyapunov exponent $\chi(v, \xi) = 0$.

---

[5] Indeed, in view of statement (8) of Proposition 10.3, the function $|Y_\xi(t)|$ is both nonincreasing and nondecreasing.

10.3. Hyperbolic properties and Lyapunov exponents

Furthermore, one can show that if for every $v \in SM$ the subspaces $E^-(v)$ and $E^+(v)$ do span the space $T_v SM$ (i.e., $T_v SM = E^-(v) \oplus E^+(v)$), then the geodesic flow is, indeed, Anosov (see [**45**]). This is the case when the curvature is strictly negative.[6] However, for manifolds of nonpositive curvature one can only expect that the geodesic flow is nonuniformly hyperbolic. To see this, consider the set

$$\Delta = \left\{ v \in SM : \limsup_{t \to \infty} \frac{1}{t} \int_0^t K(\gamma_v(s)) ds < 0 \right\}. \tag{10.13}$$

It is easy to see that this set is measurable and invariant under the flow g_t. The following result from [**89**] (see also [**91**]) shows that the Lyapunov exponents are nonzero on the set Δ.

Theorem 10.4. *For every $v \in \Delta$ we have $\chi(v, \xi) < 0$ if $\xi \in E^+(v)$ and $\chi(v, \xi) > 0$ if $\xi \in E^-(v)$.*

Proof. Let $\psi \colon \mathbb{R}^+ \to \mathbb{R}$ be a continuous function. We need the following lemma.

Lemma 10.5. *Assume that $c = \sup_{t \geq 0} |\psi(t)| < \infty$. Then:*

(1) *if $\psi(t) \leq 0$ for all $t \geq 0$ and $\tilde{\psi} > 0$, then $\overline{\psi} < 0$;*

(2) *if $\psi(t) \geq 0$ for all $t \geq 0$ and $\tilde{\psi} > 0$, then $\underline{\psi} > 0$,*

where $\overline{\psi}$ and $\underline{\psi}$ are defined by (2.15), and

$$\tilde{\psi} = \liminf_{t \to \infty} \frac{1}{t} \int_0^t \psi(s)^2 \, ds.$$

Proof of the lemma. Assume that $\psi(t) \leq 0$. Then $\overline{\psi} \leq 0$. On the other hand, if $c > 0$, then

$$-\frac{\overline{\psi}}{c} = \overline{\left|\frac{\psi}{c}\right|} \geq \widetilde{\left(\frac{\psi}{c}\right)} = \frac{\tilde{\psi}}{c^2} > 0.$$

This implies that $\overline{\psi} < 0$ and completes the proof of the first statement. The proof of the second statement is similar. $\square$

We proceed with the proof of the theorem. Fix $v \in \Delta$, $\xi \in E^+(v)$, and consider the function

$$\varphi(t) = \frac{1}{2} |Y_\xi(t)|^2.$$

---

[6]Riemannian manifolds whose geodesic flows are Anosov are said to be of *Anosov type*; see [**45**]. They are closely related to manifolds of hyperbolic type that admit a metric of negative curvature. The latter have been a subject of intensive study in differential geometry: Morse [**83**] already understood that on surfaces of negative Euler characteristic (which admit a metric of constant negative curvature) the geodesic flow in any Riemannian metric inherits to some extent the properties of the geodesic flow in a metric of negative curvature.

Using equation (10.7), we obtain
$$\frac{d^2}{dt^2}\varphi(t) = -K(t)\varphi(t) + \frac{d}{dt}|Y_\xi(t)|^2.$$

It follows from Proposition 10.3 (see statements (6) and (8)) that $\varphi(t) \ne 0$ and $\frac{d}{dt}\varphi(t) \le 0$ for all $t \ge 0$. Set
$$z(t) = (\varphi(t))^{-1}\frac{d}{dt}\varphi(t).$$

It is easy to check that the function $z(t)$ satisfies the Riccati equation
$$\frac{d}{dt}z(t) + z(t)^2 - (\varphi(t))^{-1}\frac{d}{dt}|Y_\xi(t)|^2 + K(t) = 0. \tag{10.14}$$

By Proposition 10.3,
$$\left|\frac{d}{dt}\varphi(t)\right| = \left|\frac{1}{2}\frac{d}{dt}|Y_\xi(t)|^2\right| = |Y_\xi(t)| \cdot \left|\frac{d}{dt}|Y_\xi(t)|\right|$$
$$= \|d_{g_t(v)}\pi d_v g_t \xi\| \cdot \|K d_v g_t \xi\|$$
$$\le a\|d_{g_t(v)}\pi d_v g_t \xi\|^2 = 2a\varphi(t).$$

It follows that $\sup_{t \ge 0} |z(t)| \le 2a$. Integrating the Riccati equation (10.14) on the interval $[0, t]$, we obtain that
$$z(t) - z(0) + \int_0^t z(s)^2 \, ds = \int_0^t (\varphi(s))^{-1}\frac{d}{ds}|Y_\xi(s)|^2 \, ds - \int_0^t K(s) \, ds.$$

It follows that for $v \in \Delta$ (see (10.13)) we have
$$\liminf_{t\to\infty} \frac{1}{t}\int_0^t z(s)^2 \, ds \ge \liminf_{t\to\infty} \frac{1}{t}\int_0^t (\varphi(s))^{-1}\frac{d}{ds}|Y_\xi(s)|^2 \, ds$$
$$- \limsup_{t\to\infty} \frac{1}{t}\int_0^t K(s) \, ds > 0.$$

Therefore, in view of Lemma 10.5 we conclude that
$$\limsup_{t\to\infty} \frac{1}{t}\int_0^t z(s) \, ds < 0.$$

On the other hand, using Proposition 10.3, we find that
$$\chi(v, \xi) = \limsup_{t\to\infty} \frac{1}{t}\log\|d_v g_t \xi\| = \limsup_{t\to\infty} \frac{1}{t}\log\|d_{g_t(v)}\pi d_v g_t \xi\|$$
$$= \limsup_{t\to\infty} \frac{1}{t}\log|Y_\xi(t)| = \frac{1}{2}\limsup_{t\to\infty} \frac{1}{t}\int_0^t z(s) \, ds.$$

This completes the proof of the first statement of the theorem. The second statement can be proved in a similar way. □

10.3. Hyperbolic properties and Lyapunov exponents

It follows from Theorem 10.4 that if the set Δ has positive Liouville measure, then the geodesic flow $g_t|\Delta$ is nonuniformly hyperbolic. It is, therefore, crucial to find conditions which guarantee that Δ has positive Liouville measure.

Theorem 10.6. *Let M be a smooth compact surface of nonpositive curvature $K(x)$ and genus greater than 1. Then $\mu(\Delta) > 0$. Moreover, the set Δ is open (mod 0) and is everywhere dense.*

Proof. By the Gauss–Bonnet formula, the Euler characteristic of M is
$$\frac{1}{2\pi} \int_M K(x)\, dm(x).$$
It follows from the condition of the theorem that
$$\int_M K(x)\, dm(x) < 0. \tag{10.15}$$
By the Birkhoff Ergodic Theorem, we obtain that for μ-almost every $v \in SM$ there exists the limit
$$\lim_{t \to \infty} \frac{1}{t} \int_0^t K(\pi(g_s v))\, ds = \Phi(v)$$
and that
$$\int_{SM} \Phi(v)\, d\mu(v) = \int_M K(x)\, dm(x).$$
It follows from (10.15) that $\mu(\Delta) > 0$.

We shall show that the set Δ is open (mod 0). Note that given $v \in \Delta$, there is a number t such that the curvature $K(x)$ of M at the point $x = \gamma_v(t)$ is strictly negative. Therefore, there is a disk $D(x, r)$ in M centered at x of radius r such that $K(y) < 0$ for every $y \in D(x, r)$. It follows that there is a neighborhood U of v in SM such that $\gamma_w(t) \in D(x, r)$ for every $w \in U$. We denote $\tilde{D}(x, r) = \{z \in SM : \pi(z) \in D(x, r)\}$. Note that the set $\tilde{D}(x, r)$ is open.

By Birkhoff's Ergodic Theorem, for almost every $z \in \tilde{D}(x, r)$ (with respect to the Liouville measure) the limit
$$\lim_{t \to \infty} \frac{\tilde{T}(z, t)}{t} \tag{10.16}$$
exists, where
$$\tilde{T}(z, t) = \int_0^t \chi_{\tilde{D}(x,r)}(\dot\gamma_z(\tau))\, d\tau$$
and $\chi_{\tilde{D}(x,r)}$ is the characteristic function of the set $\tilde{D}(x, r)$. In fact, using the ergodic decomposition of the measure μ, it is easy to show that the limit (10.16) is positive for almost every $z \in \tilde{D}(x, r)$. Since the set $\tilde{D}(x, r)$

is open, we obtain that the limit (10.16) is positive for almost every $w \in U$. This implies that the limit
$$\lim_{t \to \infty} \frac{T(w,t)}{t}$$
exists and is positive for almost every $w \in U$ where
$$T(w,t) = \int_0^t \chi_{D(x,r)}(\gamma_w(\tau))\,d\tau$$
and $\chi_{D(x,r)}$ is the characteristic function of the set $D(x,r)$. In view of (10.13), this implies that almost every $w \in U$ lies in Δ, and hence, the set Δ is open (mod 0). The fact that Δ is everywhere dense follows from topological transitivity of the geodesic flow g_t. □

10.4. Ergodic properties

As we saw in Section 10.3 the geodesic flow g_t on a compact surface M of nonpositive curvature and of genus greater than 1 is nonuniformly hyperbolic on the set Δ of positive Liouville measure which is defined by (10.13). Since the Liouville measure is invariant under the geodesic flow, the results of Section 9.1 apply and show that the ergodic components are of positive Liouville measure (see Theorem 9.2). In this section we show that, indeed, the geodesic flow on Δ is ergodic.

We wish to show that every ergodic component of $g_t|\Delta$ is open (mod 0). To achieve this, we construct one-dimensional foliations of SM, W^-, and W^+, such that $W^s(x) = W^-(x)$ and $W^u(x) = W^+(x)$ for almost every $x \in \Delta$ (W^- and W^+ are known as the stable and unstable horocycle foliations; see below). We then apply Theorem 9.20 to derive that the flow $g_t|\Delta$ is ergodic. In order to proceed in this direction, we need some more information on surfaces of nonpositive curvature (see [46, 92]).

We denote by H the universal Riemannian cover of M, i.e., a simply connected two-dimensional complete Riemannian manifold for which $M = H/\Gamma$ where Γ is a discrete subgroup of the group of isometries of H, isomorphic to $\pi_1(M)$. According to the Hadamard–Cartan theorem, any two points $x, y \in H$ are joined by a single geodesic which we denote by γ_{xy}. For any $x \in H$, the exponential map $\exp_x \colon \mathbb{R}^2 \to H$ is a diffeomorphism. Hence, the map
$$\varphi_x(y) = \exp_x\left(\frac{y}{1-\|y\|}\right) \tag{10.17}$$
is a homeomorphism of the open unit disk D onto H.

Two geodesics $\gamma_1(t)$ and $\gamma_2(t)$ in H are said to be *asymptotic* if
$$\sup_{t>0} \rho(\gamma_1(t), \gamma_2(t)) < \infty,$$

where ρ is the distance in H induced by the Riemannian metric. Given a point $x \in H$, there is a unique geodesic starting at x which is asymptotic to a given geodesic. The asymptoticity is an equivalence relation, and the equivalence class $\gamma(\infty)$ corresponding to a geodesic γ is called a *point at infinity*. The set of these classes is denoted by $H(\infty)$ and is called the *ideal boundary* of H. Using (10.17), one can extend the topology of the space H to $\overline{H} = H \cup H(\infty)$ so that $\overline{H}$ becomes a compact space.

The map φ_x can be extended to a homeomorphism (still denoted by φ_x) of the closed disk $\overline{D} = D \cup S^1$ onto $\overline{H}$ by the equality

$$\varphi_x(y) = \gamma_y(+\infty), \qquad y \in S^1.$$

In particular, φ_x maps S^1 homeomorphically onto $H(\infty)$.

For any two distinct points x and y on the ideal boundary there is a geodesic joining them. This geodesic is uniquely defined if the Riemannian metric is of strictly negative curvature (i.e., if inequality (10.6) is strict). Otherwise, there may exist a pair of distinct points $x, y \in H(\infty)$ which can be joined by more than one geodesic. More precisely, there exists a geodesically isometric embedding into H of an infinite strip of zero curvature which consists of geodesics joining x and y. Moreover, any two geodesics on the universal cover which are asymptotic both for $t > 0$ and for $t < 0$ (i.e., they join two distinct points on the ideal boundary) bound a flat strip. The latter means that there is a geodesically isometric embedding of a flat strip in $\mathbb{R}^2$ into the universal cover. This statement is know as the *flat strip theorem*.

The fundamental group $\pi_1(M)$ of the manifold M acts on the universal cover H by isometries. This action can be extended to the ideal boundary $H(\infty)$. Namely, if $p = \gamma_v(+\infty) \in H(\infty)$ and $\zeta \in \pi_1(M)$, then $\zeta(p)$ is the equivalence class of geodesics which are asymptotic to the geodesic $\zeta(\gamma_v(t))$.

We now describe the invariant foliations for the geodesic flow. We consider the distributions E^- and E^+ introduced in Section 10.3 (see (10.12)).

Proposition 10.7. *The distributions E^- and E^+ are integrable. Their integral manifolds form foliations of SM which are invariant under the geodesic flow g_t.*

Sketch of the proof. Consider the circle $S^1(\gamma_v(t), t)$ of radius t centered at $\gamma_v(t)$, for $t > 0$, and let $x = \pi(v)$. The intersection

$$\Gamma(t) = D(x, R) \cap S^1(\gamma_v(t), t)$$

(here $D(x, R)$ is the disk centered at x of some radius R) is a smooth convex curve passing through x. One can show that any geodesic which is orthogonal to this curve passes through the center of the circle $\gamma_v(t)$, the curvature of $\Gamma(t)$ is bounded uniformly over t, and the family of curves $\Gamma(t)$ is monotone (by inclusion) and is compact. Therefore, the sequence of curves $\Gamma(t)$ converges as $t \to \infty$ to a convex smooth curve of bounded curvature. Moreover, any geodesic which is orthogonal to this curve is asymptotic to $\gamma_v(t)$. The framing of the limit curve is a local manifold at v which is a piece of the integral manifold for the distribution E^-. The integral manifolds corresponding to the distribution E^+ can be obtained in a similar fashion. □

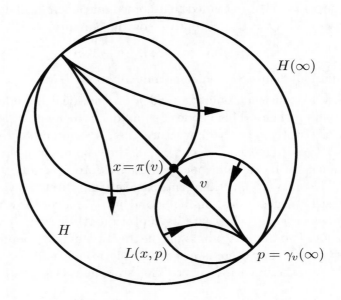

Figure 10.2. Horocycles.

We denote by W^- and W^+ the foliations of SM corresponding to the invariant distributions E^- and E^+. These foliations can be lifted from SM to SH. We denote these lifts by $\tilde{W}^-$ and $\tilde{W}^+$, respectively.

Given $x \in H$ and $p \in H(\infty)$, set
$$L(x,p) = \pi(\tilde{W}^-(v))$$
where $x = \pi(v)$ and $p = \gamma_v(\infty)$. The set $L(x,p)$ is called the *horocycle* centered at p and passing through x. See Figure 10.2.

We summarize the properties of the foliations and horocycles in the following statement.

Proposition 10.8. *The following properties hold:*

(1) *for any $x \in H$ and $p \in H(\infty)$ there exists a unique horocycle $L(x,p)$ centered at p and passing through x; it is a limit in the C^1 topology of circles $S^1(\gamma(t), t)$ as $t \to +\infty$ where γ is the unique geodesic joining x and p;*

10.4. Ergodic properties

(2) the leaf $W^-(v)$ is the framing of the horocycle $L(x,p)$ ($x = \pi(v)$ and $p = \gamma_v(+\infty)$) by orthonormal vectors which have the same direction as the vector v (i.e., they are "inside" the limit sphere). The leaf $W^+(v)$ is the framing of the horocycle $L(x,p)$ ($x = \pi(v)$ and $p = \gamma_v(-\infty) = \gamma_{-v}(+\infty)$) by orthonormal vectors which have the same direction as the vector v (i.e., they are "outside" the limit sphere);

(3) for every $\zeta \in \pi_1(M)$ we have
$$\zeta(L(x,p)) = L(\zeta(x), \zeta(p)),$$
$$d_v\zeta \tilde{W}^-(v) = \tilde{W}^-(d_v\zeta v), \quad d_v\zeta \tilde{W}^+(v) = \tilde{W}^+(d_v\zeta v);$$

(4) for every $v, w \in SH$, for which $\gamma_v(+\infty) = \gamma_w(+\infty) = p$, the geodesic $\gamma_w(t)$ intersects the horocycle $L(\pi(v), p)$ at some point.

We now state a remarkable result by Eberlein [**44**].

Proposition 10.9. *The geodesic flow g_t on any surface M of nonpositive curvature and of genus greater that 1 is topologically transitive.*

Proof. We call two points $x, y \in H(\infty)$ *dual* if for any open sets U and V, containing x and y, respectively, there exists an element $\xi \in \pi_1(M)$ such that $\xi(\bar{H} \setminus U) \subset V$. Clearly, $\xi(\bar{H} \setminus U) \subset V$ if and only if $\xi^{-1}(\bar{H} \setminus V) \subset U$. One can show that on any surface M of nonpositive curvature and of genus greater that 1 any two points $x, y \in H(\infty)$ are dual (see [**44**]).

Lemma 10.10. *If the points $x, y \in H(\infty)$ are dual, then there exists a sequence $\xi_n \in \pi_1(M)$ such that $\xi_n^{-1}(p) \to x$ and $\xi_n(p) \to y$ as $n \to \infty$ for any $p \in H$.*

Proof of the lemma. Let $\{U_n\}$ and $\{V_n\}$ be local bases for the cone topology at x and y, respectively. For each $n > 0$ there exists $\xi_n \in \pi_1(M)$ such that $\xi_n(\bar{H} \setminus U_n) \subset V_n$ and $\xi_n^{-1}(\bar{H} \setminus V_n) \subset U_n$. For any $p \in H$ and any sufficiently large n we have $p \in (\bar{H} \setminus U_n) \cap (\bar{H} \setminus V_n)$ and hence, $\xi_n(p) \in V_n$ and $\xi_n^{-1}(p) \in U_n$. The lemma follows. □

Given $v, w \in SM$, let $\tilde{v}$ and $\tilde{w}$ be their lifts to SH. Denote by $\zeta : H \to M$ the covering map. Then $d\zeta$ maps SH onto SM.

Since the points $x = \gamma_{\tilde{v}}(+\infty)$ and $y = \gamma_{\tilde{w}}(-\infty)$ are dual, by the lemma, there exists a sequence $\xi_n \in \pi_1(M)$ such that $\xi_n^{-1}(p) \to x$ and $\xi_n(p) \to y$ as $n \to \infty$ for any $p \in H$. Let $p = \pi(\tilde{v})$ and $q = \pi(\tilde{w})$ where $\pi : SM \to M$ is the projection, and set $t_n = d(\xi_n(p), q)$ and $\tilde{v}_n = \dot{\gamma}_{\xi_n(p),q}(0)$ (where $\gamma_{\xi_n(p),q}(t)$ is the geodesic connecting the points $\xi_n(p)$ and q). Then $g_{t_n}(\tilde{v}_n) = -\dot{\gamma}_{q,\xi_n(p)}(0)$.

Since $\xi_n(p) \to y$, it follows that $g_{t_n}(\tilde{v}_n) \to \tilde{w}$. Observing that $\xi_n^{-1}(q) \to x$, we obtain that $d\xi_n^{-1}\tilde{v}_n = \dot{\gamma}_{p,\xi_n^{-1}(q)}(0) \to \tilde{v}$. It follows that

$$d\zeta\tilde{v}_n = d\zeta d\xi_n^{-1}\tilde{v}_n \to d\zeta\tilde{v} = v$$

and that

$$g_{t_n}(d\zeta\tilde{v}_n) = d\zeta g_{t_n}(\tilde{v}_n) \to d\zeta\tilde{w} = w.$$

This means that for any open neighborhoods U and V of v and w, respectively, the set $g_t(U)$ meets V for arbitrarily large positive values of t, implying topological transitivity of the geodesic flow g_t. $\square$

We now state our main result.

Theorem 10.11. *Let M be a compact surface of nonpositive curvature and of genus greater than 1. Then the geodesic flow $g_t|\Delta$ is ergodic.*

Proof. By Theorem 10.6, $\mu(\Delta) > 0$ where μ is the Liouville measure. Moreover, it is open (mod 0) and is everywhere dense.

For μ-almost every $v \in \Delta$, consider one-dimensional local stable and unstable manifolds $V^s(v)$ and $V^u(v)$. We denote by $\tilde{V}^s(\tilde{v})$ and $\tilde{V}^u(\tilde{v})$ their lifts to SH (with $\tilde{v}$ being a lift of v). Given $\tilde{w} \in \tilde{V}^s(\tilde{v})$, we have

$$\rho(\pi(g_t(\tilde{v})), \pi(g_t(\tilde{w}))) \to 0 \quad \text{as } t \to \infty. \tag{10.18}$$

It follows that the geodesics $\gamma_{\tilde{v}}(t)$ and $\gamma_{\tilde{w}}(t)$ are asymptotic, and hence, $\gamma_{\tilde{v}}(+\infty) = \gamma_{\tilde{w}}(+\infty)$. We wish to show that $\tilde{w} \in \tilde{W}^-(\tilde{v})$. Assuming the contrary, consider the horocycle $L(\pi(\tilde{v}), \gamma_{\tilde{v}}(+\infty))$. Let $\tilde{z}$ be the point of intersection of the geodesic $\gamma_{\tilde{w}}(t)$ and this horocycle (such a point exists by Proposition 10.8). We have

$$\rho(\pi(g_t(\tilde{v})), \pi(g_t(\tilde{w}))) \geq \rho(\pi(\tilde{w}), \tilde{z}) > 0,$$

which contradicts (10.18). It follows that $\tilde{V}^s(\tilde{v}) \subset \tilde{W}^-(\tilde{v})$ for every $v \in \Delta$ and every lift $\tilde{v}$ of v. Arguing similarly, one can show that $\tilde{V}^u(\tilde{v}) \subset \tilde{W}^+(\tilde{v})$ for every $v \in \Delta$ and every lift $\tilde{v}$ of v. By Theorem 9.20, we conclude that every ergodic component of $g_t|\Delta$ is open (mod 0). In view of Proposition 10.9 we conclude that $g_t|\Delta$ is ergodic. $\square$

Consider the complement $\Delta^c = SM \setminus \Delta$. If $\mu(\Delta^c) = 0$, then by Theorem 10.11, the geodesic flow g_t is ergodic.

Assume now that $\mu(\Delta^c) > 0$. We claim that $K(\gamma_v(t)) = 0$ for almost every $v \in \Delta^c$ and every $t \in \mathbb{R}$. To see this observe that by Birkhoff's Ergodic Theorem, there is a set $N \subset \Delta^c$ of zero measure such that for every $v \in \Delta^c \setminus N$,

$$\lim_{t \to \infty} \frac{1}{t} \int_0^t K(\gamma_v(s)) ds = 0. \tag{10.19}$$

Let v be a Lebesgue density point of the set $\Delta^c \setminus N$. If $K(\gamma_v(t)) < 0$ for some $t \in \mathbb{R}$, then $K(\gamma_w(t)) < 0$ for every $w \in B(v,r)$ where $B(v,r)$ is a ball centered at v of some radius $r > 0$. Note that the set $B(v,r) \cap \Delta^c$ has positive measure. Applying Birkhoff's Ergodic Theorem, we obtain that for almost every $w \in B(v,r) \cap \Delta^c$,

$$\lim_{t \to 0} \frac{1}{t} l(\{t: g_t(w) \in B(v,r)\}) > 0,$$

where l is the one-dimensional Lebesgue measure. It follows that the limit in (10.19) is strictly negative, leading to a contradiction. In particular, we obtain that the Lyapunov exponent $\chi(v,\xi) = 0$ for almost every $v \in \Delta^c$ and every $\xi \in T_v SM$.

Consider the set $\Lambda \subset SM$ of flat geodesics, i.e., for $v \in \Lambda$ we have that $K(\gamma_v(t)) = 0$ for any $t \in \mathbb{R}$. It follows from what was said above that $\Lambda = \Delta^c$ (mod 0). It is conjectured that *the set Λ consists of a finite number of flat strips and a finite number of isolated closed flat geodesics*. Clearly, if this conjecture were true, the geodesic flow g_t would be ergodic.

Theorem 10.12 (Wu [112]). *Assume that the set $\{x \in M : K(x) < 0\}$ has finitely many open connected components in M. Then for every $v \in \Lambda$ the geodesic $\gamma_v(t)$ is closed. In particular, the above conjecture holds.*

Sketch of the proof. Assume that there is $v \in \Lambda$ such that the geodesic $\gamma_v(t)$ is not closed. Then one can show that there are $v_1, v_2 \in \Lambda$ not on the same orbit such that $\rho(\gamma_{v_1}(t), \gamma_{v_2}(t)) \to 0$ as $t \to \infty$. Let $\tilde{\gamma}_{v_1}$ and $\tilde{\gamma}_{v_2}$ be the lifts of geodesics γ_{v_1} and γ_{v_1} to the universal cover H. Since $\rho(\tilde{\gamma}_{v_1}(t), \tilde{\gamma}_{v_2}(t)) \to 0$ as $t \to \infty$, we obtain the ideal triangle $\triangle ABC$ with vertices $A = \tilde{\gamma}_{v_1}(-\infty)$, $B = \tilde{\gamma}_{v_2}(-\infty)$, and $C = \tilde{\gamma}_{v_1}(\infty) = \tilde{\gamma}_{v_2}(\infty)$. Since $\tilde{\gamma}_{v_1}$ and $\tilde{\gamma}_{v_2}$ are flat geodesics, they cannot intersect the lift to H of any connected component of negative curvature. Since the number of such components is finite, the radii of their inscribed circles are bounded away from zero, and since the sizes of connected components do not change when they are lifted to H, we obtain that the connected components inside $\triangle ABC$ cannot approach C. Therefore, there is $t_0 > 0$ such that the infinite triangle with vertices $\tilde{\gamma}_{v_1}(t_0)$, $\tilde{\gamma}_{v_2}(t_0)$, and C is a flat region. It follows that

$$\rho(\tilde{\gamma}_{v_1}(t), \tilde{\gamma}_{v_2}(t)) = \rho(\tilde{\gamma}_{v_1}(t_0), \tilde{\gamma}_{v_2}(t_0))$$

for any $t > t_0$, leading to a contradiction. $\square$

10.5. The entropy formula for geodesic flows

For geodesic flows on surfaces of negative curvature the entropy formula (see Section 9.3) can be transferred into a remarkable form that explicitly relates the metric entropy with the curvature of horocycles. More precisely, for

$v \in SM$ consider the horocycle $L(\pi(v), \gamma_v(+\infty))$ through the point $x = \pi(v)$, which is a submanifold in H of class C^2 if the Riemannian metric is of class C^4.

Theorem 10.13. *Let g_t be the geodesic flow on a compact surface endowed with a C^4 Riemannian metric of nonpositive curvature. Then the metric entropy of g_1 with respect to the Liouville measure μ is*

$$h_\mu(g_1) = -\int_{SM} K(v)\, d\mu(v).$$

Proof. For $v \in SM$ and $\xi \in E^+(v)$, by statement (4) of Proposition 10.3, we have for $t \geq 1$ that

$$\|d_v g_t \xi\| \leq (1+a)\|d_{g_t(v)}\pi d_v g_t \xi\|.$$

This implies that for any LP-regular point $v \in SM$ and $\xi \in E^+(v)$,

$$\begin{aligned}\chi^-(v,\xi) &= \lim_{t\to\infty} \frac{1}{t} \log \|d_v g_t \xi\| \\ &= \lim_{t\to\infty} \frac{1}{t} \log \|d_{g_t(v)}\pi d_v g_t \xi\|.\end{aligned} \quad (10.20)$$

For any $s \neq 0$ let

$$\lambda_s(v) = \frac{\|d_{g_s(v)}\pi d_v g_s \xi\|}{\|d_v \pi \xi\|}$$

(note that $\lambda_s(v)$ does not depend on the choice of the vector ξ). By the entropy formula, Birkhoff's Ergodic Theorem, and (10.20), for every $s \neq 0$,

$$\begin{aligned}h_\mu(g_s) &= -\int_{SM} \chi^-(v,\xi)\, d\mu(v) \\ &= -\int_{SM} \lim_{n\to\infty} \frac{1}{sn} \log \|d_{g_{sn}(v)}\pi d_v g_{sn} \xi\|\, d\mu(v) \\ &= -\int_{SM} \lim_{n\to\infty} \frac{1}{sn} \sum_{i=0}^{n-1} \log \lambda_{s(g_s(i+1)(v))}\, d\mu(v) \\ &= -\int_{SM} \log \lambda_s(v)\, d\mu(v).\end{aligned} \quad (10.21)$$

Since $h_\mu(g_s) = |s| h_\mu(g_1)$, we have by (10.21) that for $s > 0$,

$$\begin{aligned}h_\mu(g_1) &= \frac{1}{s} h_\mu(g_s) = \lim_{s\to 0} \frac{1}{s} h_\mu(g_s) \\ &= -\int_{SM} \lim_{s\to 0} \frac{1}{s} \log \lambda_s(v)\, d\mu(v).\end{aligned}$$

10.5. The entropy formula for geodesic flows

We shall compute the expression under the integral. For $v \in SM$, $\xi \in E^+(v)$, and $s > 0$ we have

$$\begin{aligned}|Y_\xi(s)|^2 &= |Y_\xi(0)|^2 + \int_0^s \frac{d}{du}|Y_\xi(u)|^2\, du \\ &= |Y_\xi(0)|^2 + 2\int_0^s |Y_\xi(u)| \cdot \frac{d}{du}|Y_\xi(u)|\, du.\end{aligned} \quad (10.22)$$

Let

$$a(s) = \frac{1}{s}\int_0^s |Y_\xi(u)| \cdot \frac{d}{du}|Y_\xi(u)|\, du.$$

For $s = 0$ we have

$$a(0) = 2|Y_\xi(s)| \cdot \frac{d}{ds}|Y_\xi(s)|\Big|_{s=0} = 2K(v). \quad (10.23)$$

Using (10.22) and (10.23) and noting that $|Y_\xi(0)|^2 = 1$, we obtain for $s \neq 0$,

$$\begin{aligned}\|d_{g_s(v)}\pi d_v g_s \xi\| &= \sqrt{\|d_{g_s(v)}\pi d_v g_s \xi\|^2} \\ &= \sqrt{|Y_\xi(s)|^2} = \sqrt{1+sa(s)} \\ &= 1 + \frac{1}{2}sa(s) + O(s^2).\end{aligned}$$

It follows that

$$\lim_{s\to 0}\frac{1}{s}\log \lambda_s(v) = K(v),$$

implying the desired result. $\square$

Chapter 11

Topological and Ergodic Properties of Hyperbolic Measures

In the previous chapters of the book, we studied ergodic properties of dynamical systems (diffeomorphisms and flows) with respect to smooth hyperbolic invariant measures. We have seen that for any such system every ergodic component has positive measure and up to a permutation the restriction of the system to an ergodic component is Bernoulli (see Theorems 9.2, 9.13, and 9.12). Moreover, the entropy of the system satisfies the formula (9.15). There is a larger class of invariant measures known as SRB measures after Sinai, Ruelle, and Bowen who introduced and studied these measures for Anosov diffeomorphisms and flows.

In this chapter we discuss some more general classes of hyperbolic measures. In Section 11.1 we introduce hyperbolic SRB measures and show that their ergodic properties are similar to those of smooth hyperbolic measures. We also present some rather general results on existence and uniqueness of such measures. SRB measures belong to a yet more general class of invariant measures with local product structure which as we explain is the driving force in studying ergodicity of hyperbolic measures.

In Section 11.2 we show existence and density of periodic points for general hyperbolic invariant measures. This allow us to associate to a given hyperbolic invariant measure ν a sequence of measures supported on horseshoes whose entropies approximate the entropy of ν. We then describe a more advanced approximation technique.

In Section 11.3 we discuss an important shadowing property of hyperbolic invariant measures and for diffeomorphisms on nonuniformly hyperbolic sets we outline a construction of special Markov partitions with countable number of partition elements.

Throughout this chapter we assume that f is a $C^{1+\alpha}$ diffeomorphism of a compact smooth Riemannian manifold M preserving a hyperbolic measure ν. As shown in Section 5.2.1, the set $\mathcal{E}$ given by (5.3), has full measure and is nonuniformly completely hyperbolic, so we can consider the associated families of level sets Λ_ι (see (5.28)) and regular sets Λ_ι^ℓ (see (5.29)), where $\iota = \{\lambda, \mu, \varepsilon, j\}$ is an index set (the numbers λ, μ, and ε satisfy (5.24)) and $\ell \geq 1$.

11.1. Hyperbolic measures with local product structure

We fix an index set ι and a number $\ell \geq 1$ such that the regular set Λ_ι^ℓ has positive measure. Every point $x \in \Lambda_\iota^\ell$ has local stable $V^s(x)$ and unstable $V^u(x)$ manifolds of size r_ℓ that does not depend on x (see Section 7.3.1).

11.1.1. Local product structure. We say that a subset $R \subset \Lambda_\iota^\ell$ is a *rectangle* if for any $x, y \in R$ we have $V^s(x) \cap V^u(x) \in R$. There is a number $\delta_\ell > 0$ such that if diam $R < \delta_\ell$, then the intersection consists of a single point. For $x \in R$ set $V_R^u(x) = V^u(x) \cap R$. The sets $V_R^u(x)$ corresponding to different points $x \in R$ form a partition of R. We denote by $\nu^u(x)$ the conditional measure on $V_R^u(x)$. Further, we denote by $\pi_{y,z}^s: V_R^u(y) \to V_R^u(z)$ the holonomy map generated by the family of stable manifolds $V^s(w)$ with $w \in R$ and transversals $T^1 = V_R^u(y)$ and $T^2 = V_R^u(z)$ with $y, z \in R$ (see (8.5)).

We say that the measure ν has a *local product structure* if for any rectangle R of positive measure and every $y, z \in R$ the holonomy map $\pi_{y,z}^s$ is absolutely continuous, i.e, the measure $(\pi_{y,z}^s)_* \nu^u(y)$ is absolutely continuous with respect to $\nu^u(z)$.

Theorem 11.1. *Let f be a $C^{1+\alpha}$ diffeomorphism of a compact smooth Riemannian manifold M preserving a hyperbolic measure ν which has local product structure. Then:*

(1) *the ergodic components of ν have positive measure, and hence, there are at most countably many such components;*

(2) *for every ergodic component A of positive measure there are a number $n \geq 0$ and a set $A_0 \subset A$ such that:*
  (a) *the sets $A_j = f^j(A_0)$, $0 \leq j \leq n-1$, are disjoint, their union is A, and $A_n = f^n(A_0) = A_0$;*

(b) *assume that there is $C_\ell \geq 1$ such that for any rectangle R and $y, z \in R$,*

$$C_\ell^{-1} \leq \frac{d(\pi_{y,z})_* \nu^u(y)}{\nu^u(z)} \leq C_\ell; \qquad (11.1)$$

then $f^n | A_0$ has the Bernoulli property.

Proof. The proof of the first statement can be obtained by repeating the arguments in the proof of Theorem 9.2 (see also Remark 9.7). For the proof of the second statement we refer the reader to [**2**]. □

11.1.2. SRB measures: Ergodic properties. Theorem 8.4 shows that any smooth measure has local product structure. We now introduce a larger class of invariant measures with local product structure.

A hyperbolic invariant measure ν is called a *Sinai–Ruelle–Bowen (SRB) measure* if for every regular set Λ_ι^ℓ of positive measure and almost every $x \in \Lambda_\iota^\ell$ the conditional measure $\nu^u(x)$ is equivalent to the leaf-volume $m^u(x)$ on $V^u(x)$; that is,

$$d\nu^u(x)(y) = \rho^u(x)^{-1} \rho^u(y, x) dm^u(x)(y), \qquad (11.2)$$

where the *density function* $\rho^u(y, x)$ is positive and bounded uniformly in $x \in \Lambda_\iota^\ell$ and $y \in V^u(x)$, and $\rho^u(x)$ is the normalizing factor; this function is given by (8.11) (where the superscript s should be replaced with u).

Given a Borel invariant measure ν, define its *basin of attraction* $B(\nu)$ as the set of points $x \in M$ for which the sequence of measures

$$\nu_{x,n} := \frac{1}{n} \sum_{k=0}^{n-1} \delta_{f^k(x)}$$

converges to ν in the weak*-topology. The following result describes some basic ergodic properties of SRB measures.

Theorem 11.2. *If ν is an SRB measure, then:*

(1) *ν has the ergodic properties described in Theorem 11.1;*

(2) *$\nu(B(\nu)) > 0$;*

(3) *the entropy $h_\nu(f)$ is given by the entropy formula (9.15).*

Proof. It follows from Theorem 8.2 that an SRB measure ν has local product structure. Furthermore, by Remark 8.8, the function $\rho^u(y, x)$ satisfies (8.10) (where the superscript s should be replaced with u and the map f with its inverse f^{-1}). In particular, the function $\rho^u(y, x)$ is Hölder continuous in $x \in \Lambda_\iota^\ell$ and is smooth in $y \in V^u(x)$. This implies that condition (11.1) holds, and hence, the first statement follows from Theorem 11.1. To prove the second statement choose a regular set Λ_ι^ℓ of positive ν-measure and a point $x \in \Lambda_\iota^\ell$. Consider the set $P_\ell^s(x,r)$ given by (9.2). Since ν is an SRB measure, by Theorem 8.7, the volume $m(P_\ell^s(x,r)) > 0$. Note that $P_\ell^s(x,r) \subset B(\nu)$ (mod 0) and the second statement follows. For the proof of the third statement we refer the reader to Section 3.1 in [13]. □

It follows from statement (2) of Theorem 11.2 that the diffeomorphism f may have at most countably many ergodic SRB measures. However, the situation is different in the two-dimensional case.

Theorem 11.3 (F. Rodriguez Hertz–M. A. Rodriguez Hertz–Tahzibi–Ures [98])**.** *Let f be a $C^{1+\alpha}$ topologically transitive surface diffeomorphism. Then f can have at most one SRB measure.*

Remark 11.4. In this book we only consider hyperbolic SRB measures but our definition (11.2) can easily be extended to measures which give full weight to some nonuniformly partially hyperbolic sets. For this extended class of SRB measures one can show that a Borel f-invariant measure ν is an SRB measure if and only if it satisfies the entropy formula (9.15). Furthermore, Theorem 11.3 holds for SRB measures from this extended class.

11.1.3. SRB measures: Existence. For some earlier results on construction of SRB measures we refer the reader to Section 13.3 in [13]. Here we present and discuss some more recent results on the existence of SRB measures.

1. Any smooth measure is an SRB measure. This is an immediate consequence of (11.2) and Theorems 8.2 and 8.4.

2. Theorem 1.42 is a particular case of a more general result by Climenhaga, Dolgopyat, and Pesin [39] which provides some conditions guaranteeing that a limit measure of the sequence of measures (1.12) is an SRB measure. While the statement and proof of this result are beyond the scope of this book, to illustrate the basic idea we outline here a proof of Theorem 1.19, using a "baby version" of the argument in [39].

Let Λ be a (uniformly) hyperbolic attractor for a diffeomorphism f such that $f|\Lambda$ is topologically transitive. Fix $x_0 \in \Lambda$ and consider its local unstable manifold $V^u(x_0)$. By Exercise 1.17, $V^u(x_0) \subset \Lambda$, so we can consider the leaf-volume $m^u(x_0)$ on $V^u(x_0)$, which we extend to a measure $\tilde{m}^u(x_0)$ on Λ by setting $\tilde{m}^u(x_0)(E) = m^u(x_0)(E \cap V^u(x_0))$ where $E \subset \Lambda$ is a Borel set. The evolution of $\tilde{m}^u(x_0)$ under f is the sequence of averages of the push-forward measures

$$\nu_n = \frac{1}{n} \sum_{k=0}^{n-1} f_*^k \tilde{m}^u(x_0). \tag{11.3}$$

11.1. Hyperbolic measures with local product structure

Exercise 11.5. Show that:
(1) the sequence of measures μ_n in (1.12) converges in the weak*-topology to a measure μ on Λ if and only if the sequence (11.3) converges to μ;
(2) if some subsequence ν_{n_k} converges in the weak*-topology to a measure μ on Λ, then the sequence ν_n converges to ν. *Hint:* Use the fact that Λ is uniformly hyperbolic to show that every ergodic component of an SRB measure is open (mod 0) and then apply the topological transitivity of $f|\Lambda$.

To proceed we note that for $x \in \Lambda$ its local unstable manifold $V^u(x)$ can be written as
$$V^u(x) = \exp_x\{(v, \psi_x^u(v)) : v \in B^u(r)\},$$
where $B^u(r) \subset E^u(x)$ is the ball centered at 0 of radius r and $\psi_x^u \colon B^u(r) \to E^s(x)$ is a $C^{1+\alpha}$ map satisfying $\psi_x^u(0) = 0$ and $d_0 \psi_x^u = 0$ (compare to (7.3)). The number r is the size of $V^u(x)$.

Fix $r > 0$ and let $\mathcal{R}_r$ be the space of all local unstable manifolds $V^u(x)$, $x \in \Lambda$, of size r.

Exercise 11.6. Show that $\mathcal{R}_r$ has a natural topology with respect to which it is compact.

Further, fix $L > 0$ and let $\mathcal{R}_{(r,L)}$ be the space of *standard pairs* $(V^u(x), \rho)$, where $V^u(x) \in \mathcal{R}_r$ and ρ is a $C^{1+\alpha}$ density function on $V^u(x)$ such that $\frac{1}{L} \leq \|\rho\|_{C^1} \leq L$ and the Hölder norm $|\rho|_\alpha \leq L$.

Exercise 11.7. Show that $\mathcal{R}_{(r,L)}$ is compact in the natural product topology.

A standard pair $(V^u(x), \rho)$ determines a measure $\Psi(V^u(x), \rho)$ on Λ as follows: for a Borel set $E \subset \Lambda$,
$$\Psi(V^u(x), \rho)(E) := \int_{E \cap V^u(x)} \rho \, dm^u(x).$$
Moreover, each measure η on $\mathcal{R}_{(r,L)}$ determines a measure $\Phi(\eta)$ on Λ given by
$$\Phi(\eta)(E) = \int_{\mathcal{R}_{(r,L)}} \Psi(V^u(x), \rho)(E) \, d\eta(V^u(x), \rho)$$
$$= \int_{\mathcal{R}_{(r,L)}} \int_{E \cap V^u(x)} \rho(x) \, dm^u(x) \, d\eta(V^u(x), \rho).$$
One can show that the map $\Phi \colon \mathcal{M}(\mathcal{R}_{(r,L)}) \to \mathcal{M}(M)$ is continuous; here $\mathcal{M}(M)$ and $\mathcal{M}(\mathcal{R}_{(r,L)})$ are the spaces of finite Borel measures on M and $\mathcal{R}_{(r,L)}$, respectively. In particular, $\mathcal{M}_{(r,L)} = \Phi(\mathcal{M}_{\leq 1}(\mathcal{R}_{(r),L}))$ is compact, where $\mathcal{M}_{\leq 1} \subset \mathcal{M}(M)$ is the space of measures with total weight at most 1.

Note that an invariant probability measure μ is an SRB measure if and only if $\mu \in \mathcal{M}_{(r,L)}$ for some (r, L). Note also that $m^u(x_0) \in \mathcal{M}_{(r,L)}$ for some (r, L), and hence, it suffices to show that $\mathcal{M}_{(r,L)}$ is invariant under the action of f_*. Indeed, this would imply that $f_*^k m^u(x_0) \in \mathcal{M}_{(r,L)}$ for every $k \geq 1$, and hence, $\nu_n \in \mathcal{M}_{(r,L)}$ for every $n \geq 1$. By compactness of $\mathcal{M}_{(r,L)}$, we obtain that $\mu \in \mathcal{M}_{(r,L)}$.

To show that $\mathcal{M}_{(r,L)}$ is invariant under f_*, take a standard pair $(V^u(x), \rho)$ in $\mathcal{R}_{(r,L)}$ and consider the image $f(V^u(x))$. One can show that there is a finite collection of standard pairs $\{(V^u(x_i), \rho_i) \in \mathcal{R}_{(r,L)}\}$ such that the image $f(V^u(x))$ is covered by $V^u(x_i)$ with uniformly bounded multiplicity and $f_* m^u(x)$ is a convex combination of measures $\rho_i \, dm^u(x_i)$. This implies that $\mathcal{M}_{(r,L)}$ is invariant under the action of f_* and completes the proof of the theorem.

3. Let Λ be a nonuniformly hyperbolic set of positive volume for a $C^{1+\alpha}$ diffeomorphism f. We can choose the index set ι and $\ell \geq 1$ such that the set Λ_ι^ℓ has positive volume.

We call a subset $\mathcal{P} \subset M$ *forward positively recurrent with respect to* Λ if $\mathcal{P} \subset \Lambda_\iota^\ell$ for some $\iota = \{\lambda, \mu, \varepsilon\}$ and $\ell \geq 1$ and every $x \in \mathcal{P}$ returns to Λ_ι^ℓ infinitely often in forward and backward times and with positive frequency in forward times. The set $V^s(\mathcal{P}) = \bigcup_{x \in \mathcal{P}} V^s(x)$ is called the *s-saturated set* (by local stable manifolds).

Theorem 11.8 (Ben Ovadia [18]). *Assume that there is a forward positively recurrent set $\mathcal{P}$ such that the set $V^s(\mathcal{P})$ is of positive volume. Then f possesses an SRB measure.*

Note that the set $\mathcal{P} \subset M$ contains the set $\tilde{\mathcal{E}}$ of LP-regular points with nonzero Lyapunov exponents which are both forward and backward positively recurrent with respect to Λ. While the set $\tilde{\mathcal{E}}$ may have zero volume, its s-saturated set $V^s(\tilde{\mathcal{E}})$ may have positive volume, in which case by Theorem 11.8, there is an SRB measure for f. If indeed, the set $\tilde{\mathcal{E}}$ has positive volume, then by Theorem 8.12, the SRB measure is a smooth invariant measure.

4. We consider the special case of a diffeomorphism f of a compact smooth surface M. For every $x \in M$ let $\chi_1(x) \leq \chi_2(x)$ be the Lyapunov exponents at x. It is easy to see that

$$\chi_2(x) = \limsup_{n \to \infty} \frac{1}{n} \log \|d_x f^n\|.$$

The following result provides a simple natural condition for existence of an SRB measure for f.

Theorem 11.9 (Burguet [30] and Buzzi–Crovisier–Sarig [34]). *Let f be a C^∞ diffeomorphism of a compact smooth surface M such that the set of points $x \in M$ for which $\chi_2(x) > 0$ has positive area. Then f admits an SRB measure.*

11.2. Periodic orbits and approximations by horseshoes

In this section we describe some topological properties of hyperbolic invariant measures and we begin by stating a result by Katok [62] on existence and density of hyperbolic periodic points.

Theorem 11.10 (see [13, Theorem 15.4.1]). *Given $\varepsilon > 0$ and $\ell \geq 1$ such that $\nu(\Lambda_\iota^\ell) > 0$, we have that for ν-almost every $x \in \Lambda_\iota^\ell$ there is a hyperbolic periodic point p for which $\rho(x,p) \leq \varepsilon$.*

Outline of the proof. The proof of this theorem presented in [13] is based on a powerful statement known as the Closing Lemma. Here we present a more simple and direct argument. Choose numbers $\varepsilon, \delta > 0$ and a nonperiodic Lebesgue density point $x \in \Lambda_\iota^\ell$ for ν. By the Poincaré Recurrence Theorem, for every $N > 0$ there is $n > N$ such that $f^n(x) \in \Lambda_\iota^\ell$ and $\rho(x, f^n(x)) < \delta$. If δ is sufficiently small, the transverse intersection $V^s(x) \cap V^u(f^n(x))$ consists of a single point y_1 and we let $r = r(\delta) = \rho(x, y_1)$. Consider the point $f^n(y_1) \in V^s(f^n(x))$ and note that $\rho(f^n(x), f^n(y_1)) \leq C\lambda^n e^{\varepsilon n} r$ where $C > 0$ is a constant. For sufficiently large N the distance between $f^n(x)$ and $f^n(y_1)$ becomes sufficiently small, allowing us to use Theorem 7.15 and obtain a "long" local unstable manifold $V^u(f^n(y_1))$ which transversally intersects $V^s(x)$ at a single point y_2 whose distance from x does not exceed

$$r + 2C\lambda^n e^{\varepsilon n} r \leq r + \gamma r$$

for some $0 < \gamma < 1$. Applying the above argument to the point y_2, we obtain a point $y_3 \in V^s(x)$ whose distance from x does not exceed $r + \gamma r + \gamma^2 r$. Continuing in this fashion, we obtain a sequence of points y_n converging to a point y which has a "long" local unstable manifold $V^u(y)$ and whose distance from x does not exceed $r/(1-\gamma)$.

We now consider a point z_1 which is the unique point of the transverse intersection $V^u(x) \cap V^s(f^n(x))$. Starting with the point z_1 and applying the above argument to the inverse map f^{-1}, we obtain a point $z \in V^u(x)$ which has a "long" local stable manifold $V^u(z)$ and whose distance from x does not exceed $r/(1-\beta)$ for some $0 < \beta < 1$. The intersection $V^u(y) \cap V^s(z)$ consists of a single point p which is a hyperbolic periodic point for f of period n. Choosing δ sufficiently small, we obtain that the distance between x and p does not exceed ε. $\square$

For more results on hyperbolic periodic points for f associated with hyperbolic invariant measures we refer the reader to Section 15.4 in [13].

It is easy to see that the above construction gives two hyperbolic periodic points p and q which are homoclinically related. This implies that there is a horseshoe Λ containing p and q. Indeed, we have infinitely many such horseshoes. In [**62**] Katok showed that one can use such horseshoes to approximate the entropy and the Lyapunov exponent of ν (see [**13**, Theorem 15.6.1]). We present here a more advanced approximation result by Avila, Crovisier, and Wilkinson [**8**].

Consider a C^r diffeomorphism f, $r > 1$, of a compact smooth Riemannian manifold M preserving an ergodic hyperbolic probability measure ν and let $\mathcal{V}$ be a neighborhood of ν in the space of f-invariant probability measures endowed with the weak*-topology.

Theorem 11.11. *For any $\varepsilon > 0$ there exists a horseshoe $\Lambda \subset M$ such that:*

(1) *Λ is ε-close to the support of ν in the Hausdorff distance[1];*

(2) *the topological entropy $h(f|\Lambda) > h_\nu(f) - \varepsilon$;*

(3) *every invariant probability measure supported on Λ belongs to $\mathcal{V}$;*

(4) *if $\chi_1(\nu) > \cdots > \chi_s(\nu)$, $s = s(\nu)$ are the distinct Lyapunov exponents of ν with multiplicities $k_1(\nu), \ldots, k_s(\nu)$, then there is a splitting*

$$T_\Lambda M = E_1 \oplus \cdots \oplus E_s \quad \text{with } \dim E_i = k_i(\nu), \ i = 1, \ldots s;$$

(5) *there exists $n \geq 1$ such that for each $i = 1, \ldots, s$, each $x \in \Lambda$, and each unit vector $v \in E_i(x)$ we have*

$$\exp((\chi_i - \varepsilon)n) \leq \|d_x f^n v\| \leq \exp((\chi_i + \varepsilon)n).$$

11.3. Shadowing and Markov partitions

In this section we outline a construction of special Markov partitions of a nonuniformly hyperbolic set Λ for f that allow a symbolic representation of f by a Markov shift on a countable set of states. The weak Lyapunov charts introduced in Section 5.3.3 provide an important technical tool in the construction of Markov partitions.

We first consider the case when Λ is a locally maximal hyperbolic set and we follow Bowen's approach in [**27**].

A Markov partition is a finite cover $\mathcal{R}$ of Λ whose elements are rectangles[2] (see Section 11.1 for the definition) satisfying:

(1) $R = \overline{\operatorname{int} R}$;

(2) $\operatorname{int} R_i \cap \operatorname{int} R_j = \varnothing$ for $i \neq j$;

---

[1]If X and Y are two nonempty subsets of a metric space with distance ρ, then their Hausdorff distance is given by $d_H(X, Y) = \max\{\sup_{x \in X} \inf_{y \in Y} d(x, y), \sup_{y \in Y} \inf_{x \in X} d(x, y)\}$.

[2]Despite its name, a Markov partition is actually a cover whose elements can intersect each other but only along their boundaries. While these boundaries can have a complicated topological structure, they cannot support any invariant measure of positive entropy.

11.3. Shadowing and Markov partitions

(3) if $x \in \text{int } R_i$ and $f(x) \in \text{int } R_j$, then

$$f(W^u(x, R_i)) \supseteq W^u(f(x), R_j), \; f(W^s(x, R_i)) \subseteq W^s(f(x), R_j), \quad (11.4)$$

where $W^s(x, R_i)$ is the connected component of the intersection $W^s(x) \cap R_i$ containing x and similarly for $W^u(x, R_i)$.

We describe the main steps of Bowen's construction.

1. *The alphabet $\mathcal{A}$.* Given a small number $\delta > 0$, one can choose a finite collection $\mathcal{A} = \{p_1, \ldots, p_\ell\}$ of points in Λ with the property that for any $x \in M$ there exists a point $p \in \mathcal{A}$ such that $\rho(x, p) \leq \delta$. The set $\mathcal{A}$ is the alphabet.

2. *The graph $\mathcal{G}$.* The alphabet $\mathcal{A}$ induces a directed graph with vertices p_i and edges $p_i \to p_j$ if and only if $\rho(f(p_i), p_j) \leq \delta$. Set $\mathcal{S} = \{(p, q) \in \mathcal{A}^2 : p \to q\}$ and $\mathcal{G} := (\mathcal{A}, \mathcal{S})$.

3. *The shadowing lemma.* Denote

$$\Sigma = \{u = (u_i) \in \mathcal{A}^{\mathbb{Z}} : u_i \to u_{i+1} \text{ for all } i \in \mathbb{Z}\}. \quad (11.5)$$

The elements $u \in \Sigma$ are called pseudo-orbits.[3] The shadowing lemma asserts that for every $\varepsilon > 0$ one can choose $\delta > 0$ such that the following holds: for any $u \in \Sigma$ there is a unique $x \in \Lambda$ such that for all $i \in \mathbb{Z}$ we have that $\rho(f^i(x), u_i) < \varepsilon$. We say that x shadows the pseudo-orbit u.

4. *The projection map.* Consider the map $\pi \colon \Sigma \to \Lambda$ where $\pi(u)$ is the unique point whose orbit shadows u. This map is Hölder continuous in the symbolic metric $d(u, v) = \exp(-\min\{|i| : i \in \mathbb{Z}, u_i \neq v_i\})$. In addition, π is surjective and $\pi \circ \sigma = f \circ \pi$ where $\sigma \colon \Sigma \to \Sigma$ is the left-shift.

5. *The Markov partition.* Let $[p] := \{u = (u_i) \in \Sigma : u_0 = p\}$. The collection of sets $\{\pi([p])\}_{p \in \mathcal{A}}$ forms a finite cover of M by rectangles satisfying the first condition in the definition of Markov partitions. This cover has the Markov property: if $x \in \text{int } R_i$, then there is R_j such that $f(x) \in \text{int } R_j$ and (11.4) holds. This Markov cover can be refined to obtain a cover whose elements may only intersect along their boundaries. This is the desired Markov partition. We note that the coding map π is one-to-one on the set $\pi^{-1}(Y)$ where Y is a G_δ-set given by

$$Y = M \setminus \bigcup_{n \in \mathbb{Z}} f^n \left(\bigcup_{R \in \mathcal{R}} \partial R \right) \quad (11.6)$$

and $\partial R = R \setminus \text{int } R$.

---

[3] They are "almost" orbits, since $\rho(f(u_i), u_{i+1}) \leq \delta$ for any $i \in \mathbb{Z}$.

We now outline a construction of Markov partitions for nonuniformly hyperbolic sets. For simplicity, we consider the case of a surface diffeomorphism f and we follow Sarig's work [**104**].[4] In what follows we fix an index set ι and consider the level set Λ_ι. For $x \in \Lambda_\iota$ let Ψ_x be the weak Lyapunov chart at x of size $q(x)$ and the Lyapunov change of coordinates $C(x)$ (see Theorem 5.15).

1. *Refining the weak Lyapunov chart.* One of the main steps in Sarig's construction of Markov partitions is refining the weak Lyapunov charts by replacing their square domains with smaller rectangles which have different sizes in the stable and unstable directions. It turns out that keeping track of these sizes when moving from one point to another is crucial in constructing finite-to-one symbolic codes, as we explain later.

2. *The set $\Lambda_\iota^{\#}$.* Let $\Lambda_\iota^{\#} \subset \Lambda_\iota$ be the set of points satisfying
$$\limsup_{n\to\infty}(q \circ f^n)(x) > 0, \quad \limsup_{n\to\infty}(q \circ f^{-n})(x) > 0.$$
This is an invariant set which carries every invariant probability measure which gives full weight to Λ_ι.

3. *The alphabet $\mathcal{A}$.* For $x \in \Lambda_\iota^{\#}$, let $Q(x) := c_1 \|C(x)^{-1}\|^{-c_2}$, where $c_1 > 0$ and $c_2 > 0$ are some constants. One can choose a countable set X of points in $\Lambda_\iota^{\#}$ such that for every $\ell \geq 1$ the set $X \cap \Lambda_\iota^\ell$ is finite and two discrete sets of real numbers $R^s, R^u \subset (0, \varepsilon]$ such that if $\mathcal{A}$, the alphabet, is a countable set of triplets (x, r^s, r^u) with $x \in X$, $r^s \in R^s$, $r^u \in R^u$ such that $0 < r^s, r^u \leq Q(x)$, then the following conditions hold:
(1) the set $\{(x, r^s, r^u) \in \mathcal{A}, \min\{r^s, r^u\} > t\}$ is finite for all $t > 0$ (but the upper bound may go to infinity as t goes to zero);
(2) for every $x \in \Lambda_\iota^{\#}$ there exists a sequence $(x_n, r_n^s, r_n^u)_{n \in \mathbb{Z}}$ of elements of $\mathcal{A}$ such that if $\eta_n = \min\{r_n^s, r_n^u\}$, then for every $n \in \mathbb{Z}$:
  (a) $e^{-\varepsilon} \leq \frac{Q(f^n(x))}{Q(x_n)} \leq e^{\varepsilon}$;
  (b) the three pairs $[(x_n, \eta_n)$ and $(f^n(x), \eta_n)]$, $[(f(x_n), \eta_{n+1})$ and $(x_{n+1}, \eta_{n+1})]$, $[(f^{-1}(x_{n+1}), \eta_n)$ and $(x_n, \eta_n)]$ are "close".

Here, saying that a pair $[(z_1, \zeta_1)$ and $(z_2, \zeta_2)]$ is "close" refers to a technical condition which is designed to control: (a) the closeness of the numbers ζ_1 and ζ_2 and (b) the closeness of the Lyapunov charts at z_i of size ζ_i, with respect to ζ_1 and ζ_2.

4. *The graph $\mathcal{G}$.* For $u = (x, p^s, p^u), v = (y, q^s, q^u) \in \mathcal{A}$ write $u \to v$ if:
(1) $(x, \min\{p^s, p^u\})$ and $(f^{-1}(y), \min\{p^s, p^u\})$ are "close";
(2) $(f(x), \min\{q^s, q^u\})$ and $(y, \min\{q^s, q^u\})$ are "close";
(3) the pairs (p^s, q^s) and (p^u, q^u) satisfy the so-called "greedy algorithm" condition: $q^u = \min\{p^u e^\varepsilon, Q(y)\}$ and $p^s = \min\{q^s e^\varepsilon, Q(x)\}$.

---

[4]In the multidimensional case Markov partitions were constructed by Ben Ovadia [**17**].

11.3. Shadowing and Markov partitions

This induces a directed graph structure on $\mathcal{A}$. While $\mathcal{G}$ is countable, the number of ingoing and outgoing edges at every vertex is finite.

5. *Sarig's shadowing lemma.* In the uniformly hyperbolic setting, the pseudo-orbits are sequences of points and can be viewed as sequences of charts of uniform size. In the nonuniformly hyperbolic setting, pseudo-orbits are sequences of charts of varying size. More precisely, we call a sequence of triplets $u_n = (x_n, r_n^s, r_n^u)$, $n \in \mathbb{Z}$, a pseudo-orbit if $u_n \to u_{n+1}$.

Consider the set Σ of pseudo-orbits given by (11.5). Sarig's refined shadowing lemma asserts that for all $u \in \Sigma$ there exists a unique $x \in M$ such that for every $i \in \mathbb{Z}$ we have that $d(f^i(x), x_i) < Q(x_i)$ where $u_i = (x_i, p_i^s, p_i^u)$.

6. *The projection map.* Set $\pi\colon \Sigma \to M$ where $\pi(u)$ is the unique point whose orbit shadows u. This map covers the set $\Lambda_\iota^{\#}$ and is Hölder continuous.

7. *The Markov partition.* A pseudo-orbit $v = (v_i)$ is called recurrent if there are elements v_j and w_k such that v_j appears infinitely often in the future and w_k appears infinitely often in the past. The collection

$$\mathcal{Z} = \{\pi(\{u \in [v] : v \text{ is a recurrent pseudo-orbit}\})\}_{v \in \mathcal{A}}$$

forms a countable cover of $\Lambda_\iota^{\#}$. Moreover, this cover has the "local finiteness" property: for every $Z \in \mathcal{Z}$ we have that $\operatorname{card}\{Z' \in \mathcal{Z} : Z' \cap Z \neq \varnothing\} < \infty$. The proof of this fact uses the "greedy algorithm" condition of the pseudo-orbits described above to compare the chart sizes of different pseudo-orbits which are shadowed by the same point.

This allows one to develop a procedure of refining the cover $\mathcal{Z}$ to obtain a Markov partition (in the sense of Bowen when restricted to the set Y; see (11.6)) which covers the set $\Lambda_\iota^{\#}$. An element of this partition is a fractal set whose closure may intersect the closures of infinitely many other elements of the partition. Proceeding further one can show that this Markov partition induces a new shift-space and a new coding map such that the collection of the corresponding recurrent pseudo-orbits in the new shift-space codes a set which contains $\Lambda_\iota^{\#}$ in a finite-to-one way. This is crucial in various applications of this Markov partition and the associated symbolic dynamics.

Part 2

Selected Advanced Topics

Chapter 12

Cone Techniques

As we have mentioned earlier in Chapter 5 one of the most effective ways to verify the conditions of nonuniform hyperbolicity for a diffeomorphism preserving a smooth measure ν is to show that the Lyapunov exponents are nonzero almost everywhere. While direct calculation of Lyapunov exponents is widely used in numerical studies of dynamical systems, the rigorous calculation may be difficult to carry out and the cone techniques come in handy in helping to verify that the exponents are nonzero though this does not allow us to actually compute them.

12.1. Introduction

The *cone* of size $\gamma > 0$ centered around $\mathbb{R}^{n-k}$ in the product space $\mathbb{R}^n = \mathbb{R}^k \times \mathbb{R}^{n-k}$ is defined by

$$C_\gamma = \{(v, w) \in \mathbb{R}^k \times \mathbb{R}^{n-k} : \|v\| < \gamma\|w\|\} \cup \{(0, 0)\}.$$

Let $\mathcal{A}$ be a cocycle over an invertible measurable transformation $f\colon X \to X$ preserving a Borel probability measure ν in X and let $A\colon X \to GL(n, \mathbb{R})$ be its generator. The main idea underlying the cone techniques is the following. Let $Y \subset X$ be an f-invariant subset. Assume that there exist $\gamma > 0$ and $a > 1$ such that for every $x \in Y$:

(1) *invariance*: $A(x)C_\gamma \subset C_\gamma$;

(2) *expansion*: $\|A(x)v\| \geq a\|v\|$ for every $v \in C_\gamma$.

Exercise 12.1. Show that $\chi^+(x, v) > 0$ for every $x \in Y$ and $v \in C_\gamma \setminus \{0\}$.

This implies that the $n - k$ largest values of the Lyapunov exponent are positive.

We stress that positivity of Lyapunov exponents in the above exercise holds regardless of whether the set Y has positive ν-measure or not.[1] A crucial observation made by Wojtkowski [**111**] is that in certain cases if condition (1) holds almost everywhere in X (or even on a set of positive measure), then this condition alone is sufficient to establish positivity of at least one of the Lyapunov exponents (almost everywhere or, respectively, on a set of positive measure) and no estimate on the growth of vectors inside the cone is necessary. We will present a result by Wojtkowski in this direction as well as describing another version of the cone techniques due to Burns and Katok [**63**].

We begin by stating a result of Wojtkowski establishing the positivity of the top Lyapunov exponent. Let $Q\colon \mathbb{R}^n \to \mathbb{R}$ be a nondegenerate quadratic form of type $(1, n-1)$. Without loss of generality we may assume that
$$Q(v) = v_1^2 - v_2^2 - \cdots - v_n^2.$$
Consider the cone
$$C = \{v \in \mathbb{R}^n : Q(v) > 0\} \cup \{0\}$$
and the family of matrices
$$\mathcal{F} = \{A \in GL(n, \mathbb{R}) : |\det A| = 1,\ Q(Av) > 0 \text{ for } v \in \overline{C} \setminus \{0\}\}.$$
Clearly, the cone C is preserved by the action of the matrices in $\mathcal{F}$, i.e., $AC \subset C$ for every $A \in \mathcal{F}$.

Theorem 12.2 ([**111**]). *If $\log^+\|A\| \in L^1(X, \nu)$ and the cocycle only takes values in $\mathcal{F}$, then for ν-almost every $x \in X$,*
$$\limsup_{m \to \infty} \frac{1}{m} \log\|\mathcal{A}(x, m)\| > 0.$$

We outline the proof of this result. Consider the function $\rho\colon \mathcal{F} \to \mathbb{R}_0^+$ given by
$$\rho(A) = \inf_{v \in C \setminus \{0\}} \sqrt{\frac{Q(Av)}{Q(v)}}.$$

Exercise 12.3. Show that $\|A\| \geq \rho(A)$ and $\rho(AB) \geq \rho(A)\rho(B)$.

One can also show that the function $\rho(A)$ has the crucial property
$$\rho(A) > 1 \quad \text{for } A \in \mathcal{F}. \tag{12.1}$$
It follows from Exercise 12.3 that
$$\frac{1}{m}\log\|\mathcal{A}(x,m)\| \geq \frac{1}{m}\log \rho(\mathcal{A}(x,m)) \geq \frac{1}{m}\log \prod_{i=0}^{m-1} \rho(A(f^i(x))).$$

---

[1] In fact, no invariant measure is required.

On the other hand, by Birkhoff's Ergodic Theorem and (12.1), for ν-almost every $x \in X$ we have

$$\lim_{m \to \infty} \frac{1}{m} \log \prod_{i=0}^{m-1} \rho(A(f^i(x))) > 0.$$

This yields the desired result. It should be stressed that the above argument (in particular, the bound $\rho(A) > 1$ for $A \in \mathcal{F}$) cannot be used in the case of nondegenerate quadratic forms of type $(k, n - k)$ with $k > 1$ (as noted in [**111**]).

12.2. Lyapunov functions

We now describe an approach to establishing nonzero Lyapunov exponents which is based on Lyapunov functions. It was developed by Burns and Katok in [**63**].

Consider a measurable family of cones $C = \{C_x : x \in X\}$ in $\mathbb{R}^n$ and the complementary cones

$$\widehat{C}_x = (\mathbb{R}^n \setminus \overline{C}_x) \cup \{0\}.$$

The *rank* of a cone C_x is the maximal dimension of a linear subspace $L \subset \mathbb{R}^n$ contained in C_x. We denote it by $r(C_x)$ and we have that $r(C_x) + r(\widehat{C}_x) \le n$.

A pair of complementary cones C_x and $\widehat{C}_x$ is called *complete* if $r(C_x) + r(\widehat{C}_x) = n$. We say that the family of cones C is *complete* if the pair of complementary cones $(C_x, \widehat{C}_x)$ is complete for ν-almost every $x \in X$.

Let $\mathcal{A}$ be a measurable cocycle over a measurable transformation $f \colon X \to X$ preserving a finite measure ν. We say that the family of cones C is $\mathcal{A}$-*invariant* if for ν-almost every $x \in X$,

$$A(x) C_x \subset C_{f(x)} \quad \text{and} \quad A(f^{-1}(x))^{-1} \widehat{C}_x \subset \widehat{C}_{f^{-1}(x)}. \tag{12.2}$$

Let now $Q \colon \mathbb{R}^n \to \mathbb{R}$ be a continuous function that is homogeneous of degree one (i.e., $Q(av) = aQ(v)$ for every $v \in \mathbb{R}^n$ and $a \in \mathbb{R}$) and takes both positive and negative values. The set

$$C^u(Q) := \{0\} \cup Q^{-1}(0, +\infty) \subset \mathbb{R}^n \tag{12.3}$$

is called the *positive cone of Q*, and the set

$$C^s(Q) := \{0\} \cup Q^{-1}(-\infty, 0) \subset \mathbb{R}^n \tag{12.4}$$

is called the *negative cone of Q*. The rank of $C^u(Q)$ (respectively, $C^s(Q)$) is called the *positive* (respectively, *negative*) *rank* of Q and is denoted by $r^u(Q)$ (respectively, $r^s(Q)$). Clearly, $r^u(Q) + r^s(Q) \le n$ and since Q takes both positive and negative values, we have $r^u(Q) \ge 1$ and $r^s(Q) \ge 1$. The function Q is said to be *complete* if the cones $C^u(Q)$ and $C^s(Q)$ form a complete

pair of complementary cones, i.e., if
$$r^u(Q) + r^s(Q) = n.$$
A measurable function $Q\colon X \times \mathbb{R}^n \to \mathbb{R}$ is said to be a *Lyapunov function* for the cocycle $\mathcal{A}$ (with respect to ν) if there exist positive integers r^u and r^s such that for ν-almost every $x \in X$:

(1) the function $Q_x = Q(x, \cdot)$ is continuous, homogeneous of degree one, and takes both positive and negative values;

(2) Q_x is complete, $r^u(Q_x) = r^u$, and $r^s(Q_x) = r^s$;

(3) for every $x \in \mathbb{R}^n$ we have
$$Q_{f(x)}(A(x)v) \geq Q_x(v). \tag{12.5}$$

The numbers r^u and r^s are called, respectively, the *positive* and *negative ranks* of Q.

Exercise 12.4. Using (12.5), show that if Q is a Lyapunov function, then the two families of cones
$$C^u(Q_x) = \{v \in \mathbb{R}^n : Q_x(v) > 0\} \cup \{0\},$$
$$C^s(Q_x) = \{v \in \mathbb{R}^n : Q_x(v) < 0\} \cup \{0\}$$
are $\mathcal{A}$-invariant (see (12.2)).

A Lyapunov function is said to be *eventually strict* if for ν-almost every $x \in X$ there exists $m = m(x) \in \mathbb{N}$ depending measurably on x such that for every $v \in \mathbb{R}^n \setminus \{0\}$ we have
$$Q_{f^{m(x)}}(\mathcal{A}(x,m)v) > Q_x(v), \quad Q_{f^{-m(x)}}(\mathcal{A}(x,-m)v) < Q_x(v). \tag{12.6}$$

Theorem 12.5. *Assume that* $\log^+\|A\| \in L^1(X, \nu)$ *and that there exists an eventually strict Lyapunov function for the cocycle $\mathcal{A}$. Then for ν-almost every $x \in X$ the cocycle has r^u positive and r^s negative values of the Lyapunov exponent, counted with their multiplicities.*

Proof. Without loss of generality, we may assume that the measure ν is ergodic. Indeed, let Q be an eventually strict Lyapunov function and let $(\nu_\beta)_\beta$ be an ergodic decomposition of ν. Then $\log^+\|A\| \in L^1(X, \nu_\beta)$ for almost every β and Q is an eventually strict Lyapunov function for the cocycle $\mathcal{A}$ with respect to ν_β. Moreover, if the theorem holds for ergodic measures, then it also holds for arbitrary measures. From now on we assume that ν is ergodic.

For ν-almost every $x \in X$, we will construct subspaces $D_x^u, D_x^s \subset \mathbb{R}^n$, respectively, of dimensions r^u and r^s such that for every $m \in \mathbb{Z}$,
$$\mathcal{A}(x,m)D_x^u \subset C^u(Q_{f^m(x)}) \quad \text{and} \quad \mathcal{A}(x,m)D_x^s \subset C^s(Q_{f^m(x)})$$

12.2. Lyapunov functions

and for every $v \in D_x^u \setminus \{0\}$ and $w \in D_x^s \setminus \{0\}$,

$$\lim_{m \to \infty} \frac{1}{m} \log \|\mathcal{A}(x, -m)v\| < 0 \quad \text{and} \quad \lim_{m \to \infty} \frac{1}{m} \log \|\mathcal{A}(x, m)w\| < 0.$$

This implies that $D_x^u = E^u(x)$ and $D_x^s = E^s(x)$ for ν-almost every $x \in X$. We shall establish the existence of the spaces D_x^s. The proof of the existence of the spaces D_x^u is entirely analogous.

Set $C_x = \overline{C^s(Q_x)}$ and for every $m \in \mathbb{N}$ define

$$C_{m,x} = \mathcal{A}(f^m(x), -m) C_{f^m(x)}.$$

By (12.5), we have $C_{1,x} \supset C_{2,x} \supset \cdots$. Each set $C_{m,x}$ contains a subspace of dimension r^s. By the compactness of the closed unit sphere, the intersection

$$C_{\infty,x} = \bigcap_{m=1}^{\infty} C_{m,x}$$

also contains a subspace D_x^s of dimension r^s. Fix $\varepsilon > 0$.

Exercise 12.6. Show that there exist a set $Y \subset X$ of measure $\nu(Y) > 1 - \varepsilon$ and constants $c, d > 0$ such that for every $x \in Y$ and $v \in C_{\infty,x}$,

$$c\|v\| \leq -Q_x(v) \leq d\|v\|. \tag{12.7}$$

For each $m \in \mathbb{N}$ and $x \in X$, let

$$\kappa_m(x) = \sup \left\{ \frac{Q_{f^m(x)}(\mathcal{A}(x,m)v)}{Q_x(v)} : v \in C_{\infty,x} \setminus \{0\} \right\}.$$

Since, by (12.7), the values of the function Q_x are negative, by (12.5), we have $\kappa_m(x) \in (0, 1]$. Moreover, by (12.6), for ν-almost every $x \in X$ there exists $m = m(x) \in \mathbb{N}$ such that $\kappa_m(x) < 1$. It follows from Luzin's theorem that there exist $N \in \mathbb{N}$ and a set $E \subset Y$ of measure $\nu(E) > 1 - 2\varepsilon$ such that $\kappa_N(x) < 1$ for every $x \in E$. Observe that for every $m, n \in \mathbb{N}$ and $x \in X$,

$$\kappa_{n+m}(x) \leq \kappa_n(x) \kappa_m(f^n(x)).$$

Hence, $\kappa_n(x) < 1$ for every $n \geq N$ and $x \in E$. Consider the induced map $\bar{f} \colon E \to E$ on the set E, defined (mod 0) by $\bar{f}(x) = f^{\bar{n}(x)}(x)$, where

$$\bar{n}(x) = \min\{n \in \mathbb{N} : f^n(x) \in E\}$$

is the first return time. Consider also the induced cocycle $\bar{\mathcal{A}}$ whose generator is $\bar{A}(x) = A(x, \bar{n}(x))$. For each $m \in \mathbb{N}$ and $x \in E$, let

$$\bar{\kappa}_m(x) = \sup \left\{ \frac{Q_{\bar{f}^m(x)}(\bar{\mathcal{A}}(x,m)v)}{Q_x(v)} : v \in C_{\infty,x} \setminus \{0\} \right\}.$$

We have $\bar{\kappa}_m(x) = \kappa_{\tau_m(x)}(x) \in (0, 1]$, where

$$\tau_m(x) = \sum_{i=0}^{m-1} \bar{n}(\bar{f}^i(x)).$$

Exercise 12.7. Show that for every $m, n \in \mathbb{N}$ and $x \in E$ we have

$$\bar{\kappa}_{n+m}(x) \leq \bar{\kappa}_n(x) \bar{\kappa}_m(\bar{f}^n(x)). \tag{12.8}$$

Now let

$$K_m(x) = \log \bar{\kappa}_m(x).$$

By (12.8), the sequence of functions K_m is subadditive, that is,

$$K_{m+n}(x) \leq K_n(x) + K_m(\bar{f}^n(x)). \tag{12.9}$$

Since $K_m(x) \leq 0$ (and, hence, is bounded from above), it follows from Kingman's subadditive ergodic theorem [**70**] that for ν-almost every $x \in E$ the limit

$$F(x) := \lim_{m \to \infty} \frac{K_m(x)}{m} = \lim_{m \to \infty} \frac{1}{m} \log \bar{\kappa}_m(x)$$

exists and, moreover, the function F is $\bar{f}$-invariant ν-almost everywhere in E. Since ν is ergodic, we have that

$$\begin{aligned} F(x) &= \lim_{m \to \infty} \frac{1}{m} \int_E K_m \, d\nu \\ &= \inf_{m \in \mathbb{N}} \frac{1}{m} \int_E K_m \, d\nu \\ &\leq \frac{1}{N} \int_E K_N \, d\nu < 0 \end{aligned} \tag{12.10}$$

for ν-almost every $x \in E$. Since $\tau_n(x) \geq n$, the last inequality follows from the fact that for every $x \in E$,

$$K_N(x) = \log \bar{\kappa}_N(x) = \log \kappa_{\tau_N(x)}(x) \leq \log \kappa_N(x) < 0.$$

On the other hand, it follows from (12.7) that for every $x \in E$, $m \in \mathbb{N}$, and $v \in C_{\infty, x}$,

$$\frac{c}{d} \|\bar{\mathcal{A}}(x, m) v\| \leq \bar{\kappa}_m(x) \|v\| \leq \frac{d}{c} \|\bar{\mathcal{A}}(x, m) v\|,$$

and hence,

$$\begin{aligned} \lim_{m \to \infty} \frac{1}{m} \log \|\mathcal{A}(x, m) v\| &= \lim_{m \to \infty} \frac{1}{\tau_m(x)} \log \|\bar{\mathcal{A}}(x, m) v\| \\ &\leq \lim_{m \to \infty} \frac{K_m(x)}{\tau_m(x)} = \frac{F(x)}{\tau(x)}, \end{aligned} \tag{12.11}$$

where $\tau(x) := \lim_{m \to \infty}(\tau_m(x)/m)$. By Kac's lemma and the ergodicity of the induced map $\bar{f}$ with respect to $\nu | E$, we have that for ν-almost every $x \in E$,

$$\tau(x) = \nu \left(\bigcup_{m \in \mathbb{N}} f^m(E) \right) > 0.$$

It follows from (12.10) and (12.11) that for ν-almost every $x \in E$,
$$\lim_{m \to \infty} \frac{1}{m} \log \|\mathcal{A}(x,m)v\| < 0.$$
Since $\nu(E) > 1 - 2\varepsilon$ and ε is arbitrary, the desired result follows. □

12.3. Cocycles with values in the symplectic group

Let $\mathcal{A}$ be a cocycle over an invertible measurable transformation $f \colon X \to X$ preserving a Borel probability measure ν in X and let $A \colon X \to GL(n, \mathbb{R})$ be its generator. For a homogeneous function Q of degree one and any $x \in X$ consider the two families of cones as in (12.3) and (12.4), that is,
$$C^u(Q_x) := \{0\} \cup Q_x^{-1}(0, +\infty) \subset \mathbb{R}^n$$
and
$$C^s(Q_x) := \{0\} \cup Q_x^{-1}(-\infty, 0) \subset \mathbb{R}^n.$$
If Q_x is complete and
$$A(x)C^u(Q_x) \subset C^u(Q_{f(x)}), \quad A(f^{-1}(x))^{-1}C^s(Q_x) \subset C^s(Q_{f^{-1}(x)}) \quad (12.12)$$
for ν-almost every $x \in X$, then Q is a Lyapunov function and Theorem 12.5 applies. However, there are some interesting classes of cocycles and families of cones for which condition (12.12) is satisfied but only with respect to the family of cones $C^u(Q_x)$. In this case Q may not be a Lyapunov function. One of the most interesting situations in applications when this occurs involves cocycles with values in the symplectic group $Sp(2m, \mathbb{R})$ for $m \geq 1$ and the so-called symplectic cones that we define later.

Let C be an $\mathcal{A}$-invariant measurable family of cones. We say that:

(1) C is *strict* if for ν-almost every $x \in X$,
$$A(x)\overline{C_x} \subset C_{f(x)} \quad \text{and} \quad A(f^{-1}(x))^{-1}\widehat{\overline{C_x}} \subset \widehat{C}_{f^{-1}(x)};$$

(2) C is *eventually strict* if for ν-almost every $x \in X$ there exists $m = m(x) \in \mathbb{N}$ such that
$$\mathcal{A}(x,m)\overline{C_x} \subset C_{f^m(x)} \quad \text{and} \quad \mathcal{A}(x,-m)^{-1}\widehat{\overline{C_x}} \subset \widehat{C}_{f^{-m}(x)}. \quad (12.13)$$

One can show that for cocycles with values in $Sp(2m, \mathbb{R})$ the presence of just one eventually strict invariant family of cones guarantees existence of a Lyapunov function Q for the cocycle, and hence, Theorem 12.5 applies.

We will only consider the simple case of cocycles with values in the group $SL(2, \mathbb{R})$ which is isomorphic to $Sp(2, \mathbb{R})$ and we refer the reader to [13] for the general case. We call a cone in $\mathbb{R}^n$ *connected* if its projection to the projective space $\mathbb{R}P^{n-1}$ is a connected set. A connected cone in $\mathbb{R}^2$ is simply the union of two opposite sectors bounded by two different straight lines intersecting at the origin plus the origin itself. By a linear coordinate change such a cone can always be reduced to the standard cone

$$S = \{(v, w) \in \mathbb{R}^2 : vw > 0\} \cup \{(0,0)\}.$$

Theorem 12.8. *If a cocycle with values in $SL(2, \mathbb{R})$ has an eventually strictly invariant family of connected cones $C = \{C_x : x \in X\}$, then it has an eventually strict Lyapunov function Q such that for ν-almost every $x \in X$ the function Q_x has the form*

$$Q_x(v) = \operatorname{sgn} K_x(v, v) \cdot |K_x(v, v)|^{1/2}$$

for some quadratic form K_x of signature zero, and its zero set coincides with the boundary of the cone C_x.

Proof. First assume that $C_x = S$ for ν-almost every $x \in X$. Write

$$A(x) = \begin{pmatrix} a(x) & b(x) \\ c(x) & d(x) \end{pmatrix}.$$

It follows from (12.2) that the functions a, b, c, and d are nonnegative. Since $A(x) \in SL(2, \mathbb{R})$, we have $1 = a(x)d(x) - b(x)c(x)$. For each $(u, v) \in S$ put $K(u, v) = uv$. We obtain

$$\begin{aligned} K(A(x)(u, v)) &= a(x)c(x)u^2 + b(x)d(x)v^2 + (a(x)d(x) + b(x)c(x))uv \\ &\geq (a(x)d(x) + b(x)c(x))uv \\ &\geq (a(x)d(x) - b(x)c(x))uv = K(u, v). \end{aligned}$$
(12.14)

Arguing similarly and using induction, one can show that all entries of the matrix

$$\mathcal{A}(x, m) = \begin{pmatrix} a(x, m) & b(x, m) \\ c(x, m) & d(x, m) \end{pmatrix}$$

are nonnegative. Moreover, condition (12.13) implies that there exists $m = m(x) \geq 1$ such that $b(x, m) > 0$ and $c(x, m) > 0$. By (12.14), we conclude that $K(\mathcal{A}(x, m)(u, v)) > K(u, v)$.

In the case of an arbitrary family of connected cones, let us introduce a coordinate change $L \colon X \to SL(2, \mathbb{R})$ that takes the two lines bounding the cone C_x into the coordinate axes. Then $L(x)C_x = S$. For the cocycle $B \colon X \to SL(2, \mathbb{R})$ defined by $B(x) = L(f(x))A(x)L(x)^{-1}$, the constant family of cones S is eventually strictly invariant and hence by the previous argument, the function $Q_0(x, (u, v)) = \operatorname{sgn}(uv)|uv|^{1/2}$ is an eventually strict Lyapunov function. Hence, for the original cocycle A the function $Q(x, (u, v)) = Q_0(x, L(x)(u, v))$ has the same properties. $\square$

Chapter 13

Partially Hyperbolic Diffeomorphisms with Nonzero Exponents

In this chapter we shall discuss a special class of dynamical systems which act as uniform contractions and/or uniform expansions in some directions in the tangent space while allowing nonuniform contractions and/or nonuniform expansions in some other directions, thus exhibiting hyperbolicity of a "mixed" type.

This form of hyperbolicity usually comes as a particular case of uniform partial hyperbolicity[1]. The latter is characterized by two transverse stable and unstable subspaces complemented by a central direction with weaker rates of contraction and expansion.

In general, some or all Lyapunov exponents in the central direction can be zero and the "mixed" hyperbolicity ensures that the Lyapunov exponents in the central direction are all nonzero so that the system is nonuniformly hyperbolic. In particular, a partially hyperbolic diffeomorphism with nonzero Lyapunov exponents in the central direction preserving a smooth measure can have at most countably many ergodic components of positive measure. Stronger results can be obtained if one assumes that the Lyapunov exponents in the central direction are all negative (or all positive).

---

[1] A more general setting deals with systems admitting a *dominated splitting* (see [24]) but this concept goes beyond the scope of this book.

13.1. Partial hyperbolicity

In Section 5.4 we introduced nonuniformly partially hyperbolic diffeomorphisms. We now briefly discuss the stronger notion of uniform partial hyperbolicity and we refer the reader to the book [**93**] for a more detailed exposition of uniform partial hyperbolicity theory.

A diffeomorphism f is said to be *uniformly partially hyperbolic* if there exist (1) numbers $\lambda_1 < \lambda_2 < 1 < \mu_2 < \mu_1$; (2) numbers $C > 0$ and $K > 0$; (3) subspaces $E^c(x)$, $E^s(x)$, and $E^u(x)$, for $x \in M$ such that:

(1) $d_x f E^\omega(x) = E^\omega(f(x))$ where $\omega = c$, s, or u and
$$T_x M = E^s(x) \oplus E^c(x) \oplus E^u(x);$$

(2) for $v \in E^s(x)$ and $n \geq 0$,
$$\|d_x f^n v\| \leq C \lambda_1^n \|v\|;$$

(3) for $v \in E^u(x)$ and $n \leq 0$,
$$\|d_x f^n v\| \leq C \mu_2^n \|v\|;$$

(4) for $v \in E^c(x)$ and $n \geq 0$,
$$C^{-1} \lambda_2^n \|v\| \leq \|d_x f^n v\| \leq C \mu_1^n \|v\|;$$

(5)
$$\angle(E^c(x), E^s(x)) \geq K, \quad \angle(E^c(x), E^u(x)) \geq K,$$
$$\angle(E^s(x), E^u(x)) \geq K.$$

In other words, f is uniformly partially hyperbolic if it is nonuniformly partially hyperbolic on M in the sense of Section 5.4 with the functions $\lambda_1(x)$, $\lambda_2(x)$, $\mu_1(x)$, $\mu_2(x)$, and $\varepsilon(x)$ being constant, the function $C(x)$ bounded from above, and the function $K(x)$ bounded from below.

The subspaces $E^c(x)$, $E^s(x)$, and $E^u(x)$ are called, respectively, *central*, *stable*, and *unstable*.

Exercise 13.1. Show that the subspaces $E^c(x)$, $E^s(x)$, and $E^u(x)$ depend continuously on $x \in M$.

Using Theorem 5.20, one can strengthen this result and prove that the stable and unstable subspaces depend Hölder continuously on $x \in M$. Furthermore, one can extend the Stable Manifold Theorem (see Theorem 7.1) to partially hyperbolic diffeomorphisms and show that there are local stable $V^s(x)$ and unstable $V^u(x)$ manifolds through every point $x \in M$. Due to the uniform nature of hyperbolicity they depend continuously on $x \in M$ and their sizes are uniform in x. Now using Theorem 7.16, one can construct global stable $W^s(x)$ and unstable $W^u(x)$ manifolds through points in M

13.1. Partial hyperbolicity

that generate two f-invariant transverse stable and unstable foliations of M with smooth leaves—the integrable foliations for the stable and unstable distributions.

The Absolute Continuity Theorem (see Theorem 8.2) extends to partially hyperbolic diffeomorphisms ensuring that both stable and unstable foliations possess the absolute continuity property.

Unlike the stable and unstable distributions, the central distribution may not be integrable. Indeed, there is an example due to Smale of an Anosov diffeomorphism f of a three-dimensional manifold that is a factor of a nilpotent Lie group such that the tangent bundle admits an invariant splitting $TM = E_1 \oplus E_2 \oplus E_3$ where E_3 is the unstable distribution, $E_1 \oplus E_2$ is the stable distribution, and the "weakly stable" distribution E_2 is not integrable (see [93]). One can view f as a partially hyperbolic diffeomorphism for which E_2 is the nonintegrable central distribution (and E_1 is the stable distribution). This situation is robust: any diffeomorphism g that is sufficiently closed to f in the C^1 topology is a partially hyperbolic diffeomorphism whose central distribution is not integrable. The following result by Hirsch, Pugh, and Shub [55] describes the "opposite" situation.

Theorem 13.2. *Let f be a partially hyperbolic diffeomorphism whose central distribution is integrable to a smooth foliation. Then any diffeomorphism g that is sufficiently close to f in the C^1 topology is partially hyperbolic and its central distribution is integrable to a continuous foliation with smooth leaves.*[2]

Similarly to Anosov diffeomorphisms, partially hyperbolic diffeomorphisms form an open set in the space of C^1 diffeomorphisms of M.

In studying ergodic properties of uniformly partially hyperbolic diffeomorphisms an important role is played by the accessibility property that we now introduce.

We call two points $p, q \in M$ *accessible* if they can be connected by a path which consists of finitely many segments (curves) lying in *u*nstable and *s*table manifolds of some points in M, that is, if there are points $p = z_0, z_1, \ldots, z_{\ell-1}, z_\ell = q$ such that $z_i \in W^u(z_{i-1})$ or $z_i \in W^s(z_{i-1})$ for $i = 1, \ldots, \ell$. The collection of points $z_0, z_1, \ldots, z_\ell$ is called a *us-path* connecting p and q and is denoted by $[p, q] = [z_0, z_1, \ldots, z_\ell]$.

Exercise 13.3. Show that accessibility is an equivalence relation.

The diffeomorphism f is said to have the *accessibility property* if the partition into accessibility classes is trivial, that is, if any two points $p, q \in M$

---

[2]In general, this foliation is not smooth.

are accessible. A weaker version is the *essential accessibility property*—the partition into accessibility classes is ergodic (i.e., a measurable union of equivalence classes must have zero or full measure). Finally, we call f *center bunched* if $\max\{\lambda_1, (\mu_1)^{-1}\} < \lambda_2/\mu_2$. The following result provides sufficient conditions for ergodicity of partially hyperbolic diffeomorphisms with respect to smooth measures.

Theorem 13.4 (Burns–Wilkinson [**33**]). *Let f be a C^2 diffeomorphism of a compact smooth manifold preserving a smooth measure ν. Assume that f is uniformly partially hyperbolic, essentially accessible, and center bunched. Then f is ergodic.*

The assumption that f is center bunched is technical and can be dropped if $\dim E^c = 1$ (see [**33**, **96**]).

We describe some examples of partially hyperbolic systems (see [**93**] for details and references).

1. Let M and N be compact smooth Riemannian manifolds and let $f\colon M \to M$ be a C^r Anosov diffeomorphism, $r \geq 1$. The diffeomorphism

$$F = f \times \mathrm{Id}_N \colon M \times N \to M \times N$$

is partially hyperbolic. Any sufficiently small perturbation $G \in \mathrm{Diff}^r(M \times N)$ of F is partially hyperbolic; the central distribution E^c_G is integrable and the corresponding central foliation W^c_G has compact smooth leaves, which are diffeomorphic to N. If f preserves a smooth measure ν on M, then the map F preserves the measure $\nu \times m$ where m is the Riemannian volume in N.

2. The time-t map of an Anosov flow on a compact smooth manifold M is a partially hyperbolic diffeomorphism with one-dimensional central direction. In particular, if this map is accessible, then, by Theorem 13.4, it is ergodic with respect to an invariant smooth measure. The time-t map of the geodesic flow on a negatively curved manifold is an example of such a map.

Another example is given by the special flow T^t over a C^r Anosov diffeomorphism $f\colon M \to M$ with a roof function $H(x)$ (see Section 1.2).

Theorem 13.5. *There exist $\varepsilon > 0$ and an open and dense set of C^r functions $\tilde{H}\colon M \to \mathbb{R}^+$ with $|\tilde{H}(x)| \leq 1$ such that the time-t map of the special flow T^t with the roof function*

$$H(x) = H_0 + \varepsilon \tilde{H}(x)$$

($H_0 > 0$ is a constant) is accessible (and, hence, is ergodic).

3. Let G be a compact Lie group and let $f\colon M \to M$ be a C^r Anosov diffeomorphism. Each C^r function $\varphi\colon M \to G$ defines a skew product transformation $f_\varphi\colon M \times G \to M \times G$, or G-*extension of* f, defined by the formula

$$F_\varphi(x,y) = (f(x), \varphi(x)y).$$

Left transformations are isometries of G in the bi-invariant metric, and therefore, F_φ is partially hyperbolic. If f preserves a smooth measure ν, then f_φ preserves the smooth measure $\nu \times \nu_G$ where ν_G is the (normalized) Haar measure on G.

Theorem 13.6 (Burns–Wilkinson [**32**]). *For every neighborhood* $\mathcal{U} \subset C^r(M,G)$ *of the function* φ *there exists a function* $\psi \in \mathcal{U}$ *such that the diffeomorphism* F_ψ *is accessible.*

13.2. Systems with negative central exponents

We describe an approach to the study of ergodic properties of uniformly partially hyperbolic systems preserving a smooth measure, which fits well with the spirit of this book: it takes into account the Lyapunov exponents along the central direction E^c. We consider the case when these central exponents are all negative on a subset of positive measure. The case when the central exponents are all positive on a subset of positive measure can be reduced to the previous one by switching to the inverse map.

Theorem 13.7 (Burns–Dolgopyat–Pesin [**31**]). *Let f be a C^2 diffeomorphism of a compact smooth Riemannian manifold preserving a smooth measure ν. Assume that f is uniformly partially hyperbolic, accessible, and has negative central exponents at every point x of a set A of positive measure. Then f is ergodic (indeed, has the Bernoulli property). In particular, A has full measure, and hence, f has nonzero Lyapunov exponents almost everywhere.*

Proof. Since the Lyapunov exponents in the center direction are negative, the map $f|A$ has nonzero Lyapunov exponents. We denote by $V^{sc}(x)$ the local stable manifold tangent to $E^{sc}(x)$; by Theorem 7.1, $V^{sc}(x)$ is defined for almost every $x \in A$ and its size depends measurably on x. We also denote by $m^s(x)$, $m^u(x)$, $m^{sc}(x)$ the leaf-volumes on $V^s(x)$, $V^u(x)$, and $V^{sc}(x)$, respectively.

Since the map $f|A$ has nonzero Lyapunov exponents, by Theorem 9.2, it has at most countably many ergodic components of positive measure. We shall show that every such component is open (mod 0) and, hence, so is the set A.

Since $m(A) > 0$, we can choose an index set $\iota = \{\lambda, \mu, \varepsilon, j\}$ and $\ell \geq 1$ such that the set $A^\ell_\iota = \Lambda^\ell_\iota \cap A$ has positive measure. Choose $x \in A^\ell_\iota$ and for a sufficiently small $r > 0$ consider the set $P_\ell^u(x,r)$ given by (9.4). Since the local unstable manifolds $V^u(x)$ have uniform size, we obtain that the set

$$R^u(x) = \bigcup_{y \in V^{sc}(x)} V^u(y)$$

is an open neighborhood of x which contains the set $P_\ell^u(x,r)$. Therefore, by Remark 9.7, f is ergodic on the open set $P^u(x)$ given by (9.4).

We shall show that the map $f|A$ is topologically transitive and, hence, is ergodic. In fact, we prove the following stronger statement that will help us establish that $A = M \pmod 0$.

Lemma 13.8. *Almost every orbit of f is dense.*

Proof of the lemma. It suffices to show that if U is an open set, then the orbit of almost every point enters U. To this end, let us call a point *good* if it has a neighborhood in which the orbit of almost every point enters U. We wish to show that an arbitrary point p is good. Since f is accessible, there is a *us*-path $[z_0, \ldots, z_k]$ with $z_0 \in U$ and $z_k = p$. We shall show by induction on j that each point z_j is good. This is obvious for $j = 0$.

Now suppose that z_j is good. Then z_j has a neighborhood N such that $\mathcal{O}(x) \cap U \neq \varnothing$ for almost every $x \in N$. Let S be the subset of N consisting of the points with this property that are also both forward and backward recurrent. It follows from Poincaré's Recurrence Theorem that S has full measure in N. If $x \in S$, any point $y \in W^s(x) \cup W^u(x)$ has the property that $\mathcal{O}(y) \cap U \neq \varnothing$. The absolute continuity of the foliations W^s and W^u ensures that the set

$$\bigcup_{x \in S} W^s(x) \cup W^u(x)$$

has full measure in the set

$$\bigcup_{x \in N} W^s(x) \cup W^u(x).$$

The latter is a neighborhood of z_{j+1}. Hence, z_{j+1} is good. $\square$

We shall show that $A = M \pmod 0$. Otherwise the set $B = M \setminus A$ has positive ν-measure. By Lemma 13.8, we can choose a Lebesgue density point[3] $x \in B$ whose orbit is everywhere dense. Hence, there exists $n \in \mathbb{Z}$ such that $f^n(x) \in A$. We can choose $\varepsilon > 0$ so small that $B(f^n(x), \varepsilon) \subset A$ (mod 0) and then choose $\delta > 0$ such that $f^n(B(x, \delta)) \subset B(f^n(x), \varepsilon)$. Since B is invariant and $x \in B$ is a Lebesgue density point, the set $f^n(B(x, \delta))$ contains a subset in B of positive measure. Hence, B intersects A in a set of positive measure. This contradiction implies that $A = M \pmod 0$.

---

[3] Recall that a point $x \in B$ is a Lebesgue density point if $\lim_{r \to 0} \frac{\nu(B(x,r) \cap B)}{\nu(B(x,r))} = 1$.

To show that the map f has the Bernoulli property, observe that for every $n > 0$ the map f^n is accessible and has negative central exponents. Hence, it is ergodic and the desired result follows. □

13.3. Foliations that are not absolutely continuous

As we pointed out above, the stable and unstable foliations of a partially hyperbolic diffeomorphism possess the absolute continuity property. We also mentioned that the central distribution may not be integrable. Even if it is integrable, the central foliation may not be absolutely continuous. It is indeed the case when the Lyapunov exponents in the central direction are all negative or all positive.

To better illustrate this phenomenon, consider a linear hyperbolic automorphism A of the two-dimensional torus $\mathbb{T}^2$ and the map $F = A \times \text{Id}$ of the three-dimensional torus $\mathbb{T}^3 = \mathbb{T}^2 \times S^1$. Any sufficiently small C^1 perturbation G of F is uniformly partially hyperbolic with one-dimensional central distribution. By Theorem 13.2, the latter is integrable and its integral curves form a continuous foliation W^c of M with smooth compact leaves that are diffeomorphic to the unit circle. One can show that if the perturbation G has nonzero Lyapunov exponent in the central direction, then the central foliation is not absolutely continuous: for almost every $x \in M$ the conditional measure (generated by the Riemannian volume) on the leaf $W^c(x)$ of the central foliation passing through x is atomic.

We describe a far more general version of this result which is due to Ruelle and Wilkinson [101] (see also [13]). Let (X, ν) be a probability space and let $f \colon X \to X$ be an invertible transformation, which preserves the measure ν and is ergodic with respect to ν. Also let M be a smooth compact Riemannian manifold and let $\varphi \colon X \to \text{Diff}^{1+\alpha}(M)$ be a map. Consider the skew-product transformation $F \colon X \times M \to X \times M$ given by

$$F(x, y) = (f(x), \varphi_x(y))$$

and assume it is Borel measurable. We also assume that F possesses an invariant ergodic measure μ on $X \times M$ such that $\pi_* \mu = \nu$ where $\pi \colon X \times M \to X$ is the projection.

Fix $x \in X$. Letting $\varphi_x^{(0)}$ be the identity map, define the sequence of maps $\varphi_x^{(k)}$, $k \in \mathbb{Z}$, on M by the formula

$$\varphi_x^{(k+1)} = \varphi_{f^k(x)} \circ \varphi_x^{(k)}.$$

Since the tangent bundle to M is measurably trivial, the differential of the map φ along the M direction gives rise to a cocycle

$$\mathcal{A} \colon X \times M \times \mathbb{Z} \to GL(n, \mathbb{R}), \quad n = \dim M,$$

defined by $\mathcal{A}(x,y,k) = d_y\varphi_x^{(k)}$. If $\log^+ \|d\varphi\| \in L^1(X \times M, \mu)$, then the Multiplicative Ergodic Theorem (see Theorem 4.1) and the ergodicity of μ imply that the Lyapunov exponents of this cocycle are constant μ-almost everywhere. We write the distinct values in increasing order:

$$\chi_1 < \cdots < \chi_\ell, \quad 1 \leq \ell \leq n.$$

Theorem 13.9. *Assume that for some $\gamma > 0$ the function φ satisfies*

$$\log^+ \|d\varphi\|_\gamma \in L^1(X, \nu),$$

where $\|\cdot\|_\gamma$ is the γ-Hölder norm.[4] Assume also that $\chi_\ell < 0$. Then there exists a set $S \subset X \times M$ of full measure and an integer $k \geq 1$ such that

$$\operatorname{card}(S \cap (\{x\} \times M)) = k$$

for every $(x,y) \in S$.

Proof. We need the following adaptation of the construction of regular neighborhoods in Section 4.4 to the cocycle $\mathcal{A}$.

Lemma 13.10. *There exists a set $\Lambda_0 \subset X \times M$ of full measure such that for any $\varepsilon > 0$ the following statements hold:*

(1) *there exist a measurable function $r \colon \Lambda_0 \to (0,1]$ and a collection of embeddings $\Psi_{x,y} \colon B(0, r(x,y)) \to M$ such that $\Psi_{x,y}(0) = y$ and*

$$\exp(-\varepsilon) < \frac{r(F(x,y))}{r(x,y)} < \exp(\varepsilon);$$

(2) *if*

$$\varphi_{(x,y)} = \Psi_{F(x,y)}^{-1} \circ \varphi_x \circ \Psi_{(x,y)} \colon B(0, r(x,y)) \to \mathbb{R}^n,$$

then

$$\exp(\chi_1 - \varepsilon) \leq \|d_0\varphi_{(x,y)}^{-1}\|^{-1}, \|d_0\varphi_{(x,y)}\| \leq \exp(\chi_\ell + \varepsilon);$$

(3) *the C^1 distance $d_{C^1}(\varphi_{(x,y)}, d_0\varphi_{(x,y)}) < \varepsilon$ in $B(0, r(x,y))$;*

(4) *there exist $K > 0$ and a measurable function $A \colon \Lambda_0 \to \mathbb{R}$ such that for any $z, w \in B(0, r(x,y))$,*

$$K^{-1}\rho(\Psi_{(x,y)}(z), \Psi_{(x,y)}(w)) \leq \|z - w\|$$
$$\leq A(x)\rho(\Psi_{(x,y)}(z), \Psi_{(x,y)}(w))$$

with

$$e^{-\varepsilon} < \frac{A(F(x,y))}{A(x,y)} < e^\varepsilon.$$

Exercise 13.11. Prove Lemma 13.10 applying the Reduction Theorem (see Theorem 4.10), which should be restated for the cocycle $\mathcal{A}$.

---

[4] The norm $\|d\varphi\|_\gamma$ is the smallest $C > 0$ for which $\sup_{x \in X} \sup_{y \in M} \|d_y\varphi_x(u) - d_y\varphi_x(v)\| \leq C\|u-v\|^\gamma$ for any $u, v \in T_y M$.

Decomposing μ into a system of conditional measures
$$d\mu(x,y) = d\mu_x(y)d\nu(x)$$
and using the invariance of μ with respect to F, we obtain
$$(\varphi_x)_*\mu_x = \mu_{f(x)}$$
for ν-almost every $x \in X$.

Lemma 13.12. *There exist a set $\Lambda \subset \Lambda_0$ and real numbers $R > 0$, $C > 0$, and $0 < \alpha < 1$ such that:*

(1) *$\mu(\Lambda) > 1/2$ and if $(x,y) \in \Lambda$, then $\mu_x(\Lambda_x) > 1/2$ where*
$$\Lambda_x = \{y \in M : (x,y) \in \Lambda\};$$

(2) *if $(x,y) \in \Lambda$ and $z \in M$ are such that the distance $d_M(y,z)$ between y and z in M is less than or equal to R, then for all $m \geq 0$,*
$$d_M(\varphi_x^{(m)}(y), \varphi_x^{(m)}(z)) \leq C\alpha^m d_M(y,z).$$

Exercise 13.13. Derive Lemma 13.12 from Lemma 13.10. *Hint:* Set $\alpha = e^{(\chi_\ell + \varepsilon)}$. Observe that $0 < \alpha < 1$ provided that $\varepsilon > 0$ is sufficiently small. Show that the set of points $(x,y) \in \Lambda_0$ for which statement (2) of the lemma holds for some $R > 0$ and $C > 0$ has positive measure in Λ_0.

To prove the theorem, it suffices to show that there is a set $B \subset X$ of positive ν-measure such that for any $x \in B$ the measure μ_x has an atom. To see this, for $x \in X$ let $d(x) = \sup_{y \in M} \mu_x(y)$. Clearly, the function $d(x)$ is measurable, f-invariant, and positive on B. Since f is ergodic, we have $d(x) = d > 0$ for almost every $x \in X$. Let
$$S = \{(x,y) \in X \times M : \mu_x(y) \geq d\}.$$
Note that S is F-invariant, has measure at least d, and, hence, has measure 1. The desired result follows.

To prove the above claim, let Λ be the set constructed in Lemma 13.12 and let $B = \pi(\Lambda)$. Clearly, $\nu(B) > 0$. We shall show that for any $x \in B$ the measure μ_x has an atom.

Let $\mathcal{U} = \{U_1, \ldots, U_N\}$ be a cover of M by N closed balls of radius $R/10$. For $x \in X$ set
$$m(x) = \inf \sum \operatorname{diam} U_j,$$
where the infimum is taken over all collections of closed balls $\{U_1, \ldots, U_k\}$ in M such that $k \leq N$ and $\mu_x(\bigcup_{j=1}^k U_j) \geq 1/2$. We also define the number $m = \operatorname{ess\,sup}_{x \in B} m(x)$. We will show that $m = 0$. Otherwise, there is a number $p > 0$ such that
$$C\Delta N\alpha^p < \frac{m}{2}, \tag{13.1}$$

where Δ is the diameter of M. For $x \in B$ let $U_1(x), \dots, U_{k(x)}(x)$ be those balls in the cover $\mathcal{U}$ that meet Λ_x. Since these balls cover Λ_x and $\mu_x(\Lambda_x) > 1/2$, we have that

$$\mu_x\left(\bigcup_{j=1}^{k(x)} U_j(x)\right) \geq \frac{1}{2}. \tag{13.2}$$

Taking into account that $(\varphi_x^{(i)})_*\mu_x = \mu_{f^i(x)}$, for all i we obtain

$$\mu_{f^i(x)}\left(\bigcup_{j=1}^{k(x)} \varphi_x^{(i)} U_j(x)\right) \geq \frac{1}{2}.$$

Since the balls $U_j(x)$ meet the set Λ_x and have diameter less than $R/10$, by Lemma 13.12 we obtain that

$$\text{diam}(\varphi_x^{(i)} U_j(x)) \leq C\Delta\alpha^i. \tag{13.3}$$

Let $\tau(x)$ be the first return time of the point $x \in B$ to B under the map f^p and let $B_i = \{x \in B : \tau(x) = i\}$. We have that $B = \bigcup_{i=1}^{\infty} B_i$ and since f is invertible and its inverse f^{-1} preserves ν, we also have that

$$B' = \bigcup_{i=1}^{\infty} f^{pi}(B_i) = B \pmod{0}.$$

If $z \in B'$, then $z = f^{pi}(x)$ for some $x \in B_i$ and some $i \geq 1$. It follows from the definition of $m(z)$ and inequalities (13.1), (13.2), and (13.3) that

$$m(z) \leq \sum_{j=1}^{k(x)} \text{diam}(\varphi_x^{(pi)} U_j(x)) \leq Ck(x)\Delta\alpha^{pi} \leq CN\Delta\alpha^p \leq \frac{m}{2}.$$

This implies that

$$m = \text{ess sup}_{x \in B} m(x) = \text{ess sup}_{z \in B'} m(z) \leq \frac{m}{2},$$

contradicting the assumption that $m > 0$.

We have shown that $m = 0$, and hence, $m(x) = 0$ for ν-almost every $x \in B$. For such a point x there is a sequence of closed balls $U^1(x), U^2(x), \dots$ for which

$$\lim_{i \to \infty} \text{diam}\, U^i(x) = 0$$

and $\mu_x(U^i(x)) \geq \frac{1}{2N}$ for all i. Take $z_i \in U^i(x)$. Any accumulation point of the sequence $\{z_i\}$ is an atom for μ_x. $\square$

Corollary 13.14. *Let μ be an ergodic measure for a $C^{1+\alpha}$ diffeomorphism of a compact Riemannian manifold M. If μ has all of its exponents negative, then μ is concentrated on the orbit of a periodic sink.*

13.3. Foliations that are not absolutely continuous

Proof. The result follows from Theorem 13.9 by taking $X = \{x\}$ and taking ν to be the point mass. □

Let f be a $C^{1+\alpha}$ diffeomorphism of a compact smooth Riemannian manifold M preserving a Borel probability measure μ. Assume that W is a foliation of M with smooth leaves, which is invariant under f. We say that f is W-*dissipative* if $L_W(x) \neq 0$ for μ-almost every $x \in M$, where $L_W(x)$ denotes the sum of the Lyapunov exponents of f at the point x along the subspace $T_xW(x)$.

For $x \in M$ we denote by $\operatorname{vol}(W(x))$ the leaf-volume of the leaf $W(x)$ through x. We say that the foliation W has *finite volume leaves almost everywhere* if the set of those $x \in M$ for which $\operatorname{vol}(W(x)) < \infty$ has full volume. An example of a foliation whose leaves have finite volume almost everywhere is a foliation with smooth compact leaves. If W is such a foliation, then the function $M \ni x \mapsto \operatorname{vol}(W(x))$ is well-defined (finite) but may not be bounded (see an example in [48]).

Theorem 13.15 (Hirayama–Pesin [54]). *Let f be a $C^{1+\alpha}$ diffeomorphism of a compact smooth Riemannian manifold M preserving a smooth measure μ. Also let W be an f-invariant foliation of M with smooth leaves. Assume that W has finite volume leaves almost everywhere. If f is W-dissipative, then the foliation W is not absolutely continuous in the weak sense (see Chapter 8).*

Proof. Let $A^- \subset M$ be the set of points for which $L_W(x) < 0$ and let $A^+ \subset M$ be the set of points for which $L_W(x) > 0$. They both are f-invariant and either $m(A^-) > 0$ or $m(A^+) > 0$ or both (we use here the fact that the invariant measure μ is smooth and hence equivalent to volume). Without loss of generality we may assume that $m(A^+) > 0$ (otherwise, simply replace f with f^{-1}). Then for a sufficiently small $\lambda > 0$, sufficiently large integer ℓ, and every small $\varepsilon > 0$ there exists a Borel set $A^+_{\lambda,\ell,\varepsilon} \subset A^+$ of positive μ-measure such that for every $x \in A^+_{\lambda,\ell,\varepsilon}$ and $n \geq 0$,

$$|\operatorname{Jac}(d_x f^n | T_x W(x))| \geq \ell^{-1} e^{\lambda n} e^{-\varepsilon n}. \tag{13.4}$$

Given $V > 0$, consider the set

$$Y_V = \{y \in M : \operatorname{vol}(W(y)) \leq V\}.$$

Observe that the set $A^+_{\lambda,\ell,\varepsilon}$ has positive volume and let $x \in A^+_{\lambda,\ell,\varepsilon}$ be a Lebesgue density point of m. One can choose $V > 0$ such that the set

$$R = A^+_{\lambda,\ell,\varepsilon} \cap B(x,r) \cap Y_V$$

has positive volume.

13. Partially Hyperbolic Diffeomorphisms with Nonzero Exponents

Assume on the contrary that the foliation W is absolutely continuous in the weak sense. Then for almost every $y \in R$ the set $R_y = R \cap W(y)$ has positive volume in $W(y)$. Observe that $\mu(R) > 0$ (where μ is an invariant smooth measure for f). Therefore, the trajectory of almost every point $y \in R$ returns to R infinitely often. Let y be such a point and let (n_k) be the sequence of returns to R. We also may assume that $m_{W(y)}(R_y) > 0$.

Since $f^{n_k}(y) \in R \subset Y_V$, we have that for every $k > 0$,
$$m_{W(f^{n_k}(y))}(f^{n_k}(R_y)) \leq W(f^{n_k}(y)) \leq V.$$
On the other hand, by (13.4),
$$\begin{aligned} \text{vol}(W(f^{n_k}(y))) &\geq m_{W(f^{n_k}(y))}(f^{n_k}(R_y)) \\ &= \int_{R_y} \text{Jac}(d_z f^{n_k}|T_z W(z))\, dm(z) \\ &\geq \ell^{-1} e^{(\lambda-\varepsilon)n_k} m(R_y) > V \end{aligned}$$
if n_k is sufficiently large. This yields a contradiction and completes the proof of the theorem. □

Chapter 14

More Examples of Dynamical Systems with Nonzero Lyapunov Exponents

The goal of this chapter is to present some new examples of dynamical systems with nonzero Lyapunov exponents. The corresponding constructions use more sophisticated techniques and reveal some new and subtle properties of systems with nonzero exponents.

14.1. Hyperbolic diffeomorphisms with countably many ergodic components

Consider a $C^{1+\alpha}$ diffeomorphism of a compact smooth Riemannian manifold with nonzero Lyapunov exponents with respect to a smooth invariant measure. By Theorem 9.2, it can have at most countably many ergodic components. In Section 9.1 we presented an example of such a diffeomorphism with finitely many ergodic components but the construction cannot be used to obtain a diffeomorphism with countably many ergodic components. Here we will describe a different approach due to Dolgopyat, Hu, and Pesin [**42**] to construct such a diffeomorphism, thus demonstrating that Theorem 9.2 cannot be improved. More precisely, the following statement holds.

Theorem 14.1. *There exists a C^∞ volume-preserving diffeomorphism f of the 3-torus $\mathbb{T}^3$ with nonzero Lyapunov exponents almost everywhere and with countably many ergodic components that are open* (mod 0).

Let $A\colon \mathbb{T}^2 \to \mathbb{T}^2$ be a linear hyperbolic automorphism with at least two fixed points p and p'. Consider the map $F = A \times \mathrm{Id}$ of the 3-torus $\mathbb{T}^3 = \mathbb{T}^2 \times S^1$. We will perturb F to obtain the desired diffeomorphism f.

Proposition 14.2. *Given $k \geq 2$ and $\delta > 0$, there exists a map g of the three-dimensional manifold $M = \mathbb{T}^2 \times I$ (where $I = [0,1]$) such that:*

(1) *g is a C^∞ volume-preserving diffeomorphism;*

(2) *$\|F - g\|_{C^k} \leq \delta$;*

(3) *$d^m g | \mathbb{T}^2 \times \{z\} = d^m F | \mathbb{T}^2 \times \{z\}$ for $z = 0, 1$ and all $0 \leq m < \infty$;*

(4) *g is ergodic and has nonzero Lyapunov exponents almost everywhere.*

We show how to complete the proof of the theorem by deducing it from the proposition. To this end, consider the countable collection of intervals $\{I_n\}_{n=1}^\infty$ in the circle S^1, given by

$$I_{2n} = [(n+2)^{-1}, (n+1)^{-1}], \quad I_{2n-1} = [1 - (n+1)^{-1}, 1 - (n+2)^{-1}].$$

Clearly, $\bigcup_{n=1}^\infty I_n = (0,1)$ and the interiors of I_n are pairwise disjoint. By the proposition, for each n one can construct a C^∞ volume-preserving ergodic diffeomorphism $f_n\colon \mathbb{T}^2 \times I \to \mathbb{T}^2 \times I$ satisfying:

(1) $\|F - f_n\|_{C^n} \leq e^{-n^2}$;

(2) $d^m f_n | \mathbb{T}^2 \times \{z\} = d^m F | \mathbb{T}^2 \times \{z\}$ for $z = 0, 1$ and all $0 \leq m < \infty$;

(3) f_n has nonzero Lyapunov exponents almost everywhere.

Let $L_n\colon I_n \to I$ be the affine onto map and let

$$\pi_n = (\mathrm{Id}, L_n)\colon \mathbb{T}^2 \times I_n \to \mathbb{T}^2 \times I.$$

We define the map f by setting $f | \mathbb{T}^2 \times I_n = \pi_n^{-1} \circ f_n \circ \pi_n$ for all n and $f | \mathbb{T}^2 \times \{0\} = F | \mathbb{T}^2 \times \{0\}$. Note that for every $n > 0$ and $0 \leq m \leq n$ we have

$$\|d^m F | \mathbb{T}^2 \times I_n - \pi_n^{-1} \circ d^m f_n \circ \pi_n\|_{C^n} \leq \|\pi_n^{-1} \circ (d^m F - d^m f_n) \circ \pi_n\|_{C^n}$$
$$\leq e^{-n^2} \cdot (n+1)^n \to 0$$

when $n \to \infty$. It follows that f is a C^∞ diffeomorphism of M. It is easy to see that f is volume-preserving and has nonzero Lyapunov exponents almost everywhere and that $f | \mathbb{T}^2 \times I_n$ is ergodic.

Proof of the proposition. Note that the map F is uniformly partially hyperbolic with one-dimensional center E_F^c, stable E_F^s and unstable E_F^u distributions. We obtain the desired result by arranging two C^∞ volume-preserving perturbations of F that ensure the essential accessibility property

14.1. Diffeomorphisms with countably many ergodic components

and positivity of the central Lyapunov exponent:

(1) The first perturbation is the time-t map (for a sufficiently small t) of a flow h_t generated by a vector field X, which vanishes outside a small neighborhood of a fixed point; moreover the map dh_t preserves the E_F^{uc}-planes.

(2) The second perturbation acts as small rotations in the E_F^{uc}-planes in a small neighborhood of another fixed point.

Let $\eta > 1$ and η^{-1} be the eigenvalues of A. Choose a small number $\varepsilon_0 > 0$ such that $\rho(p, p') \geq 3\varepsilon_0$ and consider the local stable and unstable one-dimensional manifolds $V^s(p)$, $V^u(p)$, $V^s(p')$, and $V^u(p')$ of "size" ε_0.

Let us choose the smallest positive number n_1 such that the intersection $A^{-n_1}(V^s(p')) \cap V^u(p) \cap B(p, \varepsilon_0)$ consists of a single point, which we denote by q_1. Similarly, let n_2 be the smallest positive number such that the intersection $A^{n_2}(V^u(p')) \cap V^s(p) \cap B(p, \varepsilon_0)$ consists of a single point, which we denote by q_2. See Figure 14.1.

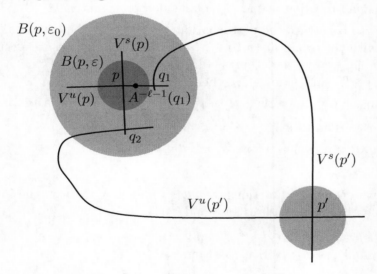

Figure 14.1

Take a sufficiently small $\varepsilon \in (0, \varepsilon_0)$ satisfying
$$\varepsilon \leq \frac{1}{2} \min\{\rho(p, q_1), \rho(p, q_2)\}.$$
There exists $\ell \geq 2$ such that (see Figure 14.1)
$$A^{-\ell}(q_1) \notin B(p, \varepsilon), \quad A^{-\ell-1}(q_1) \in B(p, \varepsilon). \tag{14.1}$$
We can choose ε so small that
$$B(p, \varepsilon) \cap (A^{-n_1}(V^s(p')) \cup A^{n_2}(V^u(p'))) = \varnothing.$$

We can further reduce ε if necessary so that for some $q \in \mathbb{T}^2$, some number $N > 0$, which will be determined later, and for every $i = 1, \ldots, N$,
$$A^i(B(q, \varepsilon)) \cap B(q, \varepsilon) = \varnothing, \quad A^i(B(q, \varepsilon)) \cap B(p, \varepsilon) = \varnothing.$$
Note that $\varepsilon = \varepsilon(N)$. Finally, we choose $\varepsilon' \in (0, \varepsilon)$ such that $A^{-\ell-1}(q_1) \in B(p, \varepsilon')$.

We define two disjoint open domains
$$\Omega_1 = B(p, \varepsilon_0) \times I \quad \text{and} \quad \Omega_2 = B^{uc}(\bar{q}, \varepsilon_0) \times B^s(\bar{q}, \varepsilon_0), \tag{14.2}$$
where $\bar{q} = (q, 1/2)$ and the sets $B^{uc}(\bar{q}, \varepsilon_0) \subset V^u(q) \times I$ and $B^s(\bar{q}, \varepsilon_0) \subset V^s(q)$ are balls of radius ε_0.

The desired map g is obtained as the result of two perturbations H_1 and H_2 of F so that $g = H_1 \circ F \circ H_2$, where each H_i is a C^∞ volume-preserving diffeomorphism of M which coincides with F outside of Ω_i, $i = 1, 2$. We construct the perturbation H_1 as a time-t map (for sufficiently small t) of a divergence-free vector field in Ω_1 and we construct the perturbation H_2 by applying a small rotation in the E_F^{uc}-plane at every point in Ω_2.

In order to construct the perturbation H_1, consider the coordinate system in Ω_1 with the origin at $(p, 0) \in M$ and the x-, y-, z-axes, respectively, to be unstable, stable, and central directions for the map F. If a point $w = (x, y, z) \in \Omega_1$ and $F(w) \in \Omega_1$, then $F(w) = (\eta x, \eta^{-1} y, z)$.

We choose a C^∞ function $\xi \colon I \to \mathbb{R}^+$ satisfying (see Figure 14.2):

(1) $\xi(z) > 0$ on $(0, 1)$;
(2) $d^i \xi(0) = d^i \xi(1) = 0$ for $i = 0, 1, \ldots, k$;
(3) $\|\xi\|_{C^k} \leq \delta$

and then two other C^∞ functions $\varphi = \varphi(x)$ and $\psi = \psi(y)$ on the interval $(-\varepsilon_0, \varepsilon_0)$ such that (see Figures 14.3 and 14.4):

(4) $\varphi(x) = \varphi_0$ for $x \in (-\varepsilon', \varepsilon')$ and $\psi(y) = \psi_0$ for $y \in (-\varepsilon', \varepsilon')$, where φ_0 and ψ_0 are positive constants;
(5) $\varphi(x) = 0$ whenever $|x| \geq \varepsilon$, $\psi(y) \geq 0$ for any y, and $\psi(y) = 0$ whenever $|y| \geq \varepsilon$;
(6) $\|\varphi\|_{C^k} \leq \delta$, $\|\psi\|_{C^k} \leq \delta$;
(7) $\int_0^{\pm \varepsilon} \varphi(s) \, ds = 0$.

We now define a vector field X in Ω_1 by the formula
$$X(x, y, z) = \left(-\psi(y) \xi'(z) \int_0^x \varphi(s) \, ds, \, 0, \, \psi(y) \xi(z) \varphi(x) \right). \tag{14.3}$$

Exercise 14.3. Show that X is a divergence-free vector field supported on $(-\varepsilon, \varepsilon) \times (-\varepsilon, \varepsilon) \times I$.

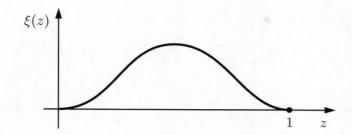

Figure 14.2. The function ξ.

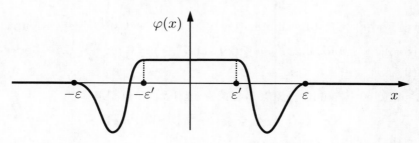

Figure 14.3. The function φ.

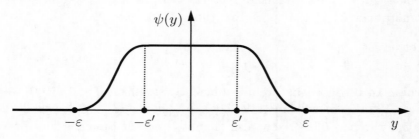

Figure 14.4. The function ψ.

We define a diffeomorphism h_t in Ω_1 to be the time-t map of the flow generated by X and we set $h_t = \mathrm{Id}$ on the complement of Ω_1. Fixing some sufficiently small $t > 0$, consider the map $H_1 = h_t$.

Exercise 14.4. Show that H_1 is a C^∞ volume-preserving diffeomorphism of M that preserves the y-coordinate, that is,
$$H_1(\mathbb{R} \times \{y\} \times \mathbb{R}) \subset \mathbb{R} \times \{y\} \times \mathbb{R}$$
for every $y \in \mathbb{R}$.

In order to construct the perturbation H_2, consider the coordinate system in Ω_2 with the origin at $(q, 1/2)$ and the x-, y-, z-axes, respectively, to be unstable, stable, and central directions. We then switch to the cylindrical coordinate system (r, θ, y), where $x = r \cos\theta$, $y = y$, and $z = r \sin\theta$.

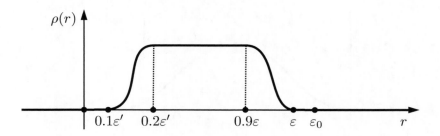

Figure 14.5. The function ρ.

We choose a C^∞ function $\rho\colon (-\varepsilon_0, \varepsilon_0) \to \mathbb{R}^+$ satisfying (see Figure 14.5):

(8) $\rho(r) > 0$ for $0.2\varepsilon' \leq r \leq 0.9\varepsilon$ and $\rho(r) = 0$ for $r \leq 0.1\varepsilon'$ and $r \geq \varepsilon$;

(9) $\|\rho\|_{C^k} \leq \delta$.

For a sufficiently small $\tau > 0$, define a map $\tilde{h}_\tau$ in Ω_2 by

$$\tilde{h}_\tau(r, \theta, y) = (r, \theta + \tau\psi(y)\rho(r), y) \tag{14.4}$$

and set $\tilde{h}_\tau = \mathrm{Id}$ on $M \setminus \Omega_2$.

Exercise 14.5. Show that the map $H_2 = \tilde{h}_\tau$ is a C^∞ volume-preserving diffeomorphism of M that preserves the y-coordinate.

Let us set
$$g = g_{t\tau} = h_t \circ F \circ \tilde{h}_\tau.$$

Exercise 14.6. Show that for all sufficiently small $t > 0$ and $\tau > 0$, the map $g_{t\tau}$ is a C^∞ diffeomorphism of M close to F in the C^1 topology and, hence, is partially hyperbolic. Show also that it is center bunched and preserves volume in M.

By Lemma 14.7 below, the map g has the essential accessibility property and hence, by Theorem 13.4, it is ergodic. It remains to explain that g has nonzero Lyapunov exponents almost everywhere.

Denote by $E^s_{t\tau}(w)$, $E^u_{t\tau}(w)$, and $E^c_{t\tau}(w)$ the stable, unstable, and central subspaces at a point $w \in M$ for the map $g_{t\tau}$. Set $\kappa_{t\tau}(w) = d_w g_{t\tau} | E^u_{t\tau}(w)$, $w \in M$. By Lemma 14.9 below, for all sufficiently small $\tau > 0$,

$$\int_M \log \kappa_{0\tau}(w)\,dw < \log \eta.$$

The subspace $E^u_{t\tau}(w)$ depends continuously on t and τ (for a fixed w) and, hence, so does $\kappa_{t\tau}$. It follows that there are $t_0 > 0$ and $\tau_0 > 0$ such that for all $0 \leq t \leq t_0$ and $0 \leq \tau \leq \tau_0$ we have

$$\int_M \log \kappa_{t\tau}(w)\,dw < \log \eta.$$

14.1. Diffeomorphisms with countably many ergodic components

Denote by $\chi^s_{t\tau}(w)$, $\chi^u_{t\tau}(w)$, and $\chi^c_{t\tau}(w)$ the Lyapunov exponents of $g_{t\tau}$ at the point $w \in M$ in the stable, unstable, and central directions, respectively (since these directions are one-dimensional the Lyapunov exponents do not depend on the vector). By ergodicity of $g_{t\tau}$ and Birkhoff's Ergodic Theorem, we have that for almost every $w \in M$,

$$\chi^u_{t\tau}(w) = \lim_{n \to \infty} \frac{1}{n} \log \prod_{i=0}^{n-1} \kappa_{t\tau}(g^i_{t\tau}(w)) = \int_M \log \kappa_{t\tau}(w)\, dw < \log \eta.$$

Since $E^s_{t\tau}(w) = E^s_{00}(w) = E^s_F(w)$ for every t and τ, we conclude that $\chi^s_{t\tau}(w) = -\log \eta$ for almost every $w \in M$. Since $g_{t\tau}$ is volume-preserving,

$$\chi^s_{t\tau}(w) + \chi^u_{t\tau}(w) + \chi^c_{t\tau}(w) = 0$$

for almost every $w \in M$. It follows that $\chi^c_{t\tau}(w) > 0$ for almost every $w \in M$, and hence, $g_{t\tau}$ has nonzero Lyapunov exponents almost everywhere. This completes the proof of the proposition. □

We now state and prove the two technical lemmas that we referred to in the proof of the theorem.

Lemma 14.7. *For every sufficiently small $t > 0$ and $\tau > 0$ the map $g_{t\tau}$ has the essential accessibility property.*

Proof of the lemma. Set $I_p = \{p\} \times (0, 1)$ (recall that p is one of the two fixed points of A; see Figure 14.1).

Exercise 14.8. Show that given a point $x \in M$, which does not lie on the boundary of M, there is a point $x' \in I_p$ such that x and x' are accessible.

It follows that in order to prove the lemma it suffices to show that any two points in I_p are accessible. Indeed, given two points $x, y \in M$, which do not lie on the boundary of M, we can find points $x', y' \in I_p$ such that the points x and x' are accessible and so are the points y and y'. If the points x', y' are accessible, the desired result follows, since accessibility is a transitive relation.

Given a point $w \in M$, denote by $\mathcal{A}(w)$ the accessibility class of w (i.e., the set of points $q \in M$ such that w and q are accessible). We use the coordinate system (x, y, z) in Ω_1 described above. Since the map h_t preserves the center leaf I_p, we have that for $z \in (0, 1)$,

$$h_t(0, 0, z) = (h_t^{(1)}(0, 0, z), h_t^{(2)}(0, 0, z), h_t^{(3)}(0, 0, z)) = (0, 0, h_t^{(3)}(0, 0, z)).$$

We claim that the accessibility property for points in I_p will follow if we show that for every $z \in (0, 1)$,

$$n\mathcal{A}(p, z) \supset \{(p, a) : a \in [(h_t^{-\ell})^{(3)}(p, z), z]\}, \tag{14.5}$$

where ℓ is chosen by (14.1). Indeed, since accessibility is a transitive relation and since $h_t^{-n}(p,z) \to (p,0)$ for any $z \in (0,1)$, condition (14.5) implies that $\mathcal{A}(p,z) \supset \{(p,a) : a \in (0,z]\}$ for all $z \in (0,1)$ and the claim follows. We now proceed with the proof of (14.5).

Denote by $V_{t\tau}^s(p)$, $V_{t\tau}^u(p)$ the local stable and unstable manifolds at p for the map $g_{t\tau}$. Let $q_1 \in V_{t\tau}^u(p)$ and $q_2 \in V_{t\tau}^s(p)$ be the two points constructed above (see Figure 14.1). The intersection $V_{t\tau}^s(q_1) \cap V_{t\tau}^u(q_2)$ is nonempty and consists of a single point q_3. We will show that for any $z_0 \in (0,1)$, there exist $z_i \in (0,1)$, $i = 1,2,3,4$, such that

$$(q_1, z_1) \in V_{t\tau}^u((p, z_0)), \quad (q_3, z_3) \in V_{t\tau}^s((q_1, z_1)),$$
$$(q_2, z_2) \in V_{t\tau}^u((q_3, z_3)), \quad (p, z_4) \in V_{t\tau}^s((q_2, z_2)),$$

and

$$z_4 \le (h_t^{-\ell})^{(3)}(p, z_0) \tag{14.6}$$

(see Figure 14.6). This will imply that $(p, z_4) \in \mathcal{A}(p, z_0)$. By continuity, we conclude that

$$\{(p,a) : a \in [z_4, z_0]\} \subset \mathcal{A}(p, z_0)$$

and (14.5) will follow.

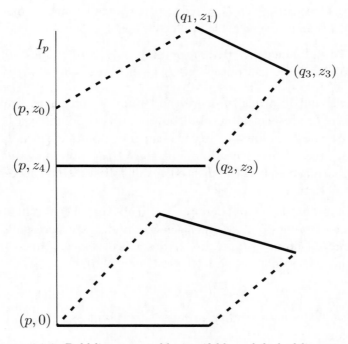

Figure 14.6. Bold lines are stable manifolds and dashed lines are unstable manifolds.

14.1. Diffeomorphisms with countably many ergodic components

Since $g_{t\tau}$ preserves the (x, z)-plane, we have $V_{t\tau}^{uc}((p, z_0)) = V_F^{uc}((p, z_0))$. Hence, there is a unique $z_1 \in (0, 1)$ such that $(q_1, z_1) \in V_{t\tau}^u((p, z_0))$. Notice that for $n \leq \ell$,

$$g_{t\tau}^{-n}(p, z_0) = (p, h_t^{-n}((p, z_0))), \quad g_{t\tau}^{-n}(q_1, z_1) = (A^{-n}q_1, z_1).$$

This is true because the points $A^{-n}q_1$, $n = 0, 1, \ldots, \ell$, lie outside the ε-neighborhood of I_p, where the perturbation map $h_t = \mathrm{Id}$. Similarly, since the points $A^{-n}q_1$, $n > \ell$, lie inside the ε'-neighborhood of I_p and the third component of h_t depends only on the z-coordinate, we have

$$g_{t\tau}^{-n}(q_1, z_1) = (A^{-n}q_1, h_t^{-n+\ell}z_1).$$

Since

$$\rho(g_{t\tau}^{-n}((p, z_0)), g_{t\tau}^{-n}((q_1, z_1))) \to 0$$

as $n \to \infty$, we have

$$\rho(h_t^{-n}((p, z_0)), h_t^{-n+\ell}((p, z_1))) \to 0$$

as $n \to \infty$. It follows that $z_1 = (h_t^{-\ell})^{(3)}((p, z_0))$.

Since $h_t = \mathrm{Id}$ outside Ω_1, the sets $A^{-n_1}V_{t\tau}^s(p')$ and $A^{n_2}V_{t\tau}^u(p')$ are pieces of horizontal lines, and hence, $z_2 = z_3 = z_1$. Since the third component of h_t does not decrease when a point moves from (q_2, z_2) to (p, z_4) along $V_{t\tau}^s(p)$, we conclude that $z_4 \leq z_3 = z_1 = (h_t^{-\ell})^3(p, z_0)$ and thus (14.6) holds, completing the proof of the lemma. $\square$

Lemma 14.9. *For any sufficiently small $\tau > 0$,*

$$\int_M \log \kappa_{0\tau}(w) dw < \log \eta.$$

Proof of the lemma. For any $w \in M$, we introduce the coordinate system in T_wM associated with the splitting $T_wM = E_F^u(w) \oplus E_F^s(w) \oplus E_F^c(w)$. Given $\tau > 0$ and $w \in M$, there exists a unique number $\alpha_\tau(w)$ such that the vector $v_\tau(w) = (1, 0, \alpha_\tau(w))^\perp$ lies in $E_{0\tau}^u(w)$ (where $\perp$ denotes the transpose). Since the map $\tilde{h}_\tau$ preserves the y-coordinate, by the definition of the function $\alpha_\tau(w)$, one can write the vector $d_w g_{0\tau}v_\tau(w)$ in the form

$$dg_{0\tau}(w)v_\tau(w) = (\bar{\kappa}_\tau(w), 0, \bar{\kappa}_\tau(w)\alpha_\tau(g_{t0}(w)))^\perp \quad (14.7)$$

for some $\bar{\kappa}_\tau(w) > 1$. Taking into account that the expansion rate of $d_w g_{0\tau}$ along its unstable direction is $\kappa_{0\tau}(w)$, we obtain that

$$\kappa_{0\tau}(w) = \bar{\kappa}_\tau(w)\frac{\sqrt{1 + \alpha_\tau(g_{0\tau}(w))^2}}{\sqrt{1 + \alpha_\tau(w)^2}}.$$

Since $E_{0\tau}^u(w)$ is close to $E_{00}^u(w)$, the function $\alpha_\tau(w)$ is uniformly bounded. Using the fact that the map $g_{0\tau}$ preserves volume, we find that

$$L_\tau = \int_M \log \kappa_{0\tau}(w) \, dw = \int_M \log \bar{\kappa}_\tau(w) \, dw. \quad (14.8)$$

Consider the map $\tilde{h}_\tau$. Since it preserves the y-coordinate, using (14.4), we can write that
$$\tilde{h}_\tau(x, y, z) = (r \cos \sigma, y, r \sin \sigma),$$
where $\sigma = \sigma(\tau, r, \theta, y) = \theta + \tau \psi(y) \rho(r)$. Therefore, the differential
$$d\tilde{h}_\tau \colon E^u_F(w) \oplus E^c_F(w) \to E^u_F(g_{0\tau}(w)) \oplus E^c_F(g_{0\tau}(w))$$
can be written in the matrix form
$$d\tilde{h}_\tau(w) = \begin{pmatrix} A(\tau, w) & B(\tau, w) \\ C(\tau, w) & D(\tau, w) \end{pmatrix}$$
$$= \begin{pmatrix} r_x \cos \sigma - r\sigma_x \sin \sigma & r_y \cos \sigma - r\sigma_y \sin \sigma \\ r_x \sin \sigma + r\sigma_x \cos \sigma & r_y \sin \sigma + r\sigma_y \cos \sigma \end{pmatrix},$$
where
$$r_x = \frac{\partial r}{\partial x} = \frac{x}{r} = \cos \theta, \quad r_z = \frac{\partial r}{\partial z} = \frac{y}{r} = \sin \theta,$$
$$\sigma_x = \frac{\partial \sigma}{\partial x} = \frac{-z}{r^2} + \frac{z}{r} \tau \tilde{\rho}_r(y, r) = \frac{\sin \theta}{r} + \tau \tilde{\rho}_r(y, r) \cos \theta,$$
$$\sigma_z = \frac{\partial \sigma}{\partial z} = \frac{x}{r^2} + \frac{x}{r} \tau \tilde{\rho}_r(y, r) = \frac{\cos \theta}{r} + \tau \tilde{\rho}_r(y, r) \sin \theta,$$
and $\tilde{\rho}(y, r) = \psi(y) \rho(r)$.

Exercise 14.10. Show that
$$A = A(\tau, w) = 1 - \tau r \tilde{\rho}_r \sin \theta \cos \theta - \frac{\tau^2 \tilde{\rho}^2}{2} - \tau^2 r \tilde{\rho} \tilde{\rho}_r \cos^2 \theta + O(\tau^3),$$
$$B = B(\tau, w) = -\tau \tilde{\rho} - \tau r \tilde{\rho}_r \sin^2 \theta - \tau^2 r \tilde{\rho} \tilde{\rho}_r \sin \theta \cos \theta + O(\tau^3),$$
$$C = C(\tau, w) = \tau \tilde{\rho} + \tau r \tilde{\rho}_r \cos^2 \theta - \tau^2 r \tilde{\rho} \tilde{\rho}_r \sin \theta \cos \theta + O(\tau^3),$$
$$D = D(\tau, w) = 1 + \tau r \tilde{\rho}_r \sin \theta \cos \theta - \frac{\tau^2 \tilde{\rho}^2}{2} - \tau^2 r \tilde{\rho} \tilde{\rho}_r \sin^2 \theta + O(\tau^3).$$

We now obtain the formula for L_τ:
$$L_\tau = \log \eta - \int_M \log(D(\tau, w) - \eta B(\tau, w) \alpha_\tau(g_{0\tau}(w))) dw. \quad (14.9)$$
Indeed, since
$$g_{0\tau} = h_0 \circ F \circ \tilde{h}_\tau = F \circ \tilde{h}_\tau,$$
we have that
$$\mathcal{D}_\tau(w) = d_w g_{0\tau}|(E^u_{0\tau}(w) \oplus E^c_{0\tau}(w)) = \begin{pmatrix} \eta A(\tau, w) & \eta B(\tau, w) \\ C(\tau, w) & D(\tau, w) \end{pmatrix}.$$
By (14.7),
$$\mathcal{D}_\tau(w) \begin{pmatrix} 1 \\ \alpha_\tau(w) \end{pmatrix} = \begin{pmatrix} \eta A(\tau, w) + \eta B(\tau, w) \alpha_\tau(w) \\ C(\tau, w) + D(\tau, w) \alpha_\tau(w) \end{pmatrix}$$
$$= \begin{pmatrix} \kappa_\tau(w) \\ \kappa_\tau(w) \alpha_\tau(g_{0\tau}(w)) \end{pmatrix}. \quad (14.10)$$

Since $\tilde{h}_\tau$ is volume-preserving, $AD - BC = 1$, and therefore,
$$A + B\alpha = \frac{1}{D} + \frac{B}{D}(C + D\alpha).$$
Comparing the components in (14.10), we obtain
$$\kappa_\tau(w) = \eta(A(\tau,w) + B(\tau,w)\alpha_\tau(w))$$
$$= \eta\left(\frac{1}{D(\tau,w)} + \frac{B(\tau,w)}{D(\tau,w)}(C(\tau,w) + D(\tau,w)\alpha_\tau(w))\right)$$
$$= \eta\left(\frac{1}{D(\tau,w)} + \frac{B(\tau,w)}{D(\tau,w)}(\kappa_\tau(w)\alpha_\tau(g_{0\tau}(w)))\right).$$
Solving for $\kappa_\tau(w)$, we get
$$\kappa_\tau(w) = \frac{\eta}{D(\tau,w) - \eta B(\tau,w)\alpha_\tau(g_{0\tau}(w))}.$$
The desired relation (14.9) follows from (14.8).

One can deduce from (14.9) by a straightforward calculation that
$$\left.\frac{dL_\tau}{d\tau}\right|_{\tau=0} = 0 \quad \text{and} \quad \left.\frac{d^2 L_\tau}{d\tau^2}\right|_{\tau=0} > 0 \qquad (14.11)$$
(we stress that the argument is not quite trivial and we refer the reader to [**13**] for details). To conclude the proof of the lemma, it remains to notice that by (14.11), $L_\tau < \log \eta$ for all sufficiently small τ. □

14.2. The Shub–Wilkinson map

We construct an example of a uniformly partially hyperbolic diffeomorphism with positive central exponents. Let A be a linear hyperbolic automorphism of the two-dimensional torus $\mathbb{T}^2$. Consider the map $F = A \times \mathrm{Id}$ of the three-dimensional torus $\mathbb{T}^3 = \mathbb{T}^2 \times S^1$ where Id is the identity map. This map preserves volume and is uniformly partially hyperbolic. It has zero Lyapunov exponent in the central direction. Any sufficiently small C^1 perturbation G of F is also uniformly partially hyperbolic with one-dimensional central direction. The following result due to Shub and Wilkinson (see [**105**] and also [**41**]) shows that this perturbation can be arranged in such a way to ensure positive Lyapunov exponents in the central direction.

Theorem 14.11. *For any $k \geq 2$ and $\delta > 0$, there exists a volume-preserving C^∞ perturbation G of F such that G is δ-close to F in the C^k topology and has positive central exponents almost everywhere.*

Proof. We follow the approach described in the previous section. Without loss of generality we may assume that the linear hyperbolic automorphism A has at least two fixed points that we denote by p and p'. For a sufficiently small ε_0 consider the two disjoint open domains Ω_1 and Ω_2 given by (14.2).

Define a diffeomorphism h_t in Ω_1 to be the time-t map of the flow generated by the vector field X given by (14.3) and set $h_t = \text{Id}$ outside Ω_1. Now define a diffeomorphism $\tilde{h}_\tau$ in Ω_2 by (14.4) and set $\tilde{h}_\tau = \text{Id}$ outside Ω_2. Finally, set $G = h_t \circ F \circ \tilde{h}_\tau$. Repeating the arguments in the proof of Proposition 14.2, one can show that for sufficiently small $t > 0$ and $\tau > 0$ the map G is a C^∞ volume-preserving diffeomorphism, which has the essential accessibility property and positive central exponents almost everywhere. The desired result follows from Theorem 13.4. $\square$

One can show that any sufficiently small C^∞ volume-preserving perturbation of G in the C^1 topology has the essential accessibility property and positive central exponents almost everywhere. Hence, it is ergodic and has the Bernoulli property. In particular, we obtain an open set in the C^1 topology in $\mathbb{T}^3$ of volume-preserving non-Anosov diffeomorphisms with nonzero Lyapunov exponents.

Chapter 15

Anosov Rigidity

In every example of volume-preserving nonuniformly hyperbolic diffeomorphisms that we have constructed in the book, the Lyapunov exponents were nonzero on a set of full volume but *not* everywhere. On the other hand, the Lyapunov exponents of an Anosov diffeomorphism are nonzero at *every* point on the manifold. This observation leads to a natural problem of whether *nonuniform hyperbolicity everywhere on a compact manifold implies uniform hyperbolicity*—a phenomenon that we call *Anosov rigidity*.

We describe two versions of the Anosov rigidity phenomenon requiring two quite different approaches. The first one deals with the situation when a diffeomorphism f is nonuniformly hyperbolic on a *compact* and *invariant* subset K, in which case the problem is to show that $f|K$ is uniformly hyperbolic. The second one requires the weaker hypothesis that Lyapunov exponents are nonzero on a set of total measure, i.e., on a set that has full measure with respect to every f-invariant Borel probability measure.

15.1. The Anosov rigidity phenomenon. I

We describe the approach to the Anosov rigidity phenomenon developed in [**52**]. It can be expressed in the following two statements. Let f be a C^1 diffeomorphism of a compact smooth Riemannian manifold M and let K be a compact invariant subset.

Theorem 15.1. *Assume that there is a continuous invariant cone family $C(x) \subset T_x M$ on K such that:*

(1) *it is invariant, i.e., $d_x f C(x) \subset C(f(x))$;*

(2) for all $x \in K$,
$$\varphi(x) = \liminf_{n\to\infty} \frac{1}{n} \min_{v \in C(x), \|v\|=1} \log \|d_x f^n v\| > 0. \tag{15.1}$$
Then there exist $c > 0$ and $\chi > 0$ such that for every $x \in K$, $v \in C(x)$, $\|v\| = 1$, and $n \in \mathbb{N}$,
$$\|d_x f^n v\| \geq c e^{\chi n}.$$
Furthermore, for every $x \in K$ there is a subspace $E(x) \subset T_x M$ such that
$$d_x f(E(x)) = E(f(x)) \quad \text{and} \quad \|d_x f^n v\| \geq c e^{\chi n}$$
for every $v \in E_x$, $\|v\| = 1$, and $n \in \mathbb{N}$.

Theorem 15.2. *Assume that there are two continuous transverse cone families $C(x)$ and $D(x)$ on K such that:*

(1) *they are invariant, i.e.,*
$$d_x f C(x) \subset C(f(x)), \quad d_x f^{-1} D(x) \subset D(f^{-1}(x));$$

(2) *for all $x \in K$,*
$$\liminf_{n\to\infty} \frac{1}{n} \min_{v \in C(x), \|v\|=1} \log \|d_x f^n v\| > 0,$$
$$\liminf_{n\to-\infty} \frac{1}{|n|} \min_{v \in D(x), \|v\|=1} \log \|d_x f^n v\| > 0.$$

Then K is a uniformly hyperbolic set for f. In particular, if $K = M$, then f is an Anosov diffeomorphism.

Exercise 15.3. Deduce Theorem 15.2 from Theorem 15.1.

Theorem 15.2 was first proved by Mañé in [**80**]. Its continuous-time version is due to Sacker and Sell [**102**]. We present a proof of Theorem 15.1 following closely the approach developed by Hasselblatt, Pesin, and Schmeling in [**52**]. It exploits some ideas from descriptive set theory.

15.1.1. Transfinite hierarchy of set filtrations. We present a set-theoretic construction providing a detailed study of representations of a compact space as a nested union of compact subsets. The main idea can be highlighted by examining the well-known proof that a positive continuous function φ on a compact space has a positive minimum: the open cover by sets $\varphi^{-1}(\frac{1}{n}, \infty)$ has a finite subcover. Attempting to extend this proof to Baire functions, one might try to cover the space with the interiors of the sets $\varphi^{-1}[\frac{1}{n}, \infty)$. If the compact set A that remains after deleting all these interiors is nonempty, then one repeats the entire process on the set A with respect to its subspace topology. We develop a transfinite process of this sort for the function $\varphi(x)$ given by (15.1). We split our construction into three steps.

15.1. The Anosov rigidity phenomenon. I

15.1.1.1. Set filtrations. Let (X, d) be a compact separable metric space. A *set filtration* of X is a collection of compact subsets $X_n \subset X$ such that:

(1) they exhaust X, i.e., $\bigcup_{n \in \mathbb{N}} X_n = X$;

(2) they are nested, i.e., $X_n \subseteq X_{n+1}$ for $n \in \mathbb{N}$;

(3) if $X_{n+1} \neq X$, then $X_n \subsetneq X_{n+1}$.

We say that X is *uniform* with respect to this filtration if $X = X_n$ for some $n \in \mathbb{N}$.

Lemma 15.4. *We have that* $X = \mathrm{Cl} \bigcup_{n \in \mathbb{N}} \mathrm{int}\, X_n$, *where* Cl *denotes closure of the corresponding set.*[1]

Proof. We need to show that given a closed ball B centered at some point x and of some radius r, the intersection $B \cap \bigcup_{n \in \mathbb{N}} \mathrm{int}\, X_n$ is nonempty. To this end, note that

$$B = B \cap X = B \cap \bigcup_{n \in \mathbb{N}} X_n = \bigcup_{n \in \mathbb{N}} B \cap X_n$$

is a complete metric space and, hence, not a countable union of sets of first category. Thus, there exists $N \in \mathbb{N}$ such that $X_N \cap B$ is of second category and, hence, not nowhere dense. This means that

$$\varnothing \neq \mathrm{int}_B(\mathrm{Cl}\, X_N) = \mathrm{int}_B X_N \subset B \cap \mathrm{int}\, X_N,$$

where int_B denotes the interior in the subspace topology of B. This implies the desired result. $\square$

The set $\Gamma = X \setminus \bigcup_{n \in \mathbb{N}} \mathrm{int}\, X_n$ is clearly compact.

Lemma 15.5. *We have that*

$$\Gamma = \{x \in X : \text{there exists } x_n \to x \text{ such that } x_n \notin X_n\}.$$

Proof. For $x \in \Gamma$ there exists a sequence $y_n \to x$ such that $y_n \notin \mathrm{int}\, X_n$. By the definition of interior, one can find $x_n \notin X_n$ such that $\rho(x_n, y_n) < 1/n$. Thus,

$$\Gamma \subset \{x \in X : \text{there exists } x_n \to x \text{ such that } x_n \notin X_n\}.$$

The reverse inclusion is clear because $x_n \notin X_n$ implies that $x_n \notin \mathrm{int}\, X_n$. $\square$

---

[1] We use different notation for closure for reasons that will be clear later (see Section 15.1.1.2).

15.1.1.2. *The hierarchy.* In view of Lemma 15.4 we wish to exhaust the set X with the interiors of sets X_n from the filtration. This leaves uncovered the compact set Γ, and we now describe how to continue this process recursively in a transfinite way.

Set $X_n^{(0)} = X_n$, $F^{(0)} = X$, and $\Gamma^{(0)} = \Gamma$. Given an ordinal β such that we already have sets $\Gamma^{(\alpha)}$ for all $\alpha < \beta$, we inductively define

$$F^{(\beta)} = \bigcap_{\alpha < \beta} \Gamma^{(\alpha)}, \quad X_n^{(\beta)} = F^{(\beta)} \cap X_n,$$

and

$$\Gamma^{(\beta)} = \mathrm{Cl}_{F^{(\beta)}} \left(\bigcup_{n \in \mathbb{N}} \mathrm{int}_{F^{(\beta)}} X_n^{(\beta)} \right) \setminus \bigcup_{n \in \mathbb{N}} \mathrm{int}_{F^{(\beta)}} X_n^{(\beta)} \subset F^{(\beta)},$$

where $\mathrm{Cl}_{F^{(\beta)}}$ denotes the closure in the subspace topology of $F^{(\beta)}$. Our next statement implies that taking the ambient closure gives the same set.

Lemma 15.6. *The sets $\Gamma^{(\beta)}$, $F^{(\beta)}$, and $X_n^{(\beta)}$ are compact.*

Proof. For $\beta = 0$ this is the compactness of Γ, X, and X_n. We proceed by induction assuming that $\Gamma^{(\alpha)}$ is compact for all $\alpha < \beta$. Then $F^{(\beta)}$ is compact because it is defined as an intersection of compact sets. Since X_n is compact, this implies compactness of $X_n^{(\beta)}$. Finally, $\Gamma^{(\beta)}$ is a closed subset of $F^{(\beta)}$, hence also compact. □

Lemma 15.7. *We have that:*

(1) *the sets $F^{(\beta)}$, $X_n^{(\beta)}$, and $\Gamma^{(\beta)}$ are nested, i.e., for $\alpha < \beta$,*

$$F^{(\beta)} \subseteq F^{(\alpha)}, \quad X_n^{(\beta)} \subseteq X_n^{(\alpha)}, \quad \Gamma^{(\beta)} \subseteq \Gamma^{(\alpha)};$$

(2) *the sets $X_n^{(\beta)}$ form a filtration of the set $F^{(\beta)}$, i.e.,*

$$\bigcup_{n \in \mathbb{N}} X_n^{(\beta)} = F^{(\beta)}, \quad X_n^{(\beta)} \subseteq X_{n+1}^{(\beta)};$$

(3) $F^{(\beta)} = \mathrm{Cl} \bigcup_{n \in \mathbb{N}} \mathrm{int}_{F^{(\beta)}} X_n^{(\beta)}$, *and hence,*

$$\Gamma^{(\beta)} = F^{(\beta)} \setminus \bigcup_{n \in \mathbb{N}} \mathrm{int}_{F^{(\beta)}} X_n^{(\beta)};$$

(4) *if $\alpha < \beta$ and $F^{(\alpha)} \neq \varnothing$, then $F^{(\beta)} \subsetneq F^{(\alpha)}$, i.e., the transfinite induction for sets $F^{(\beta)}$ can stabilize only at $\varnothing$;*

(5) $F^{(\alpha+1)} = \Gamma^{(\alpha)}$, *and hence,* $X_n^{(\alpha+1)} = \Gamma^{(\alpha)} \cap X_n$.

15.1. The Anosov rigidity phenomenon. I

Proof. (1) For the sets $F^{(\beta)}$ and $X_n^{(\beta)}$ the statement follows from the definitions and for the sets $\Gamma^{(\beta)}$ it follows from

$$\Gamma^{(\beta)} \subseteq F^{(\beta)} = \bigcap_{\tau < \beta} \Gamma^{(\tau)} \subseteq \Gamma^{(\alpha)}.$$

(2) We have that

$$F^{(\beta)} = F^{(\beta)} \cap X = F^{(\beta)} \cap \bigcup_{n \in \mathbb{N}} X_n = \bigcup_{n \in \mathbb{N}} F^{(\beta)} \cap X_n = \bigcup_{n \in \mathbb{N}} X_n^{(\beta)}$$

and

$$X_n^{(\beta)} = F^{(\beta)} \cap X_n \subset F^{(\beta)} \cap X_{n+1} = X_{n+1}^{(\beta)}.$$

(3) The result follows from (2) by applying Lemmas 15.6 and 15.4 to the set $F^{(\beta)} = \bigcup_{n \in \mathbb{N}} X_n^{(\beta)}$.

(4) The set $F^{(\alpha)} = \bigcup_{n \in \mathbb{N}} X_n^{(\alpha)}$ is nonempty, compact, and, hence, complete. Therefore, there exists $n_0 \in \mathbb{N}$ such that $X_{n_0}^{(\alpha)}$ is of second category in the induced topology of $F^{(\alpha)}$. Then $\operatorname{int} X_{n_0}^{(\alpha)} \neq \varnothing$ because $X_{n_0}^{(\alpha)}$ is compact and nonempty. It follows that

$$F^{(\beta)} = \bigcap_{\gamma < \beta} \Gamma^{(\gamma)} \subset \bigcap_{\alpha \leq \gamma < \beta} \Gamma^{(\gamma)} \subset \Gamma^{(\alpha)}$$
$$= F^{(\alpha)} \setminus \bigcup_{n \in \mathbb{N}} \operatorname{int} X_n^{(\alpha)} \subset F^{(\alpha)} \setminus \operatorname{int} X_{n_0}^{(\alpha)} \subsetneq F^{(\alpha)}.$$

(5) By statement (1) the sets $\Gamma^{(\tau)}$ are nested, and hence,

$$F^{(\alpha+1)} = \bigcap_{\tau < \alpha+1} \Gamma^{(\tau)} = \bigcap_{\tau \leq \alpha} \Gamma^{(\tau)} = \Gamma^{(\alpha)}.$$

This completes the proof of the lemma. □

15.1.1.3. *Termination of the process.* We now describe what happens in the end of the induction process. The following statement reflects the tacit assumption that $F^{(0)} = X \neq \varnothing$.

Lemma 15.8. *There is a countable ordinal ξ such that $F^{(\xi)} \neq \varnothing = F^{(\xi+1)}$.*

Proof. Since the space X is second countable, it has a countable base $\mathcal{U}$. If $F^{(\alpha)} \neq \varnothing$, then $F^{(\alpha)} \setminus F^{(\alpha+1)} \neq \varnothing$ by Lemma 15.7(4). Since this is open in the subspace topology of $F^{(\alpha)}$, there exists $\mathcal{O}_\alpha \in \mathcal{U}$ such that $\mathcal{O}_\alpha \cap F^{(\alpha)} \neq \varnothing$ and $\mathcal{O}_\alpha \cap F^{(\alpha+1)} = \varnothing$. Such sets $\mathcal{O}_\alpha$ are pairwise distinct, so there are only countably many α for which $F^{(\alpha)} \neq \varnothing$. Thus, $F^{(\alpha_0)} = \varnothing$ for a countable ordinal.

The set $\{\alpha < \omega_1 \colon F^{(\alpha)} = \varnothing\}$, where ω_1 is the first uncountable ordinal, contains α_0 and, hence, is a nonempty subset of the well-ordered set of countable ordinals. Therefore, it contains a minimal element η.

If η is a limit ordinal, i.e., is not of the form $\xi + 1$ for any ordinal ξ, then there is an increasing sequence $(\alpha_n)_{n \in \mathbb{N}}$ of ordinals such that for all $\tau < \eta$ there exists $n \in \mathbb{N}$ for which $\tau < \alpha_n < \eta$. Hence,

$$F^{(\eta)} = \bigcap_{\tau < \eta} \Gamma^{(\tau)} = \bigcap_{\tau < \eta} F^{(\tau+1)} = \bigcap_{\tau < \eta} F^{(\tau)} = \bigcap_{n \in \mathbb{N}} F^{(\alpha_n)} \neq \varnothing,$$

since $\varnothing \neq F^{(\alpha_{n+1})} \subset F^{(\alpha_n)}$. This is a contradiction; thus we can write $\eta = \xi + 1$. $\square$

Lemma 15.9. *If ξ is as in Lemma 15.8, i.e., $F^{(\xi)} \neq \varnothing = F^{(\xi+1)}$, then the following statements hold:*

(1) *$F^{(\xi)} \subset X_R$ for some $R \in \mathbb{N}$; in particular, if $\xi = 0$, then X is uniform with respect to the filtration $(X_n)_{n \in \mathbb{N}}$;*

(2) *if $\tau < \xi$, then $\bigcup_{n=1}^{\infty} \operatorname{int} X_n^{(\tau)} \subsetneq F^{(\tau)}$;*

(3) *if $\xi > 0$, then for every $\varepsilon > 0$ there exist $\tau < \xi$ and $N \in \mathbb{N}$ such that*

$$F^{(\tau)} \setminus U_\varepsilon(F^{(\xi)}) \subset \bigcup_{n=1}^{N} \operatorname{int} X_n^{(\tau)} \subset X_N^{(\tau)} \subset X_N.$$

Proof. (1) Statements (3)–(5) of Lemma 15.7 yield

$$\varnothing = F^{(\xi+1)} = \Gamma^{(\xi)} = F^{(\xi)} \setminus \bigcup_{n \in \mathbb{N}} \operatorname{int}_{F^{(\xi)}} X_n^{(\xi)}.$$

Hence, $F^{(\xi)} \subset \bigcup_{n \in \mathbb{N}} \operatorname{int}_{F^{(\xi)}} X_n^{(\xi)}$. The desired result follows since this open cover has a finite subcover.

(2) If $F^{(\tau)} = \bigcup_{n=1}^{\infty} \operatorname{int}_{F^{(\tau)}} X_n^{(\tau)}$, then by statements (3), (4), and (5) of Lemma 15.7, we have $\varnothing = \Gamma^{(\tau)} = F^{(\tau+1)}$ and $\tau \geq \xi$.

(3) We have that

$$\bigcap_{\alpha < \xi} \Gamma^{(\alpha)} = F^{(\xi)} \neq \varnothing.$$

The sets $\Gamma^{(\alpha)}$ are nested and compact, and hence, $\inf_{\alpha < \xi} d_H(F^{(\xi)}, \Gamma^{(\alpha)}) = 0$, where d_H is the Hausdorff distance. That means that there exists $\tau < \xi$ such that $\Gamma^{(\alpha)} \subseteq U_\varepsilon(F^{(\xi)})$ whenever $\tau \leq \alpha < \xi$; in particular, for $\alpha = \tau$ we have

$$F^{(\tau)} \setminus \bigcup_{n \in \mathbb{N}} \operatorname{int} X_n^{(\tau)} = \Gamma^{(\tau)} \subseteq U_\varepsilon(F^{(\xi)}),$$

15.1. The Anosov rigidity phenomenon. I

and hence,
$$F^{(\tau)} \setminus U_\varepsilon(F^{(\xi)}) \subseteq \bigcup_{n \in \mathbb{N}} \operatorname{int} X_n^{(\tau)}.$$

This is an open cover of a compact set. Hence, the claim follows. □

15.1.2. Proof of Theorem 15.1. Set
$$\varphi_n = \frac{1}{n} \min_{v \in C(x), \|v\|=1} \log \|d_x f^n v\|$$
and consider the filtration of X by the sets
$$X_n = \left\{ x \in X : \varphi_k(x) \geq \frac{1}{n} \text{ for } k \geq n \right\} \subset \left\{ x \in X : \varphi(x) \geq \frac{1}{n} \right\}.$$

The main idea of the proof is the following. Assume that we can find two compact sets $K_1 \subset K_2 \subset X$ such that the set K_1 is uniform with respect to the filtration X_n (i.e., $K_1 \subset X_N$ for some $N > 0$). Assume also that $K_2 \setminus O$ is known to be uniform whenever the set O is open and $K_1 \subset O$. If there is a uniform neighborhood U of K_1, then we conclude that $K_2 \subset U \cup K_2 \setminus U$ is uniform as well.

To implement this idea, we first show that $K_1 = F^{(\xi)}$, which is uniform by statement (1) of Lemma 15.9, has a uniform ε-neighborhood U_ε (see Lemma 15.10). This is the main step in the proof. Now we observe that if $\xi > 0$ in Lemma 15.8, then we can take $\tau < \xi$ as in statement (3) of Lemma 15.9 and conclude from the above that $K_2 = F^{(\tau)}$ is uniform. Since this implies that $F^{(\tau+1)} = \varnothing$, we conclude that $\tau + 1 > \xi$ after all, a contradiction. Consequently, $\xi = 0$, and $X = F^{(0)}$ is uniform by statement (1) of Lemma 15.9, as claimed.

Lemma 15.10. *There exist $C > 0$, $\varepsilon > 0$, and $\lambda > 1$ such that if $f^n(x) \in U_\varepsilon(F^{(\xi)})$ whenever $0 \leq n \leq K$ for some $K \in \mathbb{N}$, then*
$$\min_{v \in C(x), \|v\|=1} \|d_x f^n v\| \geq C \lambda^n \quad \text{whenever} \quad 0 \leq n \leq K.$$

Proof. By statement (1) of Lemma 15.9, there exists $R \in \mathbb{N}$ (which depends on ξ) such that $F^{(\xi)} \subset X_R$. Thus for all $n \geq R$ and $y \in F^{(\xi)}$ we have
$$\frac{1}{R} \leq \varphi_n(y) = \frac{1}{n} \min_{v \in C_y^1, \|v\|=1} \log \|d_x f^n v\|;$$
hence,
$$\min_{v \in C(y), \|v\|=1} \|d_x f^n v\| \geq e^{n/R}.$$
Now take $L \in \mathbb{N}$ so large that if $y \in F^{(\xi)}$, then
$$\min_{v \in C(y), \|v\|=1} \|d_x f^L v\| \geq 3 \max_{x \in X} \max_{v \in C(x), \|v\|=1} \|v\| = 3$$

and write $g = d_x f^L$. Note that L depends only on R and hence only on ξ. If $v \in C(y)$ and $\|v\| = 1$, then

$$\|d_x g^n v\| = \frac{\|g(g^{n-1}v/\|g^{n-1}v\|)\|}{\|(g^{n-1}v/\|g^{n-1}v\|)\|} \|g^{n-1}v\| \geq 3\|g^{n-1}v\|$$

$$\geq \cdots$$

$$\geq 3^{n-1}\|gv\| \geq 3^n.$$

Thus, for $n \in \mathbb{N}$ and $y \in F^{(\xi)}$ we have

$$\min_{v \in C(y), \|v\|=1} \|g^n(v)\| \geq 3^n.$$

If $K \leq L$, then the conclusion of Lemma 15.10 is obtained by taking

$$C \leq \min_{1 \leq n \leq L} \min_{v \in C(x), \|v\|=1} \|d_x f^n v\| \lambda^{-n},$$

where $\lambda > 1$ can be chosen arbitrarily (its particular choice will be determined below). For $K > L$ we continue as follows. For any $x \in U_\varepsilon(F^{(\xi)})$ we can choose $y \in F^{(\xi)}$ such that $\rho(x, y) < \varepsilon$. Then

$$\min_{v \in C(x), \|v\|=1} \|gv\| = \min_{v \in C(y), \|v\|=1} \|gv\| \frac{\min_{v \in C(x), \|v\|=1} \|gv\|}{\min_{v \in C(y), \|v\|=1} \|gv\|}.$$

By continuity of the cone family $C(x)$, one can choose ε so small that the last fraction is bounded from below by $2/3$.[2] This yields

$$\min_{v \in C(x), \|v\|=1} \|gv\| \geq \frac{2}{3} \min_{w \in C(y), \|w\|=1} \|gw\| \geq 2. \tag{15.2}$$

Thus, for any $n \in \mathbb{N}$ such that $nL \leq K$, by (15.2) we find that

$$\min_{v \in C(x), \|v\|=1} \|g^n v\| = \min_{v \in C(x), \|v\|=1} a\|g(g^{n-1}v)\|$$

$$\geq 2 \min_{v \in C(x), \|v\|=1} \|g^{n-1}v\|$$

$$\geq \cdots$$

$$\geq 2^{n-1} \min_{v \in C(x), \|v\|=1} \|gv\| \geq 2^n.$$

Writing $n = kL + r$, we find that for every $v \in C_x^1$,

$$\|d_x f^n v\| = \|d_x f^{kL+r} v\|$$

$$= \frac{\|d_x f^n v\|}{\|d_x f^{n-1}v\|} \cdots \frac{\|d_x f^{n-r+1}v\|}{\|d_x f^{n-r}v\|} \|g^k v\|$$

$$\geq C' 2^k = C'(2^{k/n})^n \geq C \lambda^n$$

for suitable $\lambda > 1$ and $C > 0$. $\square$

---

[2] Thus, ε depends on L and R and hence ultimately only on ξ. Note also that this is the only place where the continuity of the cone family $C(x)$ is used.

15.2. The Anosov rigidity phenomenon. II

We now conclude the proof of Theorem 15.1. Recall that we chose ξ as in Lemma 15.8, which determines R via statement (1) of Lemma 15.9, and these parameters in turn determine the choice of ε in Lemma 15.10.

Suppose that $\xi > 0$ and choose $\tau < \xi$ and N as in statement (3) of Lemma 15.9.

Consider any $x \in F^{(\tau)}$. If there is a $k_0 \in \mathbb{N}_0$ such that $f^k(x) \in U_\varepsilon(F^{(\xi)})$ for $k < k_0$ and $f^{k_0}(x) \notin U_\varepsilon(F^{(\xi)})$, then $f^{k_0}(x) \in X_N^{(\tau)}$. Thus for any $v \in C(x)$, $\|v\| = 1$, and all $n \in \mathbb{N}$ we have

$$\|d_x f^n v\| = \|d_x f^{\max(0, n-k_0)/N} d_x f^{\min(n, k_0)} v\|$$
$$\geq e^{\max(0, n-k_0)/N} \|d_x f^{\min(n, k_0)} v\|$$
$$\geq C \lambda^{\min(n, k_0)} e^{\max(0, n-k_0)/N} \geq C \gamma^n,$$

where $\gamma = \min(\lambda, e^{1/N}) > 1$. Note that the same estimate holds if $f^n(x) \in U_\varepsilon(F^{(\xi)})$ for all $n \in \mathbb{N}$, so it holds for all $x \in F^{(\tau)}$.

It is easy to check that this implies that $F^{(\tau)} \subset X_{2 \max\{1, -\log C\}/\log \gamma}$. By statement (3) of Lemma 15.7, we conclude that $F^{(\tau+1)} = \varnothing$, and hence, $\tau \geq \xi$, which is contrary to our choice of τ.

15.2. The Anosov rigidity phenomenon. II

We describe another approach to the Anosov rigidity phenomenon due to Cao [36] (see also [3, 37]).

Let f be a C^1 diffeomorphism of a compact smooth Riemannian manifold M and let $K \subset M$ be a compact invariant subset. We say that a subset $A \subset K$ has *total measure* if $\mu(A) = 1$ for any invariant Borel probability measure μ.[3]

Theorem 15.11. *Assume that there is a continuous invariant distribution $E(x) \subset T_x M$ on K such that the relation*

$$\psi(x) = \limsup_{n \to \infty} \frac{1}{n} \log \|d_x f^{-n}|E(x)\| < 0$$

holds for every x in a set A of total measure. Then there exist $c > 0$ and $\chi > 0$ such that for every $x \in K$ and $n \in \mathbb{N}$,

$$\|d_x f^{-n}|E(x)\| \leq c e^{-\chi n}.$$

Proof. We denote by $\mathcal{K}(f)$ the space of invariant measures on K, endowed with the weak*-topology.

---

[3] Of course, the set Λ itself has total measure but we are interested in the smallest such set.

Lemma 15.12. *Let φ be a continuous function on K and let $\lambda \in \mathbb{R}$. If $\int_K \varphi \, d\mu < \lambda$ for every $\mu \in \mathcal{K}(f)$, then:*

(1) *for every $x \in K$, there exists a number $n(x) > 0$ such that*
$$\frac{1}{n(x)} \sum_{i=0}^{n(x)-1} \varphi(f^i(x)) < \lambda;$$

(2) *there exists $N > 0$ such that for all $n \geq N$, we have*
$$\frac{1}{n} \sum_{i=0}^{n-1} \varphi(f^i(x)) < \lambda.$$

Proof of the lemma. Assuming the contrary, we obtain that for some $x \in K$ and every $n > 0$,
$$\frac{1}{n} \sum_{i=0}^{n-1} \varphi(f^i(x)) \geq \lambda.$$

Consider the sequence of probability measures
$$\mu_n = \frac{1}{n} \sum_{i=0}^{n-1} \delta(f^i(x)), \quad n \geq 1,$$

where $\delta(f^i(x))$ is the Dirac measure at $f^i(x)$. The set $\mathcal{K}(f)$ is compact. Let μ be an accumulation measure and let μ_{n_k} be a subsequence, which converges to μ. By the Bogolubov–Krilov theorem, the measure μ is invariant. Since the function φ is continuous, we obtain that
$$\int_K \varphi \, d\mu = \lim_{k \to \infty} \frac{1}{n_k} \sum_{i=0}^{n_k-1} \varphi(f^i(x)) \geq \lambda.$$

This contradiction proves the first statement. To show the second statement, observe that for every $x \in K$ there are numbers $n(x) > 0$ and $0 < a(x) < \lambda$ such that
$$\frac{1}{n(x)} \sum_{i=0}^{n(x)-1} \varphi(f^i(x)) < a(x).$$

Thus, by the continuity of φ, for each $x \in K$ there is a neighborhood V_x of x such that for every $y \in V_x$,
$$\frac{1}{n(x)} \sum_{i=0}^{n(x)-1} \varphi(f^i(y)) < a(x).$$

Since K is compact, there is a finite cover of K by open neighborhoods $V_{x_1}, \ldots, V_{x_p}$. Set
$$\bar{N} = \max\{n(x_1), \ldots, n(x_p)\}, \quad a = \max\{a(x_1), \ldots, a(x_p)\}$$

15.2. The Anosov rigidity phenomenon. II

and observe that $a < \lambda$. For $x \in K$ let

$$N_0(x) = 0, \quad N_1(x) = \min\{n(x_k) \colon x \in V_{x_i}, i = 1, \ldots, p\}$$

and define the sequence of functions $N_k(x)$ by

$$N_{k+1}(x) = N_k(x) + N_1(f^{N_k(x)}(x)).$$

For every $x \in K$ and $n > 0$ there exists $k > 0$ such that $N_k \leq n \leq N_{k+1}$. Letting $\alpha = \max_{x \in K} \|\varphi(x)\|$, we find that

$$\sum_{i=0}^{n-1} \varphi(f^i(x)) \leq aN_k + \alpha \bar{N} \leq an + (|a| + \alpha)\bar{N}.$$

Taking $N = \frac{2(|a|+\alpha)\bar{N}}{\lambda - a}$, we obtain the desired estimate. $\square$

For $n > 0$ consider the function on K

$$\varphi_n(x) = \log \|d_x f^{-n}|E(x)\|.$$

Since f is a C^1 diffeomorphism, the function $\varphi_n(x)$ is continuous.

Exercise 15.13. Show that the sequence of functions φ_n is subadditive, i.e., $\varphi_{m+n}(x) \leq \varphi_n(x) + \varphi_m(f^n(x))$ (see (12.9)).

By Kingman's subadditive ergodic theorem [**70**], the limit

$$\lim_{n \to \infty} \frac{\varphi_n(x)}{n} = \bar{\varphi}(x)$$

exists for almost every x and every measure $\mu \in \mathcal{K}(f)$ and the function $\bar{\varphi}$ is invariant and integrable. By the dominated convergence theorem, for every $\mu \in \mathcal{K}(f)$,

$$\lim_{n \to \infty} \int_K \frac{\varphi_n}{n} \, d\mu = \int_K \bar{\varphi} \, d\mu. \tag{15.3}$$

The assumption that the relation (15.1) holds on a set of total measure implies that $\bar{\varphi} < 0$ for almost every x and any $\mu \in \mathcal{K}(f)$. It follows that $\int_K \bar{\varphi} \, d\mu < 0$ for every invariant measure μ.

We need the following statement.

Lemma 15.14. *There exist $L > 0$ and $\lambda < 0$ such that for any $\mu \in \mathcal{K}(f)$,*

$$\frac{1}{L} \int_K \varphi_L \, d\mu < \lambda. \tag{15.4}$$

Proof of the lemma. Fix $\mu \in \mathcal{K}(f)$. Since $\int_K \bar{\varphi} \, d\mu$, by (15.3), there exists $n_\mu > 0$ such that for all $n \geq n_k$,

$$\int_K \frac{\varphi_n}{n} \, d\mu < \frac{1}{2} \int_K \bar{\varphi} \, d\mu.$$

Since $\varphi_{n_\mu}(x)/n_\mu$ is a continuous function on K, there exists an open neighborhood $O(\mu) \in \mathcal{K}(f)$ of μ such that for all $\mu' \in O(\mu)$,

$$\frac{1}{n_\mu} \int_K \varphi_{n_\mu} \, d\mu' < \frac{1}{4} \int_K \bar{\varphi} \, d\mu.$$

The collection of sets $O(\mu)$ forms an open cover of $\mathcal{K}(f)$ and since it is compact, there is a finite subcover $O(\mu_1), \ldots, O(\mu_\ell)$. Setting $n_j = n_{\mu_j}$ and $\lambda = \max\{\frac{1}{4} \int_K \bar{\varphi} \, d\mu_j\}$, we find that $\lambda < 0$ and for any $\mu \in \mathcal{K}(f)$ there is a number i such that $\mu \in O(\mu_i)$ and

$$\frac{1}{n_i} \int_K \varphi_{n_i} \, d\mu_i < \lambda.$$

Using the subadditivity of the sequence of functions φ_n repeatedly, we find that for any $k > 0$ and $x \in K$,

$$\varphi_{kn_j}(x) \leq \sum_{i=0}^{k-1} \varphi_{n_j}(f^{in_j}(x)).$$

Since the measure μ is invariant, we have that

$$\frac{1}{kn_j} \int_K \varphi_{kn_j} \, d\mu \leq \frac{1}{k} \sum_{i=0}^{k-1} \frac{1}{n_j} \int_K \varphi_{n_j}(f^{in_j}(x)) \, d\mu$$

$$= \frac{1}{k} \sum_{i=0}^{k-1} \frac{1}{n_j} \int_K \varphi_{n_j}(x) \, d\mu < \lambda.$$

Therefore, setting $L = n_1 \cdots n_\ell$, we obtain the desired inequality (15.4). $\square$

We proceed with the proof of the theorem. Since φ_L is a continuous function on K, by statement (2) of Lemma 15.12, there exists $\bar{N} > 0$ such that for all $n \geq \bar{N}$ and $x \in K$,

$$\frac{1}{n} \sum_{i=0}^{n-1} \frac{\varphi_L(f^i(x))}{L} < \lambda.$$

Using subadditivity of the sequence of functions $\varphi_n(x)$, we have that for any $x \in K$ and $k > 0$,

$$\varphi_{kL}(x) \leq \sum_{i=0}^{k-1} \varphi_L(f^{iL}(x)).$$

15.2. The Anosov rigidity phenomenon. II

It follows that for any $0 \leq j < L$,

$$\varphi_{kL}(x) \leq \varphi_j(x) + \sum_{i=0}^{k-2} \varphi_L(f^{Li+j}(x)) + \varphi_{L-j}(f^{L(k-1)+j}(x)).$$

Summing over $j = 0, \ldots, L-1$ and dividing by L, we obtain that

$$\varphi_{kL}(x) \leq \frac{1}{L}\sum_{j=0}^{L-1}\sum_{i=0}^{k-2} \varphi_L(f^{Li+j}(x)) + \frac{1}{L}\sum_{j=0}^{L-1}[\varphi_j(x) + \varphi_{L-j}(f^{L(k-1)+j}(x))]. \tag{15.5}$$

Setting $C_1 = \max_{1 \leq i \leq L} \max_{x \in K} \varphi_i(x)$, we obtain from (15.5) that

$$\varphi_{kL}(x) \leq \sum_{j=0}^{(k-1)L-1} \frac{\varphi_L(f^j(x))}{L} + 2C_1,$$

and hence, using Lemma 15.14, we find that for all k for which $L(k-1) \geq \bar{N}$,

$$\varphi_{kL}(x) \leq L(k-1)\lambda + 2C_1. \tag{15.6}$$

Fix $n \geq \bar{N} + 2L$ and write it in the form $n = kL + j$ where $0 \leq j < L$. We have that $(k-1)L = n - L + j > \bar{N}$. Again using subadditivity of the sequence of functions $\varphi_n(x)$, we have that for $x \in M$,

$$\varphi_n(x) \leq \varphi_{kL}(x) + \varphi_j(f^{kL}(x)).$$

Therefore, by (15.6), we have that

$$\varphi_n(x) \leq L(k-1)\lambda + 3C_1.$$

Since $L(k-1) < n$, we conclude that

$$\frac{1}{n}\varphi_n(x) \leq \lambda + \frac{3}{n}C_1.$$

Hence, setting $P = \max\{\bar{N} + 2L, \frac{6C_1}{-\lambda}\}$, we obtain that $\varphi_n(x) \leq \frac{\lambda}{2}$ for every $x \in K$ and every $n \geq P$. This implies that for all $x \in K$ and all $n > 0$

$$\|d_x f^{-n}|E(x)\| \leq c e^{-\chi n},$$

where

$$\chi = -\frac{\lambda}{2} > 0 \quad \text{and} \quad c = \max_{1 \leq i \leq P-1}\{\|d_x f^{-i}|E(x)\|, 1\} > 0.$$

The desired result follows. $\square$

As an immediate corollary of Theorem 15.11 one obtains a different version of the Anosov rigidity phenomenon.

Theorem 15.15. *Let $K \subset M$ be a compact f-invariant subset admitting two continuous distributions $E_1(x)$ and $E_2(x)$ on K such that $T_xM = E_1(x) \oplus E_2(x)$ and the relations*

$$\liminf_{n\to\infty} \frac{1}{n} \log \|d_x f^n | E_1(x)\| > 0, \quad \limsup_{n\to\infty} \frac{1}{n} \log \|d_x f^{-n} | E_2(x)\| < 0$$

hold for every x in a set A of total measure. Then K is a uniformly hyperbolic set for f. In particular, if $K = M$, then f is an Anosov diffeomorphism.

Theorem 15.11 is a slight modification of the result in [**36**]. One can construct an example that illustrates that the continuity requirement for the distribution $E(x)$ is essential for this theorem to hold true (see [**38**]).

Chapter 16

C^1 Pathological Behavior: Pugh's Example

We illustrate that the requirement in the Stable Manifold Theorem (see Theorem 7.1) that the diffeomorphism f is of class $C^{1+\alpha}$ for some $\alpha > 0$ is crucial. Namely, we outline a construction due to Pugh [94] of a nonuniformly hyperbolic C^1 diffeomorphism (which is not $C^{1+\alpha}$ for any $\alpha > 0$) of a four-dimensional manifold with the following property: there exists no manifold tangent to $E^s(x)$ for which (7.1) holds on some open neighborhood of x (see [13]).

Consider the sphere S^2 and denote by S_0^2 its equator and by S_-^2 and S_+^2 the southern and northern hemispheres, respectively. Let $\rho_-\colon \mathbb{R}^2 \to S_-^2$ and $\rho_+\colon \mathbb{R}^2 \to S_+^2$ be the central projections with the south pole at the origin.[1] Clearly, $\rho_+ = i \circ \rho_-$ where i is the antipodal map of S^2. Any map $f\colon \mathbb{R}^2 \to \mathbb{R}^2$ gives rise to two maps $f_\pm \colon S_\pm^2 \to S_\pm^2$ such that the diagrams

$$\begin{array}{ccc} \mathbb{R}^2 & \xrightarrow{f} & \mathbb{R}^2 \\ \rho_\pm \downarrow & & \downarrow \rho_\pm \\ S_\pm^2 & \xrightarrow{f_\pm} & S_\pm^2 \end{array}$$

---

[1] The central projection associates to every point A on the plane (that is tangent to the sphere at the south pole) the point B on the southern (respectively, northern) hemisphere that is the point of intersection of the sphere with the line passing through the center of the sphere and the point A.

are commutative, i.e., $f_\pm(x) = (\rho_\pm \circ f \circ \rho_\pm^{-1})(x)$. Thus we obtain a map $\rho_- f \cup \rho_+ f \colon S^2 \setminus S_0^2 \to S^2 \setminus S_0^2$ defined as $(\rho_- f \cup \rho_+ f)(x) = f_\pm(x)$ for $x \in S_\pm^2$. We wish to choose a map f in such a way that it can be extended to a map $\rho_\sharp f$ which is well-defined on the whole sphere S^2.

To this end, we need the following statement. Its proof while somewhat straightforward is technically involved and is omitted. We refer the reader to the proof of Lemma 4 in [**94**] for a complete argument.

Proposition 16.1. *Let $A = \begin{pmatrix} a & c \\ 0 & b \end{pmatrix}$ with $ab \neq 0$, and let $h \colon \mathbb{R} \to \mathbb{R}$ be a C^1 function with compact support. Then the map*

$$f(x, y) = (ax + cy + h(y), by)$$

lifts to a unique continuous map $\rho_\sharp f$ of S^2, which agrees with $(\rho_- f \cup \rho_+ f)(x)$ on $S^2 \setminus S_0^2$. Moreover, $\rho_\sharp f$ is a C^1 diffeomorphism whose values and derivatives at the equator S_0^2 are the same as those of $\rho_\sharp A$.

We proceed with the construction of Pugh's example. Let $g \colon (0, \infty) \to (0, \infty)$ be a smooth function such that

$$g(u) = \frac{u}{\log(1/u)} \quad \text{for } 0 < u < 1/e,$$

$g'(u) > 1$ for $u \geq 1/e$, and $g'(u) = c > 1$ is a constant for $u \geq 1$. We extend g to $\mathbb{R}$ by setting

$$g_o(u) = \begin{cases} g(u), & u > 0, \\ 0, & u = 0, \\ -g(u), & u < 0. \end{cases} \quad (16.1)$$

We have that $g_o'(u) = c > 1$ provided $|u| \geq 1$. Choose constants a and b such that $0 < a < ab < 1 < b$ and consider the maps

$$f_\pm(x, y) = (ax \pm g_o(y), by).$$

Since the function $h(y) = g_o(y) - c(y)$ has compact support, the maps $f_\pm$ satisfy the hypotheses of Proposition 16.1 and, hence, can be lifted to S^2 as $\rho_\sharp f_\pm$.

We divide S^2 into two hemispheres $H_\pm$ along the x-axis longitude L_x, that is, $H_\pm$ is the hemisphere containing the quarter sphere

$$\rho_- \{(x, y) \in \mathbb{R}^2 : \pm y > 0\}.$$

Define a map $F_S \colon S^2 \to S^2$ by

$$F_S = \begin{cases} \rho_\sharp f_+(z), & z \in H_+, \\ \rho_\sharp f_-(z), & z \in H_-. \end{cases} \quad (16.2)$$

Clearly, F_S is a C^1 diffeomorphism.

16. C^1 Pathological Behavior: Pugh's Example

Let $h\colon M \to M$ be a diffeomorphism of a compact surface M having a hyperbolic invariant set Λ on which h is topologically conjugate to the full shift on two symbols and let $T_\Lambda M = E^s \oplus E^u$ be the invariant splitting into one-dimensional stable and unstable subspaces. Then there are numbers $0 < \lambda < 1 < \mu$ such that for every $x \in \Lambda$,

$$\|d_x h v\| \leq \lambda \quad \text{for every } v \in E^s(x), \|v\| = 1,$$

$$\|d_x h v\| \geq \mu \quad \text{for every } v \in E^u(x), \|v\| = 1.$$

One can choose the map h such that the numbers λ and μ satisfy

$$\lambda < \min_{z \in S^2} \min_{u \in T_z S^2, \|u\|=1} \|d_z F_S u\|, \quad \mu > \max_{z \in S^2} \max_{u \in T_z S^2, \|u\|=1} \|d_z F_S u\|. \quad (16.3)$$

Let Λ_i, $i = 0, 1$, be the ith "cylinder", i.e., the compact set of points in Λ that corresponds to sequences of symbols with i in the initial position. Choose a smooth bump function $\mu\colon M \to [0, \pi/2]$ such that $\Lambda_0 = \mu^{-1}(0) \cap \Lambda$, $\Lambda_1 = \mu^{-1}(\pi/2) \cap \Lambda$, and $\mu^{-1}(\{0, \pi/2\})$ is a neighborhood of Λ. Now let R_θ be the rotation of S^2 by angle θ that fixes the poles.

We are now ready to define the desired map F. It acts on the four-dimensional manifold $M \times S^2$ by the formula

$$F(w, z) = \bigl(h(w), (R_{\mu(w)} \circ F_S \circ R_{-\mu(w)})(z)\bigr),$$

where R_θ is the rotation of S^2 by angle θ that fixes the poles. It is easy to see that F is a C^1 diffeomorphism and it leaves invariant the foliation $\mathcal{F}$ by 2-spheres $\{w\} \times S^2$, $w \in M$. Furthermore, by (16.3), F is normally hyperbolic to $\mathcal{F}$. This means that for every $w \in M$, $z \in S^2$ we have the splitting

$$T_{(w,z)}(M \times S^2) = E^s(w, z) \oplus T_{(w,z)}\mathcal{F} \oplus E^u(w, z)$$

into stable $E^s(w, z)$, unstable $E^u(w, z)$, and central $T_{(w,z)}\mathcal{F}$ subspaces which are invariant under $d_{(w,z)}F$. Moreover, dF contracts vectors in $E^s(w, z)$ and expands vectors in $E^u(w, z)$ with uniform rates, which are strictly bigger than the rates of contraction and expansion along $T_{(w,z)}\mathcal{F}$. The subspaces $E^s(w, z)$ and $E^u(w, z)$ are locally integrable to local stable $V^s(w, z)$ and unstable $V^u(w, z)$ manifolds.

Proposition 16.2. *There is a point $P \in M \times S^2$, which is LP-regular with nonzero Lyapunov exponents for the diffeomorphism F and for which the stable set*

$$V^{sc}(P) = \left\{ x \in M \times S^2 : \lim_{n \to \infty} \frac{1}{n} \log \rho(F^n(x), F^n(P)) < 0 \right\}$$

is not an injectively immersed submanifold tangent to $E^s(P) \oplus T_P \mathcal{F}$.

Proof. We consider another extension of the function g to $\mathbb{R}$,

$$\tilde{g}(u) = \begin{cases} g(u), & u > 0, \\ 0, & u = 0, \\ g(u), & u < 0. \end{cases}$$

Then $\tilde{g}$ is of class C^1 on $\mathbb{R}$ and of class C^∞ on $\mathbb{R} \setminus \{0\}$. Consider the C^1 maps f_S and f_T in $\mathbb{R}^2$ defined by

$$f_S(x, y) = (ax + g(y), by) \quad \text{and} \quad f_T(x, y) = (bx, g(x) + ay).$$

One can easily verify that f_S and f_T are C^1 diffeomorphisms of $\mathbb{R}^2$ onto itself. Moreover, F_S lifts f_S to S^2 (but not as $\rho_- f_S \cup \rho_+ f_S$; in fact the canonical lift $\rho_\sharp f_S$ fails to be C^1 at the equator). This follows from the fact that $f_\pm(x, y) = f_S(x, y)$ for $\pm y \geq 0$ (see (16.1)) since F_S is constructed from lifts $\rho_\sharp f_\pm$, respectively, of $f_\pm$ (see (16.2)).

Set $L_k = k(k+1)/2$ and for each $n \in \mathbb{N}$,

$$f_n = \begin{cases} f_S, & n \in \mathcal{S}, \\ f_T, & n \in \mathcal{T}, \end{cases} \qquad (16.4)$$

where

$$\mathcal{S} = \{n \in \mathbb{N} : L_{k-1} < n \leq L_k \text{ for some odd } k\},$$
$$\mathcal{T} = \{n \in \mathbb{N} : L_{k-1} < n \leq L_k \text{ for some even } k\}.$$

Notice that the sets $\mathcal{S}$ and $\mathcal{T}$ partition the set $\mathbb{N}$ of positive integers.

Since $h|\Lambda$ is topologically conjugate to the full 2-shift, there exists $p \in \Lambda$ such that $h^n(p) \in \Lambda_0$ whenever $f_n = f_S$ and $h^n(p) \in \Lambda_1$ whenever $f_n = f_T$. Let $P = (p, z_0)$ where z_0 is the south pole of S^2. Since z_0 is fixed under both F_S and R_θ, the F-orbit of P is $\{(h^n(p), z_0)\}$. We have that

$$d_{(h^n(p), z_0)} F = \begin{pmatrix} d_{h^n(p)} h & 0 \\ 0 & d_0 f_n \end{pmatrix}. \qquad (16.5)$$

Indeed, since μ is constant near Λ, F_S is a lift of f_S, and $f_T = R_{\pi/2} \circ f_S \circ R_{-\pi/2}$, one can show that the diagram

$$\begin{array}{ccc} \{h^n(p)\} \times S^2 & \xrightarrow{F} & \{h^n(p)\} \times S^2 \\ \downarrow & & \downarrow \\ S^2 & \xrightarrow{F_n} & S^2 \end{array} \qquad (16.6)$$

is commutative, where $F_n = F_S$ when $f_n = f_S$ and $F_n = F_T$ when $f_n = f_T$. It follows from (16.5) that P has one positive Lyapunov exponent corresponding to $dh|E^u$ and three negative Lyapunov exponents: one corresponds to $dh|E^s$ and the other two, being $\frac{1}{2}\log(ab)$, correspond to the products of the matrices $d_0 f_n$. Let $E^-(P)$ denote the space of vectors with negative

Lyapunov exponents. Taking into account that $d_P F^n | E_P^s$ is diagonal preserving the decomposition $E^s \oplus (x\text{-axis}) \oplus (y\text{-axis})$, we conclude that the F-orbit of P is LP-regular.

Since F is normally hyperbolic to $\mathcal{F}$, a point $(w, z) \in M \times S^2$ satisfies
$$\lim_{n \to \infty} \frac{1}{n} \log \rho(F^n(w, z), F^n(P)) < 0$$
if and only if (w, z) lies in the strongly stable manifold of some point $(p, z') \in V^{sc}(P) \cap (\{p\} \times S^2)$. This means that
$$V^{sc}(P) = V^s(V^{sc}(P) \cap (\{p\} \times S^2)).$$
By (16.6), the set $V^{sc}(P) \cap (\{p\} \times S^2)$ coincides with
$$\left\{ z \in S^2 : \lim_{n \to \infty} \frac{1}{n} \log \|(f_n \circ \cdots \circ f_0)(z)\| < 0 \right\}.$$

To complete the proof, we need the following statement. Its proof exploits the choice of maps f_n (see (16.4)) and is technically involved; we refer the reader to the proof of Theorem 1 in [**94**].

Lemma 16.3. *If $z = (x, y) \in \mathbb{R}^2$ is such that $x > 0$ and $y > 0$, then $\|f^n(z)\| \to \infty$ as $n \to \infty$.*

It follows from the lemma that $V^{sc}(P)$ is contained in $V^s(\{p\} \times S^2)$ but does not include any neighborhood of the point P, since it misses the entire first quadrant. Hence it cannot be an immersed manifold tangent to $E^-(P)$. The desired result follows. □

Despite Pugh's construction, there are some affirmative results due to Barreira and Valls [**15**] on constructing stable manifolds for diffeomorphisms with nonzero Lyapunov exponents of class C^1.

Bibliography

[1] N. Alansari, *Hyperbolic Lyapunov–Perron regular points and smooth invariant measures*, Ergodic Theory Dynam. Systems, to appear.

[2] N. Alansari, *Ergodic properties of measures with local product structure*, preprint.

[3] J. F. Alves, V. Araújo, and B. Saussol, *On the uniform hyperbolicity of some nonuniformly hyperbolic systems*, Proc. Amer. Math. Soc. **131** (2003), no. 4, 1303–1309, DOI 10.1090/S0002-9939-02-06857-0. MR1948124

[4] D. Anosov, *Tangential fields of transversal foliations in Y-systems*, Math. Notes **2** (1967), 818–823.

[5] D. Anosov, *Geodesic flows on closed Riemann manifolds with negative curvature*, Proc. Steklov Inst. Math. **90** (1969), 1–235.

[6] D. Anosov and Ya. Sinai, *Certain smooth ergodic systems*, Russian Math. Surveys **22** (1967), 103–167.

[7] L. Arnold, *Random dynamical systems*, Springer Monographs in Mathematics, Springer-Verlag, Berlin, 1998, DOI 10.1007/978-3-662-12878-7. MR1723992

[8] A. Avila, S. Crovisier, and A. Wilkinson, C^1 *density of stable ergodicity*, Adv. Math. **379** (2021), Paper No. 107496, 68, DOI 10.1016/j.aim.2020.107496. MR4198639

[9] A. Avila and R. Krikorian, *Reducibility or nonuniform hyperbolicity for quasiperiodic Schrödinger cocycles*, Ann. of Math. (2) **164** (2006), no. 3, 911–940, DOI 10.4007/annals.2006.164.911. MR2259248

[10] L. Barreira, *Lyapunov exponents*, Birkhäuser/Springer, Cham, 2017, DOI 10.1007/978-3-319-71261-1. MR3752157

[11] L. Barreira and Ya. B. Pesin, *Lyapunov exponents and smooth ergodic theory*, University Lecture Series, vol. 23, American Mathematical Society, Providence, RI, 2002, DOI 10.1090/ulect/023. MR1862379

[12] L. Barreira and Ya. Pesin, *Smooth ergodic theory and nonuniformly hyperbolic dynamics*, with an appendix by Omri Sarig, Handbook of dynamical systems, vol. 1B, Elsevier B. V., Amsterdam, 2006, pp. 57–263, DOI 10.1016/S1874-575X(06)80027-5. MR2186242

[13] L. Barreira and Ya. Pesin, *Nonuniform hyperbolicity: Dynamics of systems with nonzero Lyapunov exponents*, Encyclopedia of Mathematics and its Applications, vol. 115, Cambridge University Press, Cambridge, 2007, DOI 10.1017/CBO9781107326026. MR2348606

[14] L. Barreira, Ya. Pesin, and J. Schmeling, *Dimension and product structure of hyperbolic measures*, Ann. of Math. (2) **149** (1999), no. 3, 755–783, DOI 10.2307/121072. MR1709302

[15] L. Barreira and C. Valls, *Smoothness of invariant manifolds for nonautonomous equations*, Comm. Math. Phys. **259** (2005), no. 3, 639–677, DOI 10.1007/s00220-005-1380-z. MR2174420

[16] L. Barreira and C. Valls, *Stability of nonautonomous differential equations*, Lecture Notes in Mathematics, vol. 1926, Springer, Berlin, 2008, DOI 10.1007/978-3-540-74775-8. MR2368551

[17] S. Ben Ovadia, *Symbolic dynamics for non-uniformly hyperbolic diffeomorphisms of compact smooth manifolds*, J. Mod. Dyn. **13** (2018), 43–113, DOI 10.3934/jmd.2018013. MR3918259

[18] S. Ben Ovadia, *Hyperbolic SRB measures and the leaf condition*, Comm. Math. Phys. **387** (2021), no. 3, 1353–1404, DOI 10.1007/s00220-021-04208-6. MR4324380

[19] G. Birkhoff and G.-C. Rota, *Ordinary differential equations*, 4th ed., John Wiley & Sons, Inc., New York, 1989. MR972977

[20] J. Bochi, *Genericity of zero Lyapunov exponents*, Ergodic Theory Dynam. Systems **22** (2002), no. 6, 1667–1696, DOI 10.1017/S0143385702001165. MR1944399

[21] J. Bochi and M. Viana, *The Lyapunov exponents of generic volume-preserving and symplectic maps*, Ann. of Math. (2) **161** (2005), no. 3, 1423–1485, DOI 10.4007/annals.2005.161.1423. MR2180404

[22] C. Bonatti and L. J. Díaz, *Persistent nonhyperbolic transitive diffeomorphisms*, Ann. of Math. (2) **143** (1996), no. 2, 357–396, DOI 10.2307/2118647. MR1381990

[23] C. Bonatti, L. J. Díaz, and A. Gorodetski, *Non-hyperbolic ergodic measures with large support*, Nonlinearity **23** (2010), no. 3, 687–705, DOI 10.1088/0951-7715/23/3/015. MR2593915

[24] C. Bonatti, L. J. Díaz, and M. Viana, *Dynamics beyond uniform hyperbolicity: A global geometric and probabilistic perspective; Mathematical Physics, III*, Encyclopaedia of Mathematical Sciences, vol. 102, Springer-Verlag, Berlin, 2005. MR2105774

[25] J. Bourgain, *Positivity and continuity of the Lyapounov exponent for shifts on $\mathbb{T}^d$ with arbitrary frequency vector and real analytic potential*, J. Anal. Math. **96** (2005), 313–355, DOI 10.1007/BF02787834. MR2177191

[26] J. Bourgain and S. Jitomirskaya, *Continuity of the Lyapunov exponent for quasiperiodic operators with analytic potential*, dedicated to David Ruelle and Yasha Sinai on the occasion of their 65th birthdays, J. Statist. Phys. **108** (2002), no. 5-6, 1203–1218, DOI 10.1023/A:1019751801035. MR1933451

[27] R. Bowen, *Markov partitions for Axiom A diffeomorphisms*, Amer. J. Math. **92** (1970), 725–747, DOI 10.2307/2373370. MR277003

[28] M. Brin, *Hölder continuity of invariant distributions*, Smooth Ergodic Theory and its Applications (Seattle, WA, 1999), edited by A. Katok, R. de la Llave, Ya. Pesin, and H. Weiss, Proceedings of Symposia in Pure Mathematics, vol. 69, American Mathematical Society, Providence, RI, 2001, pp. 99–101.

[29] M. Brin and G. Stuck, *Introduction to dynamical systems*, Cambridge University Press, Cambridge, 2002, DOI 10.1017/CBO9780511755316. MR1963683

[30] D. Burguet, *SRB measures for smooth surface diffeomorphisms*, preprint.

[31] K. Burns, D. Dolgopyat, and Ya. Pesin, *Partial hyperbolicity, Lyapunov exponents and stable ergodicity*, dedicated to David Ruelle and Yasha Sinai on the occasion of their 65th birthdays, J. Statist. Phys. **108** (2002), no. 5-6, 927–942, DOI 10.1023/A:1019779128351. MR1933439

[32] K. Burns and A. Wilkinson, *Stable ergodicity of skew products* (English, with English and French summaries), Ann. Sci. École Norm. Sup. (4) **32** (1999), no. 6, 859–889, DOI 10.1016/S0012-9593(00)87721-6. MR1717580

[33] K. Burns and A. Wilkinson, *On the ergodicity of partially hyperbolic systems*, Ann. of Math. (2) **171** (2010), no. 1, 451–489, DOI 10.4007/annals.2010.171.451. MR2630044

[34] J. Buzzi, S. Crovisier, and O. Sarig, *Another proof of Burguet's existence theorem for SRB measures of C^∞ surface diffeomorphisms*, preprint.

[35] D. Bylov, R. Vinograd, D. Grobman, and V. Nemyckii, *Theory of Lyapunov Exponents and its Application to Problems of Stability*, Izdat. "Nauka", Moscow, 1966, in Russian.

[36] Y. Cao, *Non-zero Lyapunov exponents and uniform hyperbolicity*, Nonlinearity **16** (2003), no. 4, 1473–1479, DOI 10.1088/0951-7715/16/4/316. MR1986306

[37] Y. Cao, S. Luzzatto, and I. Rios, *A minimum principle for Lyapunov exponents and a higher-dimensional version of a theorem of Mañé*, Qual. Theory Dyn. Syst. **5** (2004), no. 2, 261–273, DOI 10.1007/BF02972681. MR2275440

[38] Y. Cao, S. Luzzatto, and I. Rios, *Some non-hyperbolic systems with strictly non-zero Lyapunov exponents for all invariant measures: horseshoes with internal tangencies*, Discrete Contin. Dyn. Syst. **15** (2006), no. 1, 61–71, DOI 10.3934/dcds.2006.15.61. MR2191385

[39] V. Climenhaga, D. Dolgopyat, and Ya. Pesin, *Non-stationary non-uniform hyperbolicity: SRB measures for dissipative maps*, Comm. Math. Phys. **346** (2016), no. 2, 553–602, DOI 10.1007/s00220-016-2710-z. MR3535895

[40] I. P. Cornfeld, S. V. Fomin, and Ya. G. Sinaĭ, *Ergodic theory*, translated from the Russian by A. B. Sosinskiĭ, Grundlehren der mathematischen Wissenschaften [Fundamental Principles of Mathematical Sciences], vol. 245, Springer-Verlag, New York, 1982, DOI 10.1007/978-1-4615-6927-5. MR832433

[41] D. Dolgopyat, *On dynamics of mostly contracting diffeomorphisms*, Comm. Math. Phys. **213** (2000), no. 1, 181–201, DOI 10.1007/s002200000238. MR1782146

[42] D. Dolgopyat, H. Hu, and Ya. Pesin, *An example of a smooth hyperbolic measure with countably many ergodic components*, Smooth Ergodic Theory and its Applications (Seattle, WA, 1999), edited by A. Katok, R. de la Llave, Ya. Pesin, and H. Weiss, Proceedings of Symposia in Pure Mathematics, vol. 69, American Mathematical Society, Providence, RI, 2001, pp. 102–115.

[43] D. Dolgopyat and Ya. Pesin, *Every compact manifold carries a completely hyperbolic diffeomorphism*, Ergodic Theory Dynam. Systems **22** (2002), no. 2, 409–435, DOI 10.1017/S0143385702000202. MR1898798

[44] P. Eberlein, *Geodesic flows on negatively curved manifolds. I*, Ann. of Math. (2) **95** (1972), 492–510, DOI 10.2307/1970869. MR310926

[45] P. Eberlein, *When is a geodesic flow of Anosov type? I,II*, J. Differential Geometry **8** (1973), 437–463; ibid. **8** (1973), 565–577. MR380891

[46] P. Eberlein, *Geodesic flows in manifolds of nonpositive curvature*, Smooth Ergodic Theory and its Applications (Seattle, WA, 1999), edited by A. Katok, R. de la Llave, Ya. Pesin, and H. Weiss, Proceedings of Symposia in Pure Mathematics, vol. 69, Amer. Math. Soc., Providence, RI, 2001, pp. 525–571, DOI 10.1090/pspum/069/1858545. MR1858545

[47] L. H. Eliasson, *Reducibility and point spectrum for linear quasi-periodic skew-products*, Proceedings of the International Congress of Mathematicians, Vol. II (Berlin, 1998), Doc. Math. **Extra Vol. II** (1998), 779–787. MR1648125

[48] D. B. A. Epstein, *Foliations with all leaves compact* (English, with French summary), Ann. Inst. Fourier (Grenoble) **26** (1976), no. 1, viii, 265–282. MR420652

[49] A. S. Gorodetskiĭ, *Regularity of central leaves of partially hyperbolic sets and applications* (Russian, with Russian summary), Izv. Ross. Akad. Nauk Ser. Mat. **70** (2006), no. 6, 19–44, DOI 10.1070/IM2006v070n06ABEH002340; English transl., Izv. Math. **70** (2006), no. 6, 1093–1116. MR2285025

[50] R. E. Greene and K. Shiohama, *Diffeomorphisms and volume-preserving embeddings of noncompact manifolds*, Trans. Amer. Math. Soc. **255** (1979), 403–414, DOI 10.2307/1998183. MR542888

[51] B. M. Gurevič and V. I. Oseledec, *Gibbs distributions, and the dissipativity of C-diffeomorphisms* (Russian), Dokl. Akad. Nauk SSSR **209** (1973), 1021–1023. MR0320274

[52] B. Hasselblatt, Ya. Pesin, and J. Schmeling, *Pointwise hyperbolicity implies uniform hyperbolicity*, Discrete Contin. Dyn. Syst. **34** (2014), no. 7, 2819–2827, DOI 10.3934/dcds.2014.34.2819. MR3177662

[53] G. A. Hedlund, *The dynamics of geodesic flows*, Bull. Amer. Math. Soc. **45** (1939), no. 4, 241–260, DOI 10.1090/S0002-9904-1939-06945-0. MR1563961

[54] M. Hirayama and Ya. Pesin, *Non-absolutely continuous foliations*, Israel J. Math. **160** (2007), 173–187, DOI 10.1007/s11856-007-0060-4. MR2342495

[55] M. W. Hirsch, C. C. Pugh, and M. Shub, *Invariant manifolds*, Lecture Notes in Mathematics, vol. 583, Springer-Verlag, Berlin-New York, 1977. MR0501173

[56] E. Hopf, *Statistik der geodätischen Linien in Mannigfaltigkeiten negativer Krümmung* (German), Ber. Verh. Sächs. Akad. Wiss. Leipzig Math.-Phys. Kl. **91** (1939), 261–304. MR1464

[57] H. Hu, Ya. Pesin, and A. Talitskaya, *Every compact manifold carries a hyperbolic Bernoulli flow*, Modern dynamical systems and applications, Cambridge Univ. Press, Cambridge, 2004, pp. 347–358. MR2093309

[58] M. Jakobson and G. Świątek, *One-dimensional maps*, Handbook of dynamical systems, vol. 1A, North-Holland, Amsterdam, 2002, pp. 599–664, DOI 10.1016/S1874-575X(02)80010-8. MR1928525

[59] R. A. Johnson, *The recurrent Hill's equation*, J. Differential Equations **46** (1982), no. 2, 165–193, DOI 10.1016/0022-0396(82)90114-0. MR675906

[60] R. A. Johnson and G. R. Sell, *Smoothness of spectral subbundles and reducibility of quasiperiodic linear differential systems*, J. Differential Equations **41** (1981), no. 2, 262–288, DOI 10.1016/0022-0396(81)90062-0. MR630994

[61] A. Katok, *Bernoulli diffeomorphisms on surfaces*, Ann. of Math. (2) **110** (1979), no. 3, 529–547, DOI 10.2307/1971237. MR554383

[62] A. Katok, *Lyapunov exponents, entropy and periodic orbits for diffeomorphisms*, Inst. Hautes Études Sci. Publ. Math. **51** (1980), 137–173. MR573822

[63] A. Katok and K. Burns, *Infinitesimal Lyapunov functions, invariant cone families and stochastic properties of smooth dynamical systems*, Ergodic Theory Dynam. Systems **14** (1994), 757–785.

[64] A. Katok and B. Hasselblatt, *Introduction to the modern theory of dynamical systems*, with a supplementary chapter by Katok and Leonardo Mendoza, Encyclopedia of Mathematics and its Applications, vol. 54, Cambridge University Press, Cambridge, 1995, DOI 10.1017/CBO9780511809187. MR1326374

[65] A. Katok and L. Mendoza, *Dynamical systems with nonuniformly hyperbolic behavior*, supplement in *Introduction to the modern theory of dynamical systems* by A. Katok and B. Hasselblatt, Cambridge University Press, Cambridge, 1995.

[66] A. Katok and V. Niţică, *Rigidity in higher rank abelian group actions. vol. I, Introduction and cocycle problem*, Cambridge Tracts in Mathematics, vol. 185, Cambridge University Press, Cambridge, 2011, DOI 10.1017/CBO9780511803550. MR2798364

[67] A. Katok, J.-M. Strelcyn, F. Ledrappier, and F. Przytycki, *Invariant manifolds, entropy and billiards; smooth maps with singularities*, Lecture Notes in Mathematics, vol. 1222, Springer-Verlag, Berlin, 1986, DOI 10.1007/BFb0099031. MR872698

[68] Yu. Kifer, *Ergodic theory of random transformations*, Progress in Probability and Statistics, vol. 10, Birkhäuser Boston, Inc., Boston, MA, 1986, DOI 10.1007/978-1-4684-9175-3. MR884892

[69] Yu. Kifer and P.-D. Liu, *Random dynamics*, Handbook of dynamical systems, vol. 1B, Elsevier B. V., Amsterdam, 2006, pp. 379–499, DOI 10.1016/S1874-575X(06)80030-5. MR2186245

[70] J. F. C. Kingman, *Subadditive processes*, École d'Été de Probabilités de Saint-Flour, V–1975, Lecture Notes in Math., vol. 539, Springer, Berlin, 1976, pp. 167–223. MR0438477

[71] G. Knieper, *Hyperbolic dynamics and Riemannian geometry*, Handbook of dynamical systems, vol. 1A, North-Holland, Amsterdam, 2002, pp. 453–545, DOI 10.1016/S1874-575X(02)80008-X. MR1928523

[72] R. Krikorian, *Réductibilité des systèmes produits-croisés à valeurs dans des groupes compacts* (French, with English and French summaries), Astérisque **259** (1999), vi+216. MR1732061

[73] F. Ledrappier and L.-S. Young, *The metric entropy of diffeomorphisms. I. Characterization of measures satisfying Pesin's entropy formula*, Ann. of Math. (2) **122** (1985), no. 3, 509–539, DOI 10.2307/1971328. MR819556

[74] F. Ledrappier and L.-S. Young, *The metric entropy of diffeomorphisms. II. Relations between entropy, exponents and dimension*, Ann. of Math. (2) **122** (1985), no. 3, 540–574, DOI 10.2307/1971329. MR819557

[75] P.-D. Liu and M. Qian, *Smooth ergodic theory of random dynamical systems*, Lecture Notes in Mathematics, vol. 1606, Springer-Verlag, Berlin, 1995, DOI 10.1007/BFb0094308. MR1369243

[76] S. Luzzatto, *Stochastic-like behaviour in nonuniformly expanding maps*, Handbook of dynamical systems, vol. 1B, Elsevier B. V., Amsterdam, 2006, pp. 265–326, DOI 10.1016/S1874-575X(06)80028-7. MR2186243

[77] S. Luzzatto and M. Viana, *Parameter exclusions in Hénon-like systems*, Russian Math. Surveys **58** (2003), 1053–1092.

[78] A. M. Lyapunov, *The general problem of the stability of motion*, translated from Edouard Davaux's French translation (1907) of the 1892 Russian original and edited by A. T. Fuller; with an introduction and preface by Fuller, a biography of Lyapunov by V. I. Smirnov, and a bibliography of Lyapunov's works compiled by J. F. Barrett; Lyapunov centenary issue; reprint of Internat. J. Control **55** (1992), no. 3 [MR1154209 (93e:01035)]; with a foreword by Ian Stewart, Taylor & Francis Group, London, 1992. MR1229075

[79] I. G. Malkin, *A theorem on stability in the first approximation* (Russian), Doklady Akad. Nauk SSSR (N.S.) **76** (1951), 783–784. MR0041304

[80] R. Mañé, *Quasi-Anosov diffeomorphisms and hyperbolic manifolds*, Trans. Amer. Math. Soc. **229** (1977), 351–370, DOI 10.2307/1998515. MR482849

[81] R. Mañé, *A proof of Pesin's formula*, Ergodic Theory Dynam. Systems **1** (1981), 95–102. DOI 10.1017/S0143385700001188. MR627789. Errata in **3** (1983), 159–160. DOI 10.1017/S0143385700001863. MR743033

[82] J. Milnor, *Fubini foiled: Katok's paradoxical example in measure theory*, Math. Intelligencer **19** (1997), no. 2, 30–32, DOI 10.1007/BF03024428. MR1457445

[83] M. Morse, *Instability and transitivity*, J. Math. Pures Appl. **40** (1935), 49–71.

[84] S. E. Newhouse, *Hyperbolic limit sets*, Trans. Amer. Math. Soc. **167** (1972), 125–150, DOI 10.2307/1996131. MR295388

[85] V. Oseledets, *A multiplicative ergodic theorem. Liapunov characteristic numbers for dynamical systems*, Trans. Moscow Math. Soc. **19** (1968), 197–221.

[86] O. Perron, *Die Ordnungszahlen linearer Differentialgleichungssysteme* (German), Math. Z. **31** (1930), no. 1, 748–766, DOI 10.1007/BF01246445. MR1545146

[87] Ya. Pesin, *An example of a nonergodic flow with nonzero characteristic exponents*, Func. Anal. and its Appl. **8** (1974), 263–264.

[88] Ya. Pesin, *Families of invariant manifolds corresponding to nonzero characteristic exponents*, Math. USSR-Izv. **40** (1976), 1261–1305.

[89] Ya. Pesin, *Characteristic Ljapunov exponents, and smooth ergodic theory*, Russian Math. Surveys **32** (1977), 55–114.

[90] Ya. Pesin, *A description of the π-partition of a diffeomorphism with an invariant measure*, Math. Notes **22** (1977), 506–515.

[91] Ya. Pesin, *Geodesic flows on closed Riemannian manifolds without focal points*, Math. USSR-Izv. **11** (1977), 1195–1228.

[92] Ya. Pesin, *Geodesic flows with hyperbolic behaviour of the trajectories and objects connected with them*, Russian Math. Surveys **36** (1981), 1–59.

[93] Y. B. Pesin, *Lectures on partial hyperbolicity and stable ergodicity*, Zurich Lectures in Advanced Mathematics, European Mathematical Society (EMS), Zürich, 2004, DOI 10.4171/003. MR2068774

[94] C. C. Pugh, *The $C^{1+\alpha}$ hypothesis in Pesin theory*, Inst. Hautes Études Sci. Publ. Math. **59** (1984), 143–161. MR743817

[95] C. Pugh and M. Shub, *Ergodic attractors*, Trans. Amer. Math. Soc. **312** (1989), no. 1, 1–54, DOI 10.2307/2001206. MR983869

[96] F. Rodriguez Hertz, M. A. Rodriguez Hertz, and R. Ures, *Accessibility and stable ergodicity for partially hyperbolic diffeomorphisms with 1D-center bundle*, Invent. Math. **172** (2008), no. 2, 353–381, DOI 10.1007/s00222-007-0100-z. MR2390288

[97] F. Rodriguez Hertz, M. A. Rodriguez Hertz, A. Tahzibi, and R. Ures, *New criteria for ergodicity and nonuniform hyperbolicity*, Duke Math. J. **160** (2011), no. 3, 599–629, DOI 10.1215/00127094-1444314. MR2852370

[98] F. Rodriguez Hertz, M. A. Rodriguez Hertz, A. Tahzibi, and R. Ures, *Uniqueness of SRB measures for transitive diffeomorphisms on surfaces*, Comm. Math. Phys. **306** (2011), no. 1, 35–49, DOI 10.1007/s00220-011-1275-0. MR2819418

[99] D. Ruelle, *An inequality for the entropy of differentiable maps*, Bol. Soc. Brasil. Mat. **9** (1978), no. 1, 83–87, DOI 10.1007/BF02584795. MR516310

[100] D. Ruelle, *Analycity properties of the characteristic exponents of random matrix products*, Adv. in Math. **32** (1979), no. 1, 68–80, DOI 10.1016/0001-8708(79)90029-X. MR534172

[101] D. Ruelle and A. Wilkinson, *Absolutely singular dynamical foliations*, Comm. Math. Phys. **219** (2001), no. 3, 481–487, DOI 10.1007/s002200100420. MR1838747

[102] R. J. Sacker and G. R. Sell, *Existence of dichotomies and invariant splittings for linear differential systems. I*, J. Differential Equations **15** (1974), 429–458, DOI 10.1016/0022-0396(74)90067-9. MR341458

[103] V. Sadovskaya, *Linear cocycles over hyperbolic systems*, A Vision for Dynamics in the 21st Century, Cambridge University Press, Cambridge, to appear.

[104] O. M. Sarig, *Symbolic dynamics for surface diffeomorphisms with positive entropy*, J. Amer. Math. Soc. **26** (2013), no. 2, 341–426, DOI 10.1090/S0894-0347-2012-00758-9. MR3011417

[105] M. Shub and A. Wilkinson, *Pathological foliations and removable zero exponents*, Invent. Math. **139** (2000), no. 3, 495–508, DOI 10.1007/s002229900035. MR1738057

[106] Ya. Sinai, *Dynamical systems with countably-multiple Lebesgue spectrum II*, Amer. Math. Soc. Trans. (2) **68** (1966), 34–88.

[107] S. Smale, *Differentiable dynamical systems*, Bull. Amer. Math. Soc. **73** (1967), 747–817, DOI 10.1090/S0002-9904-1967-11798-1. MR228014

[108] A. Tahzibi, *C^1-generic Pesin's entropy formula* (English, with English and French summaries), C. R. Math. Acad. Sci. Paris **335** (2002), no. 12, 1057–1062, DOI 10.1016/S1631-073X(02)02609-2. MR1955588

[109] R. F. Williams, *One-dimensional non-wandering sets*, Topology **6** (1967), 473–487, DOI 10.1016/0040-9383(67)90005-5. MR217808

[110] R. F. Williams, *Expanding attractors*, Inst. Hautes Études Sci. Publ. Math. **43** (1974), 169–203. MR348794

[111] M. Wojtkowski, *Invariant families of cones and Lyapunov exponents*, Ergodic Theory Dynam. Systems **5** (1985), no. 1, 145–161, DOI 10.1017/S0143385700002807. MR782793

Bibliography

[112] W. Wu, *On the ergodicity of geodesic flows on surfaces of nonpositive curvature* (English, with English and French summaries), Ann. Fac. Sci. Toulouse Math. (6) **24** (2015), no. 3, 625–639, DOI 10.5802/afst.1457. MR3403734

[113] L. S. Young, *Dimension, entropy and Lyapunov exponents*, Ergodic Theory Dynam. Systems **2** (1982), no. 1, 109–124, DOI 10.1017/s0143385700009615. MR684248

[114] A. Zelerowicz, *Thermodynamics of some non-uniformly hyperbolic attractors*, Nonlinearity **30** (2017), no. 7, 2612–2646, DOI 10.1088/1361-6544/aa7014. MR3670000

Index

absolute continuity, 188
   theorem, 191
absolutely continuous
   foliation in the strong sense, 188
   foliation in the weak sense, 188
   transformation, 191
accessibility property, 281
   essential –, 282
accessible points, 281
admissible submanifold, 183
Anosov
   diffeomorphism, 4, 6
   dissipative – diffeomorphism, 196
   flow, 9
asymptotic geodesics, 248
asymptotically stable solution, 149
   conditionally –, 158
attractor, 20
automorphism
   Bernoulli –, 8
   hyperbolic toral –, 4

backward
   Lyapunov exponent, 89
   regular Lyapunov exponent, 67, 79
   regular point, 121
   regular sequence of matrices, 79
basin of attraction, 259
basis
   normal –, 48
   ordered –, 48
   subordinate –, 48, 49
Bernoulli automorphism, 8

Besicovich covering lemma, 198
block
   Pesin –, 190, 212
   r-foliation –, 187
   s-foliation –, 190
   u-foliation –, 191

canonical metric, 238
Cauchy matrix, 154
center bunched diffeomorphism, 282
central space, 280
characteristic exponent, 45, 151
cocycle, 81, 82, 93
   derivative –, 82
   induced –, 84
   linear multiplicative –, 82, 93
   power –, 84
   reducible –, 93
   tempered –, 85
   triangular –, 105, 106
cocycles
   cohomologous –, 86
   equivalent –, 86
coefficient
   irregularity –, 51
   Perron –, 51
coherent filtrations, 68, 91
cohomological equation, 87
cohomologous cocycles, 86
cohomology, 85
common refinement, 223
complete
   family of cones, 273

function, 273
hyperbolicity conditions, 130
pair of cones, 273
condition
    complete hyperbolicity –, 130
    partial hyperbolicity –, 140
conditional
    entropy, 223
    measure, 208
cone, 271
    connected –, 277
    negative –, 273
    positive –, 273
    stable –, 27
    unstable –, 27
connected cone, 277
coordinate chart
    foliation –, 217
curve
    global stable –, 5
    global unstable –, 5

density function, 259
derivative cocycle, 82
diffeomorphism
    Anosov –, 4, 6
    center bunched –, 282
    flat –, 36
    hyperbolic –, 291
    structurally stable –, 9
    uniformly partially hyperbolic –, 280
    W-dissipative –, 289
Diophantine condition, 94
    recurrent –, 96
dissipative Anosov diffeomorphism, 196
distribution, 141
    Hölder continuous –, 142
    integrable –, 7
dual
    bases, 51
    Lyapunov exponents, 51
    points, 251
dynamical system
    hyperbolic –, 3
    nonuniformly hyperbolic –, 132

entropy, 224
    conditional –, 223
    formula, 222, 229
    Kolmogorov–Sinai –, 222
    lower local leaf –, 236
    measure-theoretic –, 222

metric –, 222, 224
of geodesic flow, 253
of partition, 223
upper local leaf –, 236
equivalent
    cocycles, 86
    partitions, 223
ergodic
    components, 208
    measure, 8, 10
    properties, 207
ergodicity
    local –, 216
    of smooth hyperbolic measure, 207
essential accessibility property, 282
eventually strict
    family of cones, 277
    Lyapunov function, 274
exponentially stable solution, 149
    conditionally –, 158

family of u-manifolds, 191
filtration, 47
    linear –, 47
    set –, 305
first return
    map, 84
    time, 84
flat
    diffeomorphism, 36
    strip theorem, 249
flow, 149
    Anosov –, 9
    entropy of geodesic –, 253
    geodesic –, 12, 240, 248
    special –, 9
    suspension –, 10
    topologically mixing –, 10
    topologically transitive –, 10
    weakly mixing –, 216
foliation, 7, 217
    absolutely continuous –, 188
    coordinate chart, 7, 217
    Hölder continuous –, 7
    nonabsolutely continuous –, 205
    smooth –, 7
    with finite volume leaves, 289
    with smooth leaves, 7
forward
    Lyapunov exponent, 88
    regular Lyapunov exponent, 67, 78
    regular point, 121

regular sequence of matrices, 78
function
  complete –, 273
  density –, 259
  Lyapunov –, 274
  roof –, 10
  slow-down –, 23
  tempered –, 85, 93, 114, 125

generator, 83
geodesic flow, 12, 240, 248
global
  leaf, 7, 217
  stable curve, 5
  stable manifold, 5, 7, 183, 185
  unstable curve, 5
  unstable manifold, 5, 183, 185
  weakly stable manifold, 185
  weakly unstable manifold, 185
graph transform property, 181

Hamiltonian, 26
Hölder
  constant, 142
  continuous distribution, 142
  continuous foliation, 7
  exponent, 142
holonomy map, 191
homoclinic class, 20
homoclinically related, 19
horocycle, 13, 250
horseshoe, 19
hyperbolic
  attractor, 20
  dynamical system, 3
  measure, 123
  set, 14
  toral automorphism, 4
hyperbolicity
  estimates, 6
  parameters of –, 6, 14, 130

ideal boundary, 11, 249
implicit function theorem, 173
inclination lemma, 181
independent partitions, 223
index, 19
  rational – set, 211
  set, 133
induced
  cocycle, 84
  transformation, 84

inner product
  Lyapunov –, 136
  strong Lyapunov –, 110
  weak Lyapunov –, 136
integrability condition, 87
integrable distribution, 7
invariant
  family of cones, 273
  splitting, 6
irregularity coefficient, 51

Jacobian, 191

Katok map, 22, 25, 27
kernel
  Pesin tempering –, 111
  tempering –, 114
Kolmogorov–Sinai entropy, 222

lamination, 184
leaf
  global –, 7, 217
  local –, 7, 217
  lower – pointwise dimension, 235
  lower local – entropy, 236
  upper – pointwise dimension, 235
  upper local – entropy, 236
  volume, 187
level set, 133
limit solution
  negative –, 243
  positive –, 243
linear
  extension, 82, 83
  filtration, 47
  skew product, 83
local
  ergodicity, 216
  leaf, 7, 217
  manifold theory, 163
  product structure, 258
  s-canonical – transversal, 191
  smooth submanifold, 159, 164
  stable manifold, 7, 148, 185
  transversal, 190
  u-canonical – transversal, 190
  unstable manifold, 7, 178, 185
locally maximal set, 19
lower
  leaf pointwise dimension, 235
  local leaf entropy, 236
  pointwise dimension, 234

LP-regular
  Lyapunov exponent, 69, 79
  point, 91, 121
Lyapunov
  backward – exponent, 89
  change of coordinates, 111
  characteristic exponent, 45
  chart, 137
  eventually strict – function, 274
  exponent, 45, 54
  forward – exponent, 88
  function, 274
  inner product, 110, 136
  LP-regular – exponent, 69, 79
  Lyapunov–Perron regular – exponent, 69, 79
  metric, 17
  norm, 137
  spectrum, 47, 120, 123
  stability theorem, 153
  stability theory, 147
  strong – chart, 118
  strong – inner product, 110
  strong – norm, 111
  value of – exponent, 46
  weak – inner product, 136
  weal – norm, 137
Lyapunov–Perron regular
  Lyapunov exponent, 69, 79
  point, 91, 121

manifold
  global stable –, 5, 7
  global unstable –, 5
  local stable –, 7
  local unstable –, 7, 178
  of nonpositive curvature, 240
map
  absolutely continuous –, 191
  holonomy –, 191
  Katok –, 22, 25
  nonuniformly hyperbolic –, 130
  nonuniformly partially hyperbolic –, 140
  uniformly hyperbolic –, 14
measurable
  lamination, 184
  vector bundle, 83
measure
  conditional –, 208
  ergodic –, 8, 10
  hyperbolic –, 123

Sinai–Ruelle–Bowen –, 22
smooth –, xiv, 207
SRB –, 22
measure-theoretic entropy, 222
metric
  canonical –, 238
  entropy, 222, 224
  Lyapunov –, 17
multiplicative ergodic theorem, 97, 105, 122

negative
  cone, 273
  limit solution, 243
  rank, 273, 274
neighborhood
  regular –, 118
nonabsolutely continuous foliation, 205
nonpositive curvature, 240, 242
nonuniform
  hyperbolicity, xv, 129
  hyperbolicity theory, 119, 158, 163
nonuniformly
  hyperbolic dynamical system, 132
  hyperbolic map, 130
  hyperbolic set, 130, 131
  partially hyperbolic map, 140
  partially hyperbolic set, 140
nonzero Lyapunov exponents, 120, 123
  diffeomorphism with –, 33
  flow with –, 39
norm
  Lyapunov –, 137
  strong Lyapunov –, 111
normal basis, 48
normalization property, 45

order of perturbation, 151
ordered basis, 48
Oseledets
  decomposition, 69, 91
  subspace, 91
Oseledets–Pesin reduction theorem, 111

parameters of hyperbolicity, 6, 14, 130
partial hyperbolicity conditions, 140
partitions
  equivalent –, 223
  independent –, 223
Perron coefficient, 51
perturbation, 151
  order of –, 151

Index

Pesin
    block, 190, 212
    set, 133
    tempering kernel, 111
point
    at infinity, 249
    backward regular –, 121
    forward regular –, 90, 121
    LP-regular –, 91, 121
    Lyapunov–Perron regular –, 91, 121
pointwise dimension, 234
    lower –, 234
    lower leaf –, 235
    stable –, 235
    unstable –, 235
    upper –, 234
    upper leaf –, 235
positive
    cone, 273
    limit solution, 243
    rank, 273, 274
power cocycle, 84
property
    accessibility –, 281
    essential accessibility –, 282

q-foliation, 217
    with smooth leaves, 217

r-admissible submanifold, 183
r-foliation block, 187
r-local transversal, 187
rank
    negative –, 273, 274
    positive –, 273, 274
rational index set, 211
rectangle, 258
recurrent Diophantine condition, 96
reducible cocycle, 93
refinement, 223
    common –, 223
regular
    backward –, 67
    backward – Lyapunov exponent, 79
    backward – point, 90
    forward –, 67, 78
    forward – point, 90
    Lyapunov–Perron – point, 91
    neighborhood, 115, 118
    pair of Lyapunov exponents, 53
    set, 133
return
    map, 84
    time, 84
roof function, 10

s-canonical local transversal, 191
s-foliation block, 190
s-manifold, 190
sequence of matrices
    backward regular –, 79
    forward regular –, 78
set
    filtration, 305
    hyperbolic –, 14
    index –, 133
    level –, 133
    locally maximal –, 19
    nonuniformly hyperbolic –, 130, 131
    nonuniformly partially hyperbolic –, 140
    Pesin –, 133
    regular –, 133
    uniformly completely hyperbolic –, 14
Sinai–Ruelle–Bowen measure, 22, 259
slow-down, 44
    function, 23
    procedure, 23
Smale–Williams solenoid, 20
    slow-down of –, 43
smooth
    ergodic theory, xiv, 207, 208, 222
    foliation, 7
    measure, xiv, 207
solenoid, 20
    slow-down of –, 43
solution
    asymptotically stable –, 149, 158
    conditionally stable –, 158
    exponentially stable –, 149, 158
    stable –, 149
    unstable –, 149
space
    central –, 280
    stable –, 280
    unstable –, 280
special flow, 9
spectral decomposition theorem, 215
spectrum
    Lyapunov –, 47, 123
SRB measure, 22, 259
stability theory, 147
stable
    conditionally – solution, 158

cone, 27
   global – curve, 5
   global – manifold, 5, 7, 183, 185
   global weakly – manifold, 185
   local – manifold, 7, 148, 185
   manifold theorem, 164
   manifold theorem for flows, 185
   pointwise dimension, 235
   solution, 149
   subspace, 5, 6, 14, 124, 130, 148, 280
strict family of cones, 277
strong
   Lyapunov chart, 118
   Lyapunov inner product, 110
   Lyapunov norm, 111
structurally stable diffeomorphism, 9
submanifold
   local smooth –, 159, 164
   r-admissible –, 183
subordinate basis, 48, 49
subspace
   Oseledets –, 91
   stable –, 5, 6, 14, 124, 130, 148
   unstable –, 5, 6, 14, 124, 130
suspension flow, 10
symplectic group, 277
system of variational equations, 28, 150

tempered
   cocycle, 85
   equivalence, 85
   function, 85, 93, 114, 125
tempering kernel, 111, 114
   lemma, 114
theorem
   absolute continuity –, 191
   flat strip –, 249
   implicit function –, 173
   Lyapunov stability –, 153
   multiplicative ergodic –, 97, 105, 122
   reduction –, 110, 111
   spectral decomposition –, 215
   stable manifold –, 164
   stable manifold – for flows, 185
theory
   local manifold –, 163
   Lyapunov stability –, 147
   nonuniform hyperbolicity –, 119, 158, 163
   stability –, 147
topologically
   mixing flow, 10

transitive flow, 10
transitive map, 220
total measure, 311
transitive
   topologically – map, 220
transversal, 187
   local –, 190
   s-canonical local –, 191
   u-canonical local –, 190
transverse subspaces, 143
trapping region, 20
triangular cocycle, 105, 106

u-canonical local transversal, 190
u-foliation block, 191
uniformly
   completely hyperbolic set, 14
   hyperbolic map, 14
   partially hyperbolic diffeomorphism, 280
unstable
   cone, 27
   global – curve, 5
   global – manifold, 5, 183, 185
   global weakly – manifold, 185
   local – manifold, 7, 178, 185
   pointwise dimension, 235
   solution, 149
   subspace, 5, 6, 14, 124, 130, 280
upper
   leaf pointwise dimension, 235
   local leaf entropy, 236
   pointwise dimension, 234

variational equations, 28, 148, 150
vector bundle, 83

W-dissipative diffeomorphism, 289
weak
   Lyapunov chart, 137
   Lyapunov inner product, 136
   Lyapunov norm, 137
weak*-topology, 22
weakly
   mixing flow, 216
   stable foliation, 9
   unstable foliation, 9

Selected Published Titles in This Series

231 **Luís Barreira and Yakov Pesin,** Introduction to Smooth Ergodic Theory, Second Edition, 2023
228 **Henk Bruin,** Topological and Ergodic Theory of Symbolic Dynamics, 2022
227 **William M. Goldman,** Geometric Structures on Manifolds, 2022
226 **Milivoje Lukić,** A First Course in Spectral Theory, 2022
225 **Jacob Bedrossian and Vlad Vicol,** The Mathematical Analysis of the Incompressible Euler and Navier-Stokes Equations, 2022
224 **Ben Krause,** Discrete Analogues in Harmonic Analysis, 2022
223 **Volodymyr Nekrashevych,** Groups and Topological Dynamics, 2022
222 **Michael Artin,** Algebraic Geometry, 2022
221 **David Damanik and Jake Fillman,** One-Dimensional Ergodic Schrödinger Operators, 2022
220 **Isaac Goldbring,** Ultrafilters Throughout Mathematics, 2022
219 **Michael Joswig,** Essentials of Tropical Combinatorics, 2021
218 **Riccardo Benedetti,** Lectures on Differential Topology, 2021
217 **Marius Crainic, Rui Loja Fernandes, and Ioan Mărcuţ,** Lectures on Poisson Geometry, 2021
216 **Brian Osserman,** A Concise Introduction to Algebraic Varieties, 2021
215 **Tai-Ping Liu,** Shock Waves, 2021
214 **Ioannis Karatzas and Constantinos Kardaras,** Portfolio Theory and Arbitrage, 2021
213 **Hung Vinh Tran,** Hamilton–Jacobi Equations, 2021
212 **Marcelo Viana and José M. Espinar,** Differential Equations, 2021
211 **Mateusz Michałek and Bernd Sturmfels,** Invitation to Nonlinear Algebra, 2021
210 **Bruce E. Sagan,** Combinatorics: The Art of Counting, 2020
209 **Jessica S. Purcell,** Hyperbolic Knot Theory, 2020
208 **Vicente Muñoz, Ángel González-Prieto, and Juan Ángel Rojo,** Geometry and Topology of Manifolds, 2020
207 **Dmitry N. Kozlov,** Organized Collapse: An Introduction to Discrete Morse Theory, 2020
206 **Ben Andrews, Bennett Chow, Christine Guenther, and Mat Langford,** Extrinsic Geometric Flows, 2020
205 **Mikhail Shubin,** Invitation to Partial Differential Equations, 2020
204 **Sarah J. Witherspoon,** Hochschild Cohomology for Algebras, 2019
203 **Dimitris Koukoulopoulos,** The Distribution of Prime Numbers, 2019
202 **Michael E. Taylor,** Introduction to Complex Analysis, 2019
201 **Dan A. Lee,** Geometric Relativity, 2019
200 **Semyon Dyatlov and Maciej Zworski,** Mathematical Theory of Scattering Resonances, 2019
199 **Weinan E, Tiejun Li, and Eric Vanden-Eijnden,** Applied Stochastic Analysis, 2019
198 **Robert L. Benedetto,** Dynamics in One Non-Archimedean Variable, 2019
197 **Walter Craig,** A Course on Partial Differential Equations, 2018
196 **Martin Stynes and David Stynes,** Convection-Diffusion Problems, 2018
195 **Matthias Beck and Raman Sanyal,** Combinatorial Reciprocity Theorems, 2018
194 **Seth Sullivant,** Algebraic Statistics, 2018
193 **Martin Lorenz,** A Tour of Representation Theory, 2018
192 **Tai-Peng Tsai,** Lectures on Navier-Stokes Equations, 2018
191 **Theo Bühler and Dietmar A. Salamon,** Functional Analysis, 2018

For a complete list of titles in this series, visit the
AMS Bookstore at www.ams.org/bookstore/gsmseries/.